Methodologies and Applications of Computational Statistics for Machine Intelligence

Debabrata Samanta
Christ University (Deemed), India

Raghavendra Rao Althar
QMS, First American India, Bangalore, India

Sabyasachi Pramanik
Haldia Institute of Technology, India

Soumi Dutta
Institute of Engineering and Management, Kolkata, India

A volume in the Advances in Systems Analysis,
Software Engineering, and High Performance
Computing (ASASEHPC) Book Series

Published in the United States of America by
IGI Global
Engineering Science Reference (an imprint of IGI Global)
701 E. Chocolate Avenue
Hershey PA, USA 17033
Tel: 717-533-8845
Fax: 717-533-8661
E-mail: cust@igi-global.com
Web site: http://www.igi-global.com

Library of Congress Cataloging-in-Publication Data

Names: Samanta, Debabrata, 1987- editor. | Althar, Raghavendra Rao, 1981-
    editor. | Pramanik, Sabyasachi, 1978- editor. | Dutta, Soumi, 1986-
    editor.
Title: Methodologies and applications of computational statistics for
    machine intelligence / Debabrata Samanta, Raghavendra Rao Althar,
    Sabyasachi Pramanik, and Soumi Dutta, editors.
Description: Hershey, PA : Engineering Science Reference (an imprint of IGI
    Global), 2021. | Includes bibliographical references and index. |
    Summary: "This book delves into computational statistics that focus on
    devising an efficient methodology to obtain quantitative solutions for
    problems that are devised quantitatively and brings together
    computational capability and statistical advanced thought processes to
    solve some of the problems encountered in the field"-- Provided by
    publisher.
Identifiers: LCCN 2021012380 (print) | LCCN 2021012381 (ebook) | ISBN
    9781799877011 (hardcover) | ISBN 9781799877028 (paperback) | ISBN
    9781799877035 (ebook)
Subjects: LCSH: Machine learning--Statistical methods.
Classification: LCC Q325.5 .M39 2021 (print) | LCC Q325.5 (ebook) | DDC
    006.3/1--dc23
LC record available at https://lccn.loc.gov/2021012380
LC ebook record available at https://lccn.loc.gov/2021012381

This book is published in the IGI Global book series Advances in Systems Analysis, Software Engineering, and High Performance Computing (ASASEHPC) (ISSN: 2327-3453; eISSN: 2327-3461)

British Cataloguing in Publication Data
A Cataloguing in Publication record for this book is available from the British Library.

All work contributed to this book is new, previously-unpublished material. The views expressed in this book are those of the authors, but not necessarily of the publisher.

For electronic access to this publication, please contact: eresources@igi-global.com.

Advances in Systems Analysis, Software Engineering, and High Performance Computing (ASASEHPC) Book Series

Vijayan Sugumaran
Oakland University, USA

ISSN:2327-3453
EISSN:2327-3461

MISSION

The theory and practice of computing applications and distributed systems has emerged as one of the key areas of research driving innovations in business, engineering, and science. The fields of software engineering, systems analysis, and high performance computing offer a wide range of applications and solutions in solving computational problems for any modern organization.

The **Advances in Systems Analysis, Software Engineering, and High Performance Computing (ASASEHPC) Book Series** brings together research in the areas of distributed computing, systems and software engineering, high performance computing, and service science. This collection of publications is useful for academics, researchers, and practitioners seeking the latest practices and knowledge in this field.

COVERAGE

- Engineering Environments
- Virtual Data Systems
- Distributed Cloud Computing
- Parallel Architectures
- Metadata and Semantic Web
- Software Engineering
- Computer System Analysis
- Computer Graphics
- Enterprise Information Systems
- Computer Networking

IGI Global is currently accepting manuscripts for publication within this series. To submit a proposal for a volume in this series, please contact our Acquisition Editors at Acquisitions@igi-global.com or visit: http://www.igi-global.com/publish/.

Titles in this Series

For a list of additional titles in this series, please visit: www.igi-global.com/book-series

Handbook of Research on Software Quality Innovation in Interactive Systems
Francisco Vicente Cipolla-Ficarra (Latin Association of Human-Computer Interaction, Spain & International Association of Interactive Communication, Italy)
Engineering Science Reference • ©2021 • 501pp • H/C (ISBN: 9781799870104) • US $295.00

Handbook of Research on Methodologies and Applications of Supercomputing
Veljko Milutinović (Indiana University, Bloomington, USA) and Miloš Kotlar (University of Belgrade, Serbia)
Engineering Science Reference • ©2021 • 393pp • H/C (ISBN: 9781799871569) • US $345.00

MATLAB® With Applications in Mechanics and Tribology
Leonid Burstein (Independent Researcher, Israel)
Engineering Science Reference • ©2021 • 368pp • H/C (ISBN: 9781799870784) • US $195.00

Advancements in Fuzzy Reliability Theory
Akshay Kumar (Graphic Era Hill University, India) Mangey Ram (Department of Mathematics, Graphic Era (Deemed to be University), India) and Om Prakash Yadav (North Dakota State University, USA)
Engineering Science Reference • ©2021 • 322pp • H/C (ISBN: 9781799875642) • US $245.00

Impacts and Challenges of Cloud Business Intelligence
Shadi Aljawarneh (Jordan University of Science and Technology, Jordan) and Manisha Malhotra (Chandigarh University, India)
Business Science Reference • ©2021 • 263pp • H/C (ISBN: 9781799850403) • US $195.00

Handbook of Research on Modeling, Analysis, and Control of Complex Systems
Ahmad Taher Azar (Faculty of Computers and Artificial Intelligence, Benha University, Benha, Egypt & College of Computer and Information Sciences, Prince Sultan University, Riyadh, Saudi Arabia) and Nashwa Ahmad Kamal (Faculty of Engineering, Cairo University, Giza, Egypt)
Engineering Science Reference • ©2021 • 685pp • H/C (ISBN: 9781799857884) • US $295.00

Artificial Intelligence Paradigms for Smart Cyber-Physical Systems
Ashish Kumar Luhach (The PNG University of Technology, Papua New Guinea) and Atilla Elçi (Hasan Kalyoncu University, Turkey)
Engineering Science Reference • ©2021 • 392pp • H/C (ISBN: 9781799851011) • US $225.00

IGI Global
PUBLISHER of TIMELY KNOWLEDGE

701 East Chocolate Avenue, Hershey, PA 17033, USA
Tel: 717-533-8845 x100 • Fax: 717-533-8661
E-Mail: cust@igi-global.com • www.igi-global.com

The key strength of this book is that it provides a missing link between existing research in this field and future trends. The book is well-organized so that readers can quickly understand the various aspects of multimedia information analysis as they relate to various types of Computational Statistics for Machine Intelligence. It comes highly recommended.

Dr. Suplab Podder, *MBA (Mkt), MBA (HRM), M.Com., Ph.D*
Department of Management and Commerce, Dayananda Sagar College of Arts, Science and Commerce, India

Dr. Debabrata Samanta would like to dedicate this book to his parents Mr. Dulal Chandra Samanta and Mrs. Ambujini Samanta, his elder sister Mrs. Tanusree Samanta, who have encouraged him all the way, and to his beloved daughter Ms. Aditri Samanta.

Mr. Raghavendra Rao Althar would like to dedicate this book to his parents Mr. Heriyanna and Mrs. Gulabi and his soulmate Jyothi and son Ramateja.

Dr. Sabyasachi Pramanik would like to thank everyone who contributed to this project, particularly the writers and reviewers who participated in the review process. This book would not have been possible without their support. I thank the Almighty for the opportunity to research such an important topic at this time, and I thank each of the writers for their individual contributions. I am grateful to all of the chapter writers who dedicated their time and experience to this book from all over the world. I appreciate all of the peer reviewers' helpful suggestions for improving the consistency, coherence, and content of the chapters. Some authors volunteered to be referees, and I am grateful for their time and effort. I'd like to express my gratitude to the IGI-Global production team, who have provided the highest quality of support and have been incredibly supportive and inspiring during this research process; this publication would not have been possible without their perseverance. A good book publication is the product of the efforts of many people, not just those credited as editor or author. I'd like to dedicate this book to my parents and wife, who have always been an inspiration and support to me.

Dr. Soumi Dutta would like to dedicate this book to her parents and daughter Adriti.

Editorial Advisory Board

Table of Contents

Detailed Table of Contents

Human action recognition (HAR) is one of most significant research topics, and it has attracted the concentration of many researchers. Automatic HAR system is applied in several fields like visual surveillance, data retrieval, healthcare, etc. Based on this inspiration, in this chapter, the authors propose a new HAR model that considers an image as input and analyses and exposes the action present in it. Under the analysis phase, they implement two different feature extraction methods with the help of rotation invariant Gabor filter and edge adaptive wavelet filter. For every action image, a new vector called as composite feature vector is formulated and then subjected to dimensionality reduction through principal component analysis (PCA). Finally, the authors employ the most popular supervised machine learning algorithm (i.e., support vector machine [SVM]) for classification. Simulation is done over two standard datasets; they are KTH and Weizmann, and the performance is measured through an accuracy metric.

Chapter 2

Microblogging, where millions of users exchange messages to share their opinions on different trending and non-trending topics, is one of the popular communication media in recent times. Several researchers are concentrating on these data due to a huge source of information exchanges in online social media. In platforms such as Twitter, dataset-generated lacks coherence, and manually extracting meaning or knowledge from them proves to be painstakingly difficult. It opens up the challenges to the researchers for knowledge extraction driven by a summarization approach. Therefore, automated summary generation tools are recommended to get a meaningful summary out of a given topic becomes crucial in the age of big data. In this work, an unsupervised, extractive summarization model has been proposed. For categorization of data, k-means algorithm has been used, and based on scoring of each document in the corpus, summarization model is designed. The proposed methodology achieves an improved outcome over existing methods, such as lexical rank, sum basic, LSA, etc. evaluated by rouge tool.

Chapter 3

Ankita Mandal, Institute of Engineering and Management, Kolkata, India
Soumi Dutta, Institute of Engineering and Management, Kolkata, India
Sabyasachi Pramanik, Haldia Institute of Technology, India

In the present research work, the use of geometrical figures have been made for the calculation of the value of pi. Instead of circle and square, ellipse and rectangle had been used to derive the value of pi. Ellipse can be considered as an extension of a circle where it had been stretched in two dimensions in unequal manner giving rise to the concept of major axis and minor axis. These axes are considered as the length and breadth of the considered rectangle. The ellipse has been considered within the rectangle and some random points are generated to see the position occurrence of the generated points. If the point lies within the ellipse, then the specific counter is incremented; otherwise, the counter for the rectangle is incremented.

Chapter 4

Venkat Narayana Rao T., Sreenidhi Institute of Science and Technology, Hyderabad, India
Manogna Thumukunta, Sreenidhi Institute of Science and Technology, Hyderabad, India
Muralidhar Kurni, Anantha Lakshmi Institute of Technology and Sciences, Ananthapuram, India
Saritha K., Sri Venkateswara Degree and PG College, Ananthapuram, India

Artificial intelligence and automation are believed by many to be the new age of industrial revolution. Machine learning is an artificial intelligence section that recognizes patterns from vast amounts of data and projects useful information. Prediction, as an application of machine learning, has been sought after by all kinds of industries. Predictive models with higher efficiencies have proven effective in reducing market risks, predicting natural disasters, indicating health risks, and predicting stock values. The quality of decision making through these algorithms has left a lasting impression on several businesses and is bound to alter how the world looks at analytics. This chapter includes an introduction to machine learning and prediction using machine learning. It also sheds light on its approach and its applications.

Chapter 5

Raghavendra Rao Althar, First American India, India & Christ University (Deemed), India
Debabrata Samanta, Christ University (Deemed), India

The chapter focuses on exploring the work done for applying data science for software engineering, focusing on secured software systems development. With requirements management being the first stage of the life cycle, all the approaches that can help security mindset right at the beginning are explored. By exploring the work done in this area, various key themes of security and its data sources are explored, which will mark the setup of base for advanced exploration of the better approaches to make software systems mature. Based on the assessments of some of the work done in this area, possible prospects are explored. This exploration also helps to emphasize the key challenges that are causing trouble for the software development community. The work also explores the possible collaboration across machine learning, deep learning, and natural language processing approaches. The work helps to throw light on critical dimensions of software development where security plays a key role.

Chapter 6

Exploratory data analysis is a technique to analyze data sets in order to summarize the main characteristics of them using quantitative and visual aspects. The chapter starts with the introduction of exploratory data analysis. It discusses the conventional view of it and describes the main limitations of it. It explores the features of quantitative and visual exploratory data analysis in detail. It deals with the statistical techniques relevant to EDA. It also emphasizes the main visual techniques to represent the data in an efficient way. R has extraordinary capabilities to deal with quantitative and visual aspects to summarize the main characteristics of the data set. The chapter provides the practical exposure of various plotting systems using R. Finally, the chapter deals with current research and future trends of the EDA.

Chapter 7

The financial time series have a high frequency and the difference between their observations is not regular. Therefore, continuous models can be used instead of discrete-time series models. The purpose of this chapter is to define Lévy-driven continuous autoregressive moving average (CARMA) models and their applications. The CARMA model is an explicit solution to stochastic differential equations, and also, it is analogue to the discrete ARMA models. In order to form a basis for CARMA processes, the structures of discrete-time processes models are examined. Then stochastic differential equations, Lévy processes, compound Poisson processes, and variance gamma processes are defined. Finally, the parameter estimation of CARMA(2,1) is discussed as an example. The most common method for the parameter estimation of the CARMA process is the pseudo maximum likelihood estimation (PMLE) method by mapping the ARMA coefficients to the corresponding estimates of the CARMA coefficients. Furthermore, a simulation study and a real data application are given as examples.

Chapter 8

Direct signal, clutter, and multipath echoes are received along with surveillance signal in passive bistatic radars. These signals degrade the target detection capability of the radar processing algorithm and thus require additional processing to achieve a decent performance. Different clutter and multipath cancellation algorithms are devised for removal of unwanted signals. These algorithms require different computational complexity to provide different level of clutter cancellation. This chapter reviews different clutter cancellation techniques and compares their performance based on the computational complexity. This performance comparison allows understanding the computation load put up by different clutter cancellation techniques and ultimately the response rate of the radar system while maintaining decent target detection.

Chapter 9

Yibeltal Meslie, Mekdela Amba University, Ethiopia
Wegayehu Enbeyle, Department of Statistics, Mizan-Tepi University, Ethiopia
Binay Kumar Pandey, Govind Ballabh Pant University of Agriculture and Technology, Pantnagar, India
Sabyasachi Pramanik, Haldia Institute of Technology, India
Digvijay Pandey, Department of Technical Education, Institute of Engineering and Technology, Lucknow, India
Pankaj Dadeech, Swami Keshvanand Institute of Technology, Management, and Gramothan (SKIT), Jaipur, India
Assaye Belay, Department of Statistics, Mizan-Tepi University, Ethiopia
Ashwini Saini, RPS Group of Institution, Haryana, India

COVID-19 is likely to pose a significant threat to healthcare, especially for disadvantaged populations due to the inadequate condition of public health services with people's lack of financial ways to obtain healthcare. The primary intention of such research was to investigate trend analysis for total daily confirmed cases with new corona virus (i.e., COVID-19) in the countries of Africa and Asia. The study utilized the daily recorded time series observed for two weeks (52 observations) in which the data is obtained from the world health organization (WHO) and world meter website. Univariate ARIMA models were employed. STATA 14.2 and Minitab 14 statistical software were used for the analysis at 5% significance level for testing hypothesis. Throughout time frame studied, because all four series are non-stationary at level, they became static after the first variation. The result revealed the appropriate time series model (ARIMA) for Ethiopia, Pakistan, India, and Nigeria were Moving Average order 2, ARIMA(1, 1, 1), ARIMA(2, 1, 1), and ARIMA (1, 1, 2), respectively.

Chapter 10

Raghavendra Rao Althar, First American India, India & Christ University (Deemed), India
Debabrata Samanta, Christ University (Deemed), India

This chapter focuses on knowledge graphs application in software engineering. It starts with a general exploration of artificial intelligence for software engineering and then funnels down to the area where knowledge graphs can be a good fit. The focus is to put together work done in this area and call out key learning and future aspirations. The knowledge management system's architecture, specific application of the knowledge graph in software engineering like automation of test case creation and aspiring to build a continuous learning system are explored. Understanding the semantics of the knowledge, developing an intelligent development environment, defect prediction with network analysis, and clustering of the graph data are exciting explorations.

Chapter 11

Nagadevi Darapureddy, Chaitanya Bharathi Institute of Technology, Hyderabad, India
Muralidhar Kurni, Anantha Lakshmi Institute of Technology and Sciences, Ananthapuramu, India
Saritha K., Sri Venkateswara Degree and PG College, Ananthapuram, India

Artificial intelligence (AI) refers to science-generating devices with functions like reasoning, thinking, learning, and planning. A robot is an intelligent artificial machine capable of sensing and interacting with its environment utilizing integrated sensors or computer vision. In the present day, AI has become a more familiar presence in robotic resolutions, introducing flexibility and learning capabilities. A robot with AI provides new opportunities for industries to produce work safer, save valuable time, and increase productivity. Economic impact assessment and awareness of the social, legal, and ethical problems of robotics and AI are essential to optimize the advantages of these innovations while minimizing adverse effects. The impact of AI and robots affects healthcare, manufacturing, transport, and jobs in logistics, security, retail, agri-food, and construction. The chapter outlines the vision of AI, robot's timeline, highlighting robot's limitations, hence embedding AI to robotic real-world applications to get an optimized solution.

Chapter 12

M. R. Sundara Kumar, Sona College of Technology, Salem, India
S. Sankar, Sona College of Technology, Salem, India
Vinay Kumar Nassa, South Point Group of Institutions, Sonepat, India
Digvijay Pandey, Institute of Engineering and Technology, Lucknow, India
Binay Kumar Pandey, College of Technology, Govind Ballabh Pant University of Agriculture and Technology, India
Wegayehu Enbeyle, Department of Statistics, Mizan-Tepi University, Ethiopia

In this digital world, a set of information about the real-world entities is collected and stored in a common place for extraction. When the information generated has no meaning, it will convert into meaningful information with a set of rules. Those data have to be converted from one form to another form based on the attributes where it was generated. Storing these data with huge volume in one place and retrieving from the repository reveals complications. To overcome the problem of extraction, a set of rules and algorithms was framed by the standards and researchers. Mining the data from the repository by certain principles is called data mining. It has a lot of algorithms and rules for extraction from the data warehouses. But when the data is stored under a common structure on the repository, the values derived from that huge volume are complicated. Computing statistical data using data mining provides the exact information about the real-world applications like population, weather report, and probability of occurrences.

Preface

Computational statistics is concerned with developing a system for obtaining quantitative solutions to problems that are formulated quantitatively. To solve some of the problems efficiently, this topic combines analytical capacity and statistical sophisticated thought processes. Optimization techniques in statistical inference, expectation maximization algorithms, and Monte Carlo simulation are just a few of the fascinating aspects of the topic. Simulation and graphical analysis advancements have also accelerated the field of mathematical analytics. Financial applications, such as risk management and derivative pricing, biological applications, such as bioinformatics and computational biology, and computer network security applications that affect people's lives all use computational statistics. With such high-impact areas, it's critical to delve deeper into the topic and examine the main areas as well as their recent development. This book serves as a primer on how to put recent developments in computational statistics to use. Topics in computational statistics, methodologies, and implementations will be covered in the book. The impact of computational arithmetic on computational statistics is investigated, as well as numerical algorithms in statistical application software. The book examines the fundamentals of computer systems, which serve as the foundation for algorithms, as well as the mathematical techniques that underpin simulation techniques. It also covers linear algebra and its importance in optimization techniques, the evolution of optimization techniques, and the most efficient use of computer resources. The role of statistical graphics in data analysis is also looked into. Additionally, computationally expensive statistical approaches are investigated. Computational inference and the role of the computer model in experiment design are also discussed. The book also covers Bayesian analysis, survival analysis, and data mining in computational statistics, as well as groundbreaking computational statistics applications.

Chapter 1 focuses on Human Action Recognition is one of the most important research topics that has drawn the attention of a large number of researchers. Automatic Human Action Recognition systems are widely used in a variety of areas, including visual surveillance, medical assistance, data retrieval, human-computer interaction (HCI), and health care, among others. In this paper, we propose a new Human Action Recognition model based on this inspiration, which takes an image as input and analyses and exposes the action present in it. We use the Rotation Invariant Gabor filter and the Edge adaptive Wavelet filter in the analysis process to introduce two separate feature extraction methods. For each action image, a new vector called the Composite Feature Vector is created, which is then subjected to Principal Component Analysis to reduce dimensionality (PCA). Finally, for classification, we use the most common supervised machine learning algorithm, Support Vector Machine (SVM). Simulation tests are conducted on three standard action datasets: the UCF 11 action dataset, the KTH dataset, and the Weizmann dataset, with performance metrics such as Recall, Precision, F-Score, False Negative Rate, and False Positive Rate used to assess performance. Furthermore, we conducted a comparative analysis by varying the features, with accuracy and false alarm rate being used to assess results.

Chapter 2 deep dives Microblogging has become one of the most common forms of communication in recent years, with millions of users exchanging messages to express their thoughts on various trending and non-trending topics. Owing to the vast amount of information exchanged in online social media, many researchers are focusing on these data. The dataset created on platforms such as Twitter lacks coherence, making manual extraction of sense or information extremely difficult. It presents researchers with new challenges in terms of information extraction using a summarization approach. In the age of big data, automated summary generation tools are recommended for getting a meaningful summary out of a given subject. An unsupervised, extractive Summarization model is proposed in this paper. The K-Means algorithm was used to categorise data, and a summarization model was created based on the scoring of each document in the corpus. The proposed methodology outperforms current methods such as Lexical Rank, Sum Basic, LSA, and others, according to Rouge Tool.

Chapter 3 intends the value of pi is considered to be one of the most important numbers in today's world, and it is often used in mathematical and scientific calculations. Pi is an irrational number with a non-recurring decimal part, according to its characteristics. It has a transcendental quality to it, which justifies its statistical randomness. For the measurement, various methodologies have been used since ancient times. Computers' computing ability has been used for quicker and more precise calculations as technology has progressed. Finding the ratio between the area of a circle and the area of a square with the circle inscribed inside it is one of the most popular methods for determining the value of pi. Such an approach confines the computation to normal surfaces with the least amount of parameter variance. For the purpose of calculating the value of pi, geometrical figures were used in the current study. The value of pi was calculated using an ellipse and a rectangle instead of a circle and a square. Ellipse can be thought of as a circle that has been extended in two dimensions unequally, giving rise to the concepts of major axis and minor axis. These axes reflect the length and width of the rectangle under consideration. The ellipse has been considered inside the rectangle, and some random points have been developed to see how the generated points are distributed. The specific counter is incremented if the point is inside the ellipse; otherwise, the rectangle's counter is incremented. In order to put the above-mentioned principle into practise, Monte Carlo Simulation is used. Monte Carlo Simulation is a statistical method for calculating the likelihood of unknown occurrences. The Monte Carlo method is one of the best mathematical methods to use since it uses the instantiation of random numbers to find the likelihood of points lying inside the ellipse. The probability is calculated using the ratio of the ellipse's area to the rectangle's area, where the value of pi is four times the probability. This entire model was created with Scilab, an open source, free programme that can be used for numerical computations in a variety of applications such as numerical problem optimization, signal processing, image processing and enhancement, statistical model formulation and analysis, and so on. Scilab has become a very common forum among students and academicians for various simulation methodologies due to the free licenced availability of the programme in an open platform. The overall application of Scilab in locating an effective solution using the proposed methodology is outlined in this chapter. It has been discovered that increasing the number of simulation iterations increases the precision of the model. When the simulation is run 8,000 times, the value of pi comes out to be 3.1395 with a precision of around 99.79 percent when compared to the normal value of 3.14159.

In Chapter 4, with the rapid growth of the internet, there is an ever-increasing need for automated and effective solutions to a variety of problems. In the early twenty-first century, a major change occurred, resulting in a reliance on Artificial Intelligence and its applications. Many people agree that artificial intelligence and automation were the catalysts for the modern industrial revolution. Artificial Intelligence has many divisions, each with functionalities that are used in a variety of industries. Machine Learning is a branch of Artificial Intelligence that recognises trends in large quantities of data and projects useful data. This data can then be used to make critical choices. Machine learning has made its way into almost every area imaginable, including science, politics, industry, medicine, and research. It can be used for a variety of tasks, including making forecasts, classifying data, collecting relevant data, and calculating probabilities. Prediction, as a Machine Learning application, has been sought after by a variety of industries. Prediction algorithms have been used to solve long-standing problems. Predictive models have given answers to ambiguity, making for a more informed approach to it. Higher-efficiency predictive models have been shown to be effective in minimizing market risks, predicting natural disasters, identifying health risks, and predicting stock prices, among other things. The high quality of decision-making enabled by these algorithms has left an indelible mark on many companies and is set to change the way the world views analytics. A predictive model collects data, cleans it, and trains it using a sample data collection, during which the algorithm recognizes trends in the data and learns to predict outcomes based on the user's requirements. This chapter provides an overview of Machine Learning, as well as predictions based on Machine Learning, as well as its approach and implementations.

The paper examines the work that has been done to apply data science to software engineering, with an emphasis on the creation of stable software systems. Since requirements management is the first stage of the life cycle, all methods that can aid in the development of a security mindset are investigated. Various main themes of security and its data sources are discussed as a result of the work done in this field, which will serve as a foundation for further exploration of better ways to mature software systems. Possible futures are investigated based on evaluations of some of the work done in this field. This investigation also aids in highlighting the major issues plaguing the software development industry. Work also looks at how machine learning, deep learning, and natural language processing approaches could work together. Work contributes to shedding light on crucial aspects of software development in which security plays a critical role in this Chapter 5.

Exploratory data analysis is a method of analysing data sets in order to summarise their key characteristics using both quantitative and visual elements. Exploratory Data Analysis is introduced first in this chapter. It addresses the popular perception of it as well as its major drawbacks. It delves into the specifics of quantitative and visual exploratory data analysis. It is concerned with statistical techniques that are applicable to EDA. It also emphasises the most important visual techniques for effectively representing data. R has exceptional skills for dealing with quantitative and visual dimensions of data sets to summarise key characteristics. The chapter gives you a hands-on introduction to different plotting systems that you can use with R. Finally, the chapter discusses recent EDA research and future trends, in Chapter 6.

In Chapter 7, the financial time series have a high frequency and the difference between their observations is not regular. Therefore, continuous models can be used instead of discrete-time series models. The purpose of this chapter is to define Lévy - Driven Continuous Autoregressive Moving Average (CARMA) models and their applications. The CARMA model is an explicit solution to stochastic differential equations and also, it is analog to the discrete ARMA models. In order to form a basis for CARMA processes, the structures of discrete-time processes models are examined, then stochastic differential equations, Lévy processes, Compound Poisson processes, and Variance Gamma processes are defined. Finally, the parameter estimation of CARMA (2, 1) is discussed as an example. The most common method for the parameter estimation of the CARMA process is the Pseudo Maximum Likelihood Estimation (PMLE) method by mapping the ARMA coefficients to the corresponding estimates of the CARMA coefficients. Furthermore, a simulation study and a real data application are given as examples.

Chapter 8 takes in passive bi-static radars, direct signal, debris, and multipath echoes are received alongside surveillance signal. These signals degrade the radar processing algorithm's target detection capability, necessitating additional processing to achieve acceptable results. For the elimination of unnecessary signals, various clutter and multipath cancellation algorithms have been invented. To provide different levels of clutter cancellation, these algorithms need different levels of computational complexity. This article compares the efficiency of various clutter cancellation strategies based on computational complexity. This comparison of results allows for a better understanding of the computational load imposed by various clutter cancellation techniques and, as a result, the radar system's response rate while maintaining good target detection.

Chapter 9 throws Due to the insufficient state of public health facilities and people's lack of financial means to access health care, COVID-19 is likely to pose a major threat to healthcare, especially for vulnerable communities. The primary goal of this type of study was to look at trend analysis for total daily reported cases of the new corona virus, COVID-19, in African and Asian countries. The data was collected from the world health organisation (WHO) and the world metre website, and the analysis used a daily documented time series observed for two weeks (52 observations). ARIMA models with only one variable were used. The study was conducted using STATA 14.2 and Minitab 14 statistical tools with a 5% significance level for hypothesis testing. Since all four series are non-stationary at the level, they became static after the first variation over the time period studied. Moving Average order 2, ARIMA(1, 1, 1), ARIMA(2, 1, 1), and ARIMA(1, 1, 2) were found to be the acceptable time series models (ARIMA) for Ethiopia, Pakistan, India, and Nigeria, respectively. For Ethiopia, Pakistan, India, and Nigeria, the Autoregressive Integrated Moving Average models of total daily verified case (s) data from COVID-19 were fitted, and all of the underlined assumptions were checked and found to be correct. Following that, forecasts for the next two weeks (14 period/days) were made for the entire region.

In Chapter 10, this chapter focuses on the use of information graphs in software engineering. It begins with a broad overview of Artificial Intelligence for software engineering before narrowing down on the areas where information graphs can be useful. The aim is to bring together previous work in this area and highlight important takeaways and potential goals. The design of a knowledge management system, as well as basic applications of the knowledge graph in software engineering such as test case automation and the need to create a continuous learning system, is investigated. Explorations include understanding the semantics of information, designing an intelligent development environment, defect prediction using network analysis, and clustering graph data. Other areas of research include pattern recognition in the context of the information graph, code property graph exploration for software vulnerabilities, and dependency graphs for software vulnerabilities. Multiple potential research areas are highlighted as a result of the work performed in these areas, which can include directions to select areas that can bring a lot of value to software engineering.

Chapter 11 intends Artificial intelligence (AI) refers to science-based systems that perform tasks such as reasoning, thought, learning, and planning, implying that computers can think and behave. A robot is an intelligent artificial machine that uses integrated sensors or computer vision to sense and communicate with its environment. AI provides the Robot with computer vision, allowing it to navigate, hear, and measure the appropriate response. A robot is a semi-autonomous or autonomous system that is controlled and groomed by humans when performing a task autonomously or partially autonomously.

AI has become a more popular presence in robotic resolutions in recent years, adding versatility and learning capabilities. Robotics and artificial intelligence (AI) are a powerful mix for automating factory tasks. Healthcare, manufacturing, transportation, and jobs in logistics, defence, retail, agri-food, and construction are all affected by AI and robots. Robots have made their way into our professional and personal lives in a variety of areas, including manufacturing (assembly), medicine (assisting surgeons), music (conducting orchestras), radiology (detection of cancer), entertainment (including stand-up), restaurants (gourmet meals), military preparation, and many others. To maximize the benefits of these technologies while mitigating their negative effects, economic impact assessments and understanding of the social, legal, and ethical issues surrounding Robotics and AI are needed. A robot with artificial intelligence opens up new possibilities for industries to generate work in a safer, more efficient, and more productive manner. There has been a lot of research into using AI to improve Robot and process efficiency, expand Robot functionality, and save businesses money. The central areas of the locus are extending their picking abilities to deal with items that aren't solid or in static AI study positions in robots. Expand robot autonomy to allow them to function effectively in non-standard conditions (such as rugged terrain), allowing them to be controlled by oral commands and indications. AI is being used in research to allow robots to make decisions based on video presentations and unusual tests. Small-to-medium-sized businesses can embrace robots as programming time and expense are reduced, allowing them to be more profitable. This chapter discusses AI's vision, the timeline of robots, and the drawbacks of robots, with the aim of embedding AI into robotic real-world applications to achieve an integrated solution.

Chapter 12 focuses a collection of information about real-world individuals is gathered and stored in a common location in this digital world for extraction. When the data is generated, it has no value, but it can be transformed into meaningful data using a set of rules. Those records must be translated from one format to another based on the attributes that were used to create them. Although storing these large volumes of data in one location, extracting them from the repository presents challenges. The standards and researchers framed a set of rules and algorithms to solve the problem of extraction. Data mining is the process of extracting data from a repository using certain concepts. For extracting data from data warehouses, it has a lot of algorithms and rules. When the data is stored in a repository with a similar structure, however, calculating values from such a large volume becomes more difficult. Because of its typical format and numerical figures, statistical data can be manipulated from the archive, causing disputes. Data mining is used to compute statistical data, and offers precise details about real-world applications such as population, weather, and likelihood of events, among others. Data mining uses a variety of algorithms and techniques to deal with complex statistical statistics. With Knowledge Data Discovery, complex equations and mathematical problems in statistical data can be solved using data mining principles (KDD). The size of the data stored in the archive has generally varied depending on the form of information stored as knowledge from discovery. It has a number of problems and difficulties to categorise under various principles. Statistics is the concept of archiving exact data from a database and using formulas and equations to generate numerical values as output. The statistics' main goal is to collect data from the archive using conventional methods. However, it is a little sluggish, and the current data has been augmented with duplication and noisy data. To prevent such disputes, data mining in computational statistics is used to identify and categorise their attributes. The challenges and issues faced by data mining in computational statistics are discussed in this chapter, along with potential solutions. The solutions of numerical statistical data by its algorithms and rules to balance the user input data and output are known as data mining concepts. Furthermore, when a large amount of data is managed in the repository, data mining can be used to solve problems in the statistical field. This chapter discusses the various forms of statistics and data mining problems that must be addressed simultaneously in order to maintain a massive repository of real-world data that can be accessed in one location and reliably retrieved by the user at any time. When data mining concepts are used to collect statistical data, reliability, precision, and availability play a big role.

The aim of the book is to present a wide range of emerging methods for conflict resolution in organisations in order to inspire top-level managers to resolve organisational disputes. Once the above-mentioned goal is met, it will be possible to create a welcoming environment in various organisations, resulting in improved results and throughputs.

Debabrata Samanta
Christ University (Deemed), India

Raghavendra Rao Althar
QMS, First American India, Bangalore, India

Sabyasachi Pramanik
Haldia Institute of Technology, India

Soumi Dutta
Institute of Engineering and Management, Kolkata, India

Acknowledgment

First and foremost, the editors would like to thank almighty God for keeping us healthy when working on the edited book *Methodologies and Applications of Computational Statistics for Machine Intelligence*. The editors wish to express their gratitude to all of the writers for their contributions. The editors would also like to express their heartfelt gratitude to all members of the editorial advisory board for their unwavering support and helpful advice during the editing process. We also want to thank all of the reviewers for their thoughtful and timely assistance in the review process. Thanks also go out to all of my coworkers and friends for their unwavering encouragement and assistance. The editors would like to express their gratitude to IGI Global for allowing us to publish with them.

The editors would like to dedicate this book to those deceased and infected due to the ongoing COVID-19 pandemic.

Chapter 1

Hybrid Feature Vector–Assisted Action Representation for Human Action Recognition Using Support Vector Machines

L. Nirmala Devi
Osmania University, India

A.Nageswar Rao
Osmania University, India

ABSTRACT

Human action recognition (HAR) is one of most significant research topics, and it has attracted the concentration of many researchers. Automatic HAR system is applied in several fields like visual surveillance, data retrieval, healthcare, etc. Based on this inspiration, in this chapter, the authors propose a new HAR model that considers an image as input and analyses and exposes the action present in it. Under the analysis phase, they implement two different feature extraction methods with the help of rotation invariant Gabor filter and edge adaptive wavelet filter. For every action image, a new vector called as composite feature vector is formulated and then subjected to dimensionality reduction through principal component analysis (PCA). Finally, the authors employ the most popular supervised machine learning algorithm (i.e., support vector machine [SVM]) for classification. Simulation is done over two standard datasets; they are KTH and Weizmann, and the performance is measured through an accuracy metric.

I. INTRODUCTION

Human Action Recognition (HAR)has attained a great research interest in recent years due to its wide variety applicability in different scenarios like Visual Surveillance, Gesture Recognition, Human Robot Interactions (HRI), Human -computer Interactions (HCI), analysis of behavior content based data retrieval etc. HAR is a very significant and an inspiring topic of research. One of the most popular problems with

DOI: 10.4018/978-1-7998-7701-1.ch001

HAR is that the similar action may perform in several ways, even by similar person. One more serious issue in the HAR is the similarity of poses those appears in the same manner from multiple viewpoints. Hence the major problem is to determine an appropriate action representation method which is more discriminative (thus one can discriminate different actions) as well as generalized (thus the same actions can detected, even performed by different types of moving people and acquired from any view point) (M. Keestra et al. (2015)).

In earlier, the HAR is accomplished with the help of external peripherals like Gyroscopes, Accelerometers and Magnetometers etc. However, there is need of a physical interaction between sensors and human beings those were doing action. Thus, the main problem rises at the restricted movements due to the presence of connected wires and complex device settings (M. Jain et al. (2013)). To solve these issues, the new HAR has come into picture which is based on video sensors. In video based HAR, the activities are supervised with the help of video cameras. The main aim of HAR is to make the system to identify action by modelling the action through some mathematical computational algorithms. Similar to Human Visual System (HVS) these mathematical algorithms must produce a label after analyzing the entire or partial action from the video. In the computer vision oriented research, the development is done in the direction of developing such type of computational algorithms such that we can gain a great understanding about the actions of human beings from digital videos and images.

Due to the great significance of HAR in several real world applications, it has gained a lot of research interest and so many researchers concentrated in this direction. It has been studied from past decades and it is still a very challenging issue in real time applications. One of the most popular problems with HAR is that the similar action may perform in several ways, even by similar person. Another important issue is that the same human pose appears quite different when observed from different viewpoints. Hence the major problem is to determine an appropriate action representation method which is more discriminative (thus one can discriminate different actions) as well as generalized (thus the same actions can detected, even performed by different types of moving people and acquired from any view point) (Manish Khare et al., (2017)).

This chapter proposes a new HAR method for recognizing human actions from videos. Towards the recognition of human action, we have modeled an automatic system that takes an image as input, analyses it through signal processing approaches and then exposes the action present in it. At this contribution, we have proposed a new feature extraction method by combining the features of two individual filters; they are Gabor filter and wavelet filter. For dimensionality reduction, Principal Component Analysis (PCA) is employed and for classification, the Support Vector Machine (SVM) is used.

The remaining chapter structure is organized as follows; the literature survey on HAR is explored in section II. Further the details of developed method for HAR are explored in section III. The particulars of experimental analysis are explored in section IV and the conclusions are issued in the last section V.

II. LITERATURE SURVEY

Based on the theme of methodology employed for HAR, the earler developed methods are categorized into two categories, they are the methods focused on action representation and the other is the method focused on classiification. In the former category, the main focus in to extract the features from an action video. The extracted features have to represent an action video in such a manner it can explore the motion movements of actions. A most effective action representation features can assure more discriminative information between different actions such that the classification algorithm gains a perfect clarity about the sequences of an action (Michael W. Davidson et al. (2015)). Next, in the classification phase, the main intention is made on the complexity reduction and robustness provision. A classifier must be designed in such a way it can derive an appropriate results for different types of actions. Then only the classifier is called as robust classifier. Further, the framework employed for action recognition also has to focus over computational burden and has to select an appropriate classifier who has fewer burdens with more recognition efficiency.

FE is an imperative aspect in the HAR model. For a given action video, we can see that they are differed with so many issues like motion speed, visual appearance, movement of cameras, pose variations etc., which makes the extraction of efficient features much complex (P A. Dhulekar et al. (2018)). Depends on the mode of features extraction, they are categorized into two categories. They are; Global Features and Local Features. Global methods are employed to extract the motion information from the whole body of human actions. "Motion History Image (MHI)" and "Motion Energy Image (MEI)" are the two common approaches those alleviates the information of an action with respect to its time of occurance and position of occurance. Spatio-Time Interest Points (STIPs) and the Motion Trajectories (MT) are the two most popular methods under the category of local representation. These methods are robust in the case of varying appearances and translations.

Due to the flexibility in the ease of implementation of STIPs, P. Dollar et al. (2005) proposed to determine two different STIPs. The first method employed 2-D Gabor filter and its main intention is determine the speed and shape of moving foreground in every frame of an action video. Next, the second method employed STIP clouds to extract the features at different orientations as well as at different scales.

MT is one more way to handle the long the long-duration STIPs. With this inspiration, Raju S, S.et al. (2004) developed an enhance MT based method for action representation. This method considered the motion of camera which was obtained by a new feature matching method employed through the successive frames through the SURF descriptor and dense OFVs. This prediction is also having an advantage in the removal of camera motion effect in OFVs based methods.

The researchers (A. Bobick et al. 2001) developed an innovative MT form for action recognition through which we can capture the temporal discriminative relationships between actions. In this method, the MTs are computed with the help of STIPs and termed as cuboids. These features are extracted by performing a matching criterion between SIFT descriptors of consecutive frames of an action video. Then the MT points are represented in the form of Bag of visual words (BoVW) and then processed through SVM algorithm for classification. However, the main drawback with cuboid features is that for a particular span of frames, the detected interest points may or may not exist at the same spatial location within the temporal bounds of cuboid.

Christian Schuldt et al. (2004) employed two action descriptors such as (1) Distance Mean Histogram (DMH) of Gradients and (2) Segmented block of mean image (SBMI). For action analysis followed by classification, this approach employed the most popular random forest algorithm. For experimental validation, they have used two datasets namely ATM action video dataset and HMDB action video dataset. For the analysis purpose, they had taken the help of block based segmentation and then for ach block they have measured the mean which can give an overall description about the features present in that block (Yuan Shen et al. (2010)).

Considering the gradient of motion, (D. K.Vishwakarma et al. 2015) proposed an action recognition based on Accumulated Motion Image (AMI) in which the histograms are built based on the energy distributions. After the evaluation of AMI, Discrete Fourier Transform (DFT) is employed and mean and variance are measured. Finally the Dynamic Time Wrapping (DTW) is employed for training.

The Gradient descriptor is much effective in the extraction of edge structure or the appearance of an image because it is computed from the local distribution gradient. However, the gradient based algorithms are sensitive to noise. To overcome this problem, (D. Weinland et al. 2006) proposed to extract the gradient features along with difference of Gaussian kernel features. "Spatial distribution gradients (SDG)" are computed at several resolution levels of sub-images, resulting different shape variations. Further these Difference of Gaussian features are combined with STIPs and created a fused vector for every action class. SDGs are measured at several resolution levels of sub-images of AEI, resulting in the complete shape variations of human silhouette during the activity. Due to translational, rotation and scale characteristics of STIPs, the vocabulary of DoG-based STIPs are constructed through vector quantization which will become a unique for every class of activity (VikasTripathi et al. (2017)).

Spatio-Time Interest Points (STIPs) and the Motion Trajectories (MT) are the two most popular methods under the category of local representation. These methods are robust in the case of varying appearances and translations. (D.K. Vishwakarma et al. (2019)) proposed a system for detection and recognition of movements in human where each point if represented as a function corresponding to its location spatially in the sequence of images.This feature has been for the detection of movements in the overall human body Gdyczynski, et al. (2014). S Sabari Raju et al. (2005) used temporal features in spatial domain for recognizing behaviours and actions of individuals. Due to the flexibility in the ease of implementation of STIPs, Bregonzio et al proposed to determine two different STIPs. The first method employed 2-D Gabor filter and its main intention is determine the speed and shape of moving foreground in every frame of an action video. Next, the second method employed STIP clouds to extract the features at different orientations as well as at different scales. But the STIPs have efficiency in the action video of short time but not of the videos of longer duration.

MT is one more way to handle the long the long-duration STIPs. With this inspiration, H. A. Moghaddam et al. (2019) developed an enhance MT based method for action representation. This method considered the motion of camera which was obtained by a new feature matching method employed through the successive frames through the SURF descriptor and dense OFVs. This prediction is also having an advantage in the removal of camera motion effect in OFVs based methods. H. Bay et al. (2008) used micro CT report images to understand the trabecular alignment within bone stem cells using Gabor filters and edge enhancing algorithms based on Gaussian methods. Gabor filters have also been implemented in extracting texts from digital documents H. Wang et al., (2013), localization and extraction of coloured images with complexities H. Wang, et al., (2011).

H. Wang, et al., (2013) developed an innovative MT form for action recognition through which we can capture the temporal discriminative relationships between actions. In this method, the MTs are computed with the help of STIPs and termed as cuboids. These features are extracted by performing a matching criterion between SIFT descriptors of consecutive frames of an action video H. Wang, et al., (2015). Then the MT points are represented in the form of Bag of visual words (BoVW) and then processed through SVM algorithm for classification. However, the main drawback with cuboid features is that for a particular span of frames, the detected interest points may or may not exist at the same spatial location within the temporal bounds of cuboid (V. Thanikachalam et al. (2012)).

Considering the gradient of motion, Hamim A. Abdul Azim et al., (2015) proposed an action recognition based on Accumulated Motion Image (AMI) in which the histograms are built based on the energy distributions. After the evaluation of AMI, Discrete Fourier Transform (DFT) is employed and mean and variance are measured. Finally the Dynamic Time Wrapping (DTW) is employed for training.

Jin Jiang et al. (2014) used the "Wavelet Packet Transform (WPT)" to extract the features and the actions are classified through SVM classifier. Further, to determine the optimal parameters of kernel values in SVM, they employed an "Improved Adaptive Genetic Algorithm (IAGA)".

Jingen Liu et al., (2009) studied the advantages of Ridge let Transform (RT) and Gabor Filter and then they combined the features extracted by both filers to perform HAR. The main reason behind the effectiveness of RT is its characteristics of dependency on the property of orientation and the main reason behind the effectiveness of Gabor filter is its orientation and scale invariance.

K. Ruben Raju et al. (2020) employed DWT for feature extraction from action frames at multiple resolutions. Here the DWT is employed to decompose the action frame into different frequency bands. The DWT is more operative in the provision of multi-resolution features. However, the main problem with the DWT is its shift invariance property for action images.

Larson, et al. (2007) employed two action descriptors such as (1) Distance Mean Histogram (DMH) of Gradients and (2) Segmented block of mean image (SBMI). For action analysis followed by classification, this approach employed the most popular random forest algorithm. For experimental validation, they have used two datasets namely ATM action video dataset and HMDB action video dataset. For the analysis purpose, they had taken the help of block based segmentation and then for ach block they have measured the mean which can give an overall description about the features present in that block.

Lena Gorelick et al. (2007) proposed a new feature extraction technique comprised of HOG. For classification, it employed two different algorithms such as K-Nearest neighbor and SVM. Six different experiments are conducted on the basis of combination of feature extractors and classifiers

Lyons, M.et al. (1998) focused over the wavelet coefficients correlation and applied a new WT technique called as "Spatio-temporal wavelet correlogram (SWTC)" to extract the features form an action frame in action video. SWTC has an advantage of multi-scale and multi-resolution properties and also measures the correlation between the obtained wavelet coefficients.

M. Bregonzio et al. (2009) employed Gabor filter along with wavelet filter for the extraction of features from action videos. After the extraction of Gabor and wavelet features, they constructed a composite feature vector and it was fed to SVM classifier for the recognition of action. However, the major disadvantage with wavelet transform its lack of shift invariance, means for scaled and shifted features, it won't have effective recognition performance.

III. PROPOSED METHOD

3.1. Overview

Here, we explore the entire particulars of developed HAR system. The entire system is executed in two stages. They are Training and Testing. At the training stage, the developed system is subjected to training through a huge action videos while in the second stage; the system which is exposed for testing. In the both phases, we don't refer the raw action videos for processing. Initially, the action videos are subjected to the Feature Extraction over which the characteristics of an action are characterized such that the system can understand. At the feature extraction phase, we employed two different types of features and they are extracted through two different methods, they are Gabor filer and Wavelet filter. After extracting

Figure 1. Block diagram of proposed action recognition system

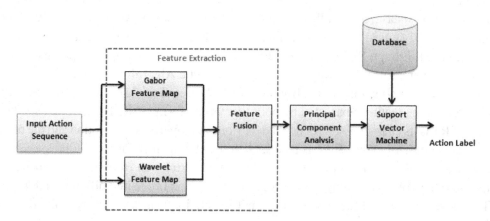

the features from action videos, they are trained to the system through machine learning algorithm. The block diagram of proposed HAR method is shown in Figure 1.

In the developed method for Feature Extraction, we have employed totally two different types of transformation filters; they are Wavelet Filter and Gabor Filter. The main aim of Gabor Features is to exploit the scale and rotational invariant features while the main aim of Wavelet transformation filter is to exploit the multi-resolution features. The major contribution of this method is the construction of composite feature vector by which different actions can be discriminated much effectively. Unlike the existing methods those focused only on the single FE methods, this method combined two different filters such that the obtained Fused Feature Vector is more descriptive and ensures a more discrimination between different actions eve under scaling, rotation and multi-resolution environments. Moreover, we also employed a more popular dimensionality reduction method, PCA to reduce the size of Fused Feature Vector.

3.2 Feature Extraction

Here the proposed contribution is a hybrid Feature Extraction method which was employed with the help of two different signal processing filters. They are; Gabor filters (GF) and Wavelet filters. For a given action video, the first filter tries to find the scale as well as rotational invariant features while the next filter tries to determine the features with multi-resolution in nature. The second filter is employed for totally five levels thereby the features at different resolutions can be extracted successfully. Finally every action video is represented with a Fused Feature Vector which is formed by the combination of filter maps derived through these two filters.

3.2.1. Gabor Features

Gabor Filter is one of best prevalent filter which has a widespread application sin image oriented aspects. A 2-D Gabor filter can be viewed as a sinusoidal signal of specific frequency and orientation by a Gaussian wave. The Gabor filter has real and imaginary components representing orthogonal directions. These two forms may form a complex number or they can also be used individually. However, here we have represented real, complex and imaginary terms, as shown below in the equations (1),(2) and (3) respectively;

Real:

$$G\left(x,y,\lambda,\theta,\sigma,\gamma\right) = \exp\left(-\frac{X^2 + \gamma^2 Y^2}{2\sigma^2}\right)\cos\left(\frac{2\pi}{\lambda}X\right) \tag{1}$$

Imaginary

$$G\left(x,y,\lambda,\theta,\sigma,\gamma\right) = \exp\left(-\frac{X^2 + \gamma^2 Y^2}{2\sigma^2}\right)\sin\left(\frac{2\pi}{\lambda}X\right) \tag{2}$$

Complex:

$$G\left(x,y,\lambda,\theta,\sigma,\gamma\right) = \exp\left(-\frac{X^2 + \gamma^2 Y^2}{2\sigma^2}\right)\exp\left(i\left(\frac{2\pi}{\lambda}X\right)\right) \tag{3}$$

Where

$$X = x\cos\theta + y\sin\theta \tag{4}$$

$$Y = -x\sin\theta + y\cos\theta \tag{5}$$

The equations (4) and (5) refers to the translated points (X,Y) Where $\left(x,y\right)$ is position relative to pixel coordinates of the key frame.

In the above equations

λ = wavelength of sinusoidal component
θ = Orientation of Gabor function
σ = standard deviation of Gaussian envelope
γ = spatial aspect ratio

These four parameters are called control parameters that control the size and shape of Gabor function. Gabor filter is more advantageous because it helps in the representation of an action image in different orientations, means it study and analyzes the action image in different angles. In this contribution, we have employed the Gabor filter at six different scales and at eight different orientations. Let's

$$S = \left\{5\times 5, 7\times 7, 9\times 9, 11\times 11, 13\times 13, 15\times 15\right\}$$

and

$$\theta = \left\{ 0^0, 45^0, 90^0, 135^0, 180^0, 225^0, 270^0, 315^0 \right\},$$

the max-pooling procedure for key feature maps extraction at different orientations is represented in equation (6) as;

Figure 2. (a) Original Action Image (b) Scaled image at the scale of 5×5

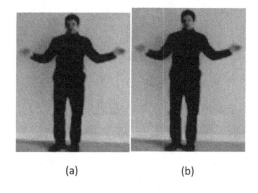

(a)   (b)

Figure 3. Gabor feature maps of a scaled action image at an orientation of (a) 0^0, (b) 45^0, (c) 90^0, (d) 135^0, (e) 180^0, (f) 225^0, (g) 270^0, and (h) 315^0

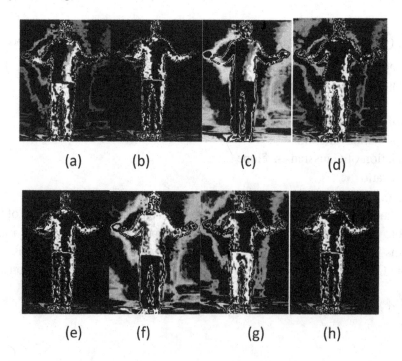

(a) (b) (c) (d)

(e) (f) (g) (h)

$$F_{max,i} = \max_{\substack{i=1\,to\,length(\theta) \\ j=1\,to\,length(S)}} (x,y)\left\{f_{S_i}\left(x,y,\theta_j\right)\right\}$$

(6)

Where

f_{S_i} = feature map at i_{th} orientation and

θ_j = the j_{th} scale.

Figure 2(a) shows the original action image which has been scaled to 5 x 5 in Figure 2(b).

Figure 3 represents the Gabor feature maps of a scaled action image at an different orientation of (a) 0^0, (b) 45^0, (c) 90^0, (d) 135^0, (e) 180^0, (f) 225^0, (g) 270^0, and (h) 315^0 respectively.

3.2.2. Complex Wavelet (Contour) Features

The Wavelet features are more important in determining the pose of Human action. A complete and effective contour itself provides efficient information about the Human action. DWT is a most popular and effective WT that represents an action image by assuming the linear relations with basic functions. For a given action image, the DWT decomposes it into four sub bands. They are namely;

1. Approximations (CA),
2. Verticals (CV),
3. Horizontals (CH), and
4. Diagonals (CD).

The main advantage with the wavelet function is the extraction of edge information from the image. Figure 4 shows the block schematic of DWT and the sub bands obtained after the decomposition of an action image through DWT is demonstrated in Figure 5. Figure 5 (a) shows the image of sub bands derived after the decomposition of an image with an action of Wave hand and the next Figure 5 (b) the DWT bands of boxing actions. For both action images, we have applied the three level decomposition and hence we have obtained totally eight sub bands.

IV. SIMULATION EXPERIMENTS

For the simulation purpose, we used the MATLAB software followed by image processing toolbox and statistics toolbox. For input data purpose, we have referred two action datasets; they are Weizmann action dataset and KTH action dataset. The entire simulation is done with the help of some predefined commands those are available in MATLAB and at some instances we have coded through statistical toolbox commands. Since there is an availability of multiple toolboxes in MATLAB, we have used it for simulation and realized the entire concept. To do the simulation, initially we need to prepare the dataset, means we have to make ready the actions videos. Under these settings, we focus on the provision of number of training and testing action videos. In general, the training is employed through 70%

Figure 4. Schematic of 2D DWT

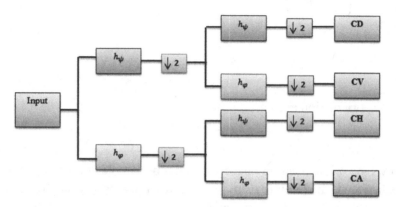

Figure 5. Action image after decomposition through DWT (a) Hand Wave action and (b) Boxing action

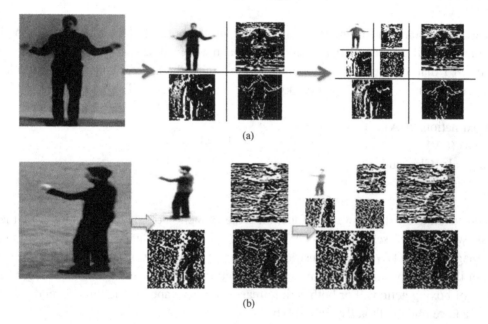

dataset and testing is employed over the 30% dataset if the available dataset is of 100%. Here also we employed the same process.

4.1 Results on Weizmann and KTH Datasets

Here in the present section, we initially explain the details of KTH dataset settings and then the details of Weizmann dataset. Under this dataset, we have considered totally six different actions. They are namely;

1. Boxing,
2. Jumping,
3. Running,

4. Walking,
5. Handclapping and
6. Hand wave.

Figure 6 shows some examples of KTH dataset and Figure 7 shows some examples of Weizmann dataset. The total number of action categories present in the Weizmann dataset are ten; they are;

1. Jump forward on two legs,
2. Jumping jack,
3. Skipping,
4. Running,
5. Walking,
6. Bending,
7. Wave sigle hand,
8. Handwaving,
9. Gallopsideways,
10. Jump in the same location with two legs.

The results obtained after the simulation of developed approach over KTH dataset is shown in table 1 and table 2. Next the results obtained after the simulation of developed approach over KTH dataset is shown in table 3.

Figure 6. KTH action samples

Figure 7. Weizmann action samples

Table 1 shows the confusion matrix of the KTH dataset. The values shown in this table are obtained after the simulation of multiple runs over the KTH dataset. The proposed method is simulated through KTH dataset for several times and at every simulation the recognized results are noticed. Finally all these results are averaged and they are represented in the form of confusion matrix in Table 1. From these values, the performance is measured and the performance metrics are shown in Table 2. From this table, we can see that the maximum recall is observed at Walking action while the minimum recall is observed at running action.Next the maximum precision is observed at Hand waving action while the minimum precision is observed at jogging action. The maximum F-Score is noticed at Hand waving action and minimum F-score is noticed at Jogging action. The next two performance metrics such as FNR and FDR are simply inversely related to the recall and PPV and hence the values are obtained by subtracting them from 100. Thus the minimum FNR is observed at walking action and maximum FNR is observed at running action. Next the minimum FDR is observed at hand waving action while the maximum FDR is observed at Jogging action.

Table 1. Confusion matrix for over KTH dataset

	Walk	Jump	Run	Box	Hand Clap	Hand Wave	Total
Walk	**188**	13	9	0	0	0	210
Jump	12	**97**	11	0	0	0	120
Run	17	18	**145**	0	0	0	180
Box	0	0	0	**216**	20	4	240
Hand Clap	0	2	4	18	**154**	2	180
Hand wave	0	8	8	4	9	**211**	240
Total	217	138	177	238	183	217	**1170**

Table 2. Performance Metrics for the simulation of proposed approach over KTH dataset

Action / Metric	Recall (%)	PPV (%)	F-Score (%)	FDR (%)	FNR (%)
Walking	90.2541	90.4312	90.3426	9.5688	9.7459
Jogging	87.5021	75.0012	80.7708	24.9988	12.4979
Running	83.8956	83.4363	83.6653	16.5637	16.1044
Boxing	90.1245	87.8023	88.9482	12.1977	9.8755
Hand Clapping	86.6745	85.7114	86.1903	14.2886	13.3255
Hand Waving	85.8396	97.1245	91.1340	2.8755	14.1604

Table 3 shows the confusion matrix of the Weizmann dataset. This matrix is constructed based on the values obtained after the simulation of proposed action recognition model over the Weizmann dataset for three times cross validation. In the first validation, we have considered the first three subjects (S1, S2 and S3) for testing and the remaining six subjects (S4, S5, S6, S7, S8, and S9) for training. In the 2nd validation, the second three subjects (S4, S5 and S6) are considered for testing and the remaining six subjects (S1, S2, S3, S7, S8, and S9) for training. In the 3rd validation phase, the last three subjects

Table 3. Confusion matrix after the simulation of proposed action recognition model over Weizmann Dataset

	A1	A2	A3	A4	A5	A6	A7	A8	A9	A10	Total
A1	**248**	4	3	5	4	0	4	2	0	0	270
A2	0	**344**	10	10	5	0	0	0	6	30	405
A3	0	9	**117**	9	0	0	0	0	0	0	135
A4	0	14	14	**232**	3	0	3	0	4	0	270
A5	0	0	0	0	**117**	5	10	3	0	0	135
A6	0	2	2	6	10	**108**	0	7	0	0	135
A7	0	3	3	3	15	0	**101**	10	0	0	135
A8	3	4	3	0	10	8	5	**237**	0	0	270
A9	3	10	8	10	0	0	0	0	**364**	10	405
A10	0	10	4	3	0	0	0	0	10	**243**	270
Total	254	400	164	278	164	121	123	259	384	283	**2430**

(S7, S8 and S9) are considered for testing and the remaining six subjects (S1, S2, S3, S4, S5, and S6) for training. After each cross validation, we have constructed a confusion matrix based on the recognized actions from the given input actions. The confusion matrix shown in Table 3 is an average of three confusion matrix obtained at individual validations. Based on the results observed in Table 3, the performance is measured and the performance metrics are shown in Table 4. From this table, we can see that the maximum recall is observed at Jump forward on two legs action while the minimum recall is observed at wave single hand action. Next the maximum precision is observed at Jump forward on two legs action while the minimum precision is observed at two actions namely skipping and walking. The maximum F-Score is noticed at Jump forward on two legs action and minimum F-score is noticed at two actions namely skipping and walking. The next two performance metrics such as FNR and FDR

Table 4. Performance Metrics for the simulation of proposed approach over Weizmann dataset

Action / Metric	Recall (%)	PPV (%)	F-Score (%)	FDR (%)	FNR (%)
Jump forward on two legs	91.8518	97.6377	94.6564	8.1481	2.3622
Jumping jack	84.9382	86.0000	85.4658	15.0610	14.0000
Skipping	86.6666	71.3414	78.2608	13.3330	28.6585
Running	85.9259	83.4532	84.6715	14.0740	16.5467
Walking	86.6666	71.3414	78.2608	13.333	28.6585
Bending	80.0000	89.2561	84.3750	20.0000	10.7438
Wave single hand	74.8148	82.1138	78.2945	25.1851	17.8861
Hand waving	87.7777	91.5057	89.6030	12.2222	8.49420
Gallop side ways	89.8765	94.7916	92.2686	10.1234	5.20833
Jump in the same location with two legs	90.0000	85.8657	87.8842	10.0000	14.1342

Figure 8. Comparison between existing and prosposed approaches through individual average Accuracy of different actions of KTH Dataset

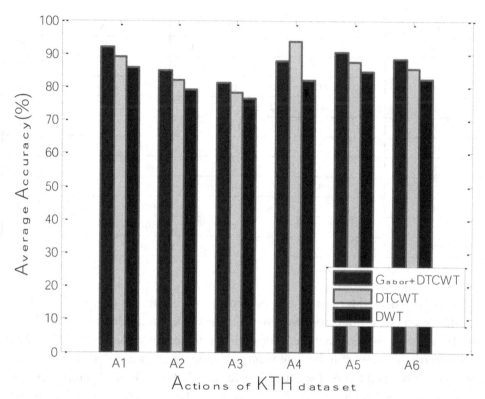

are simply inversely related to the recall and PPV and hence the values are obtained by subtracting them from 100. Thus the minimum FNR is observed at Jump forward on two legs action and maximum FNR is observed at Wave single hand action. Next the minimum FDR is observed at Jump forward on two legs action while the maximum FDR is observed at two actions namely skipping and walking.

On an average, for KTH actions, the proposed recognition model gained an average acuracy of 87.4888% while for the existing models, it is observed as 85.6631% and 81.7075% for DTCWT and DWT respectively graphically shown in Figure 8. The average of average accuracy of the actions presnet in weizmann action datatset is observed as 85.8519%, 83.1369% and 80.1336% for Gabor + DTCWT, DTCWT and DWT respectively shown in Figure 9.

4.2 Results on UCF 11 Action Dataset

UCF 11 is a very challenging dataset because it has so many variations in pose, scaling, viewpoints, cluttered backgrounds, illuminations, appearance and camera motion. For every action category, the videos are grouped into 25 groups and every group is composed of more than four clips. The grouping is done based on the observation of similar characteristics of action videos. The common features those considered for grouping are viewpoint similarity, background similarity, and same actor and so on. All the Videos present in this dataset are in the .MPEG4 format. The name of actions present in this dataset is

Figure 9. Comparison between existing and prosposed approaches through individual average Accuracy of different actions of Weizmann Dataset

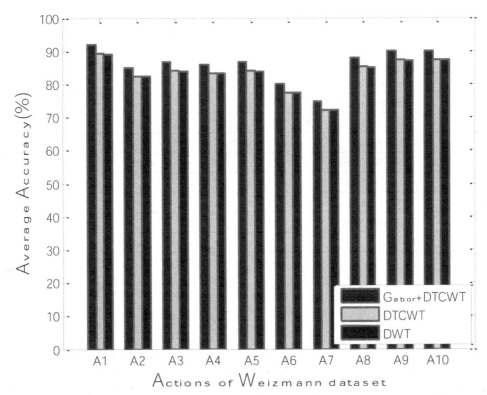

1.   Horseback riding (HR),
2.   Golf swinging (GS),
3.   Diving (DV),
4.   Biking/cycling (B/C),
5.   Basketball shooting (BS),
6.   Swinging (SW),
7.   Soccer juggling (SJ),
8.   Trampoline jumping (TJ),
9.   Tennis swinging (TS),
10. Walking with dog (WWD), and
11. Volleyball spiking (VS).

After the grouping of actions with similar characteristic into 25 groups, we have processed them for training and testing. For every action category, the Training and testing settings are done as shown in Table 5

During the simulation experiments, we have conducted three case studies by varying the K values of K-NN algorithm. Every simulation is completed in in two stages; they are

Table 5. UCF 11 Dataset Settings

S. No	Action Name	Training	Test	Total
1	Horse Riding	139	59	198
2	Golf Swing	99	43	142
3	Diving	109	47	156
4	Biking / Cycling	102	43	145
5	Basketball shooting	99	42	141
6	Swinging	132	57	189
7	Soccer Juggling	109	47	156
8	Trampoline Jumping	83	33	116
9	Tennis Swing	117	50	167
10	Walk with dog	86	37	123
11	Volleyball Spiking	81	35	116
Total		**1156**	**493**	**1649**

1. Training and
2. Testing.

In the first stage different actions are trained according to the details described under dataset settings section. For every action video, initially we have extracted the three features such as Gaussian, and Gabor features. After extraction of the features, we have constricted a Fused Feature Vector. After the determination of optimal and discriminative features from every action video, they are subjected to classification through K-NN algorithm. After the completion of training of different actions, then we employ testing process in which the action videos are processed for testing one-by-one. After the completion of simulation, the obtained results are shown below;

Table 6 shows the results of recognized actions both correctly and falsely. The values demonstrated in the Table 6 are the results obtained at the simulation of K-NN algorithm. The numbers depicted at along the diagonal are called as TPs, means the total number of correctly recognized actions. In a row, the summation of all values except the highlighted value determines the FN value, means the total number of incorrectly recognized actions. In the same manner, in the column, the summation of all values except the TP determines the FP, means incorrectly recognized actions. Based on these values, we can calculate the above specified performance metrics such as False Positive Rate, False Negative Rate, F-Score, precision and detection rate or recall.

The performance metrics shown in Table 7 are derived from the results shown in confusion matrix in Table 6. From this table, we can see that the maximum Recall rate is observed for Golf swinging action and it is approximately 81.3953% while the minimum recall is observed for Soccer Juggling and it is approximately 48.9361%. Next the maximum precision is observed at Soccer Juggling and it is approximately 88.4615% while the minimum precision is observed at Biking or Cycling and it is approximately 46.0317%. The next metric is F-score which is nothing but the harmonic mean of recall and precision. The maximum F-score is observed at Golf swinging and it is approximately observed as 80.4597% while the minimum F-score is observed at Basketball shooting and it is approximately 53.1645%. The

Table 6. Confusion Matrix of the Different Action Classes

	WWD	VS	TJ	TS	SW	SJ	HR	GS	DV	B/C	BS	Total
WWD	**26**	3	0	3	0	0	0	0	3	2	0	**37**
VS	4	**24**	0	5	0	0	0	0	0	2	0	**35**
TJ	0	2	**25**	0	4	1	1	0	0	0	0	**33**
TS	2	1	0	**38**	0	0	0	0	0	3	6	**50**
SW	0	6	10	0	**30**	0	0	0	0	8	3	**57**
SJ	0	0	0	4	6	**23**	2	0	2	8	2	**47**
HR	0	5	3	2	0	0	**40**	0	0	9	0	**59**
GS	0	1	1	0	2	0	0	**35**	0	0	4	**43**
DV	0	5	0	0	2	2	0	0	**36**	1	1	**47**
B/C	1	0	1	0	1	0	5	3	3	**29**	0	**43**
BS	3	3	0	5	2	0	0	6	1	1	**21**	**42**
Total	36	50	40	57	47	26	48	44	45	63	37	**493**

Table 7. Performance Analysis of Proposed Approach

Action	Recall	Precision	F-Score	FNR	FPR
WWD	70.2702	72.2222	71.2328	29.7297	27.7777
VS	68.5714	48.0000	56.4705	31.4285	52.0000
TJ	75.7575	62.5000	68.4931	24.2424	37.5000
TS	76.0000	66.6666	71.0280	24.0000	33.3333
SW	52.6315	63.8297	57.6923	47.3684	36.1702
SJ	48.9361	88.4615	63.0136	51.0638	11.5384
HR	67.7966	83.3333	74.7663	32.2033	16.6666
GS	81.3953	79.5454	80.4597	18.6046	20.4545
DV	76.5957	80.0000	78.2608	23.4042	20.0000
B/C	67.4418	46.0317	54.7169	32.5581	53.9682
BS	50.0000	56.7567	53.1645	50.0000	43.2432

further performance metrics such FNR and FPR are just follows inverse relation with recall and preci-sion respectively. Thus the maximum recall gained action have minimum FNR and maximum precision gained action minimum FPR.

For the effectiveness alleviation of proposed approach, we conduct a simulation with different meth-ods as well with differed set of action videos. At this simulation, we divided the entire dataset into three groups and simulation is accomplished for three times, for every simulation we used one set. At each simulation set, we employed three sub simulations with three different set of features, they are Wavelet features, complex wavelet features and Gabor with complex wavelet features. Thus totally we conduct nine simulation studies and at each simulation we measured the recall rate of proposed approach and the observed recall is demonstrated in Figure 10. From this simulation we observed that there is an impact of

Figure 10. Recall at different training sets

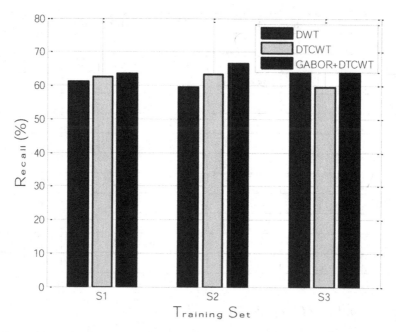

Figure 11. average accuracy of different actions of UCF 11 dataset

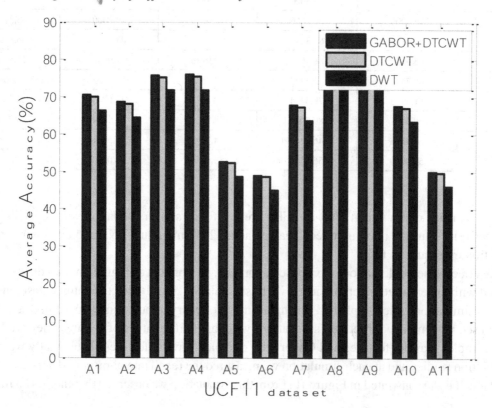

Figure 12. average False Alarm Rate of different actions of UCF 11 dataset

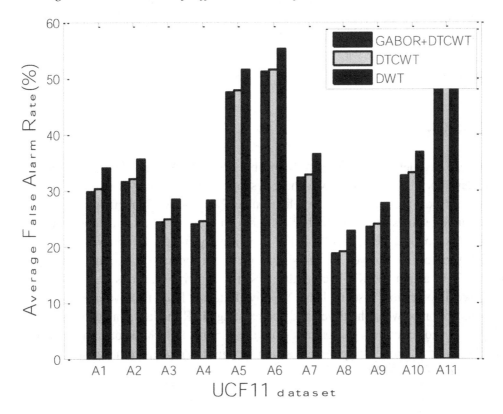

the action videos as well as features on the action recognition. Form this Figure 10; we can see that for all set of simulation, the fused features have gained maximum recall compared to the individual features. On an average, the average accuracy obtained is observed as 65.4171% for Gabor with complex wavelet features while it is of only 63.2314% and 61.4457% for complex wavelet features and wavelet features.

Figure 11shows the accuracy results of different actions of UCF 11 action dataset. The values plotted in this figure are obtained after averaging the accuracy in three simulation studies. The three simulation studies vary with set of action videos trained. For three different set of action videos, we extract initially intensity features only and they are used for action recognition through K-NN. In the next three simulation studies, we extract initially orientational features only and they are used for action recognition through K-NN. In the last three simulation studies, extract both features and fused them. Then the fused vector is processed to K-NN for action recognition. From the above figure, we can observe that the maximum accuracy is obtained for golf swinging and it is approximately 78.6635%. This is obtained when the fused feature vector is used. Next the maximum accuracy is gained for Diving action when only orientational features are used as features and the maximum accuracy is gained for Tennis Swinging when only intensity features are used as features. On an average, the overall accuracy gained at fused feature is observed as 68.5689% while for orientational features and intensity features it is observed as 64.2315% and 62.3369% respectively.

Figure 12 shows the average False Alarm Rate (FAR) results of different actions of UCF 11 action dataset. Generally the False Alarm Rate is measured as the average of False Negative Rate and False Positive Rate. The evaluation of FAR is similar to the process of accuracy evaluation. Instead of considering the two metrics such as False Negative Rate and False Positive Rate we calculated one measured by combining those two. The False Alarm Rate is indirectly shown the false recognition capability of proposed method. On an average, the overall False Alarm Rate gained at fused feature is observed as 33.1458% while for complex wavelet features and wavelet features it is observed as 35.7688% and 33.6348% respectively.

V. CONCLUSION

In this chapter, we have demonstrated our first contribution towards the development of an effective HAR system. The developed HAR model is two phase system in which initially the action videos are trained and then tested. In both phases, we have employed a newly proposed FE method to derive the effective and discriminative features through which the system can recognize human actions more accurately. The FE method is a hybrid method, collected of two different filters; they are wavelet filter and Gabor filter. SVM algorithm is employed for classification. Experimental validation is done with the help Weizmann, UCF 11 action and KTH action datasets and the performance is measured through several metrics like Recall, precision, F-score, False Negative Rate and False positive rate. Further we also did a detailed comparison between the developed and existing HAR models. On an average, the Accuracy obtained by proposed action recognition model on the simulation of KTH dataset is 88.5624%. On an average, the Accuracy obtained by proposed action recognition model on the simulation of Weizmann dataset is observed as 87.4512%.

REFERENCES

Azim & Hemayed. (2015). Human action recognition using trajectory-based representation. *Egyptial Informatics Journal, 16*(2), 187-198.

Bay, H., Ess, A., Tuytelaars, T., & Van Gool, L. (2008). Speeded-up robust features (SURF). *Computer Vision and Image Understanding, 110*(3), 346–359. doi:10.1016/j.cviu.2007.09.014

Bobick, A., & Davis, J. (2001). The recognition of human movement using temporal templates. *IEEE Transactions on Pattern Analysis and Machine Intelligence, 23*(3), 257–267. doi:10.1109/34.910878

Bregonzio, Gong, & Xiang. (2009). *Recognizing action as clouds of space-time interest points.* CVPR.

Davidson, M. W., & Abramowitz, M. (2015). *Molecular Expressions Microscopy Primer: Digital Image Processing – Difference of Gaussians Edge Enhancement Algorithm.* Olympus America Inc., and Florida State University.

Dhulekar, P. A., & Gandhe, S. T. (2018). Action recognition based on Histogram of Oriented gradients and Spatio-temporal interest points. *IJET, 7*(4), 2153–2160. doi:10.14419/ijet.v7i4.17274

Dollar, Rabaud, Cottrell, & Belongie. (2005). Behavior recognition via sparse Spatio-temporal features. *ICCV VS-PETS.*

Gdyczynski, C. M., Manbachi, A., Hashemi, S. M., Lashkari, B., & Cobbold, R. S. C. (2014). On estimating the directionality distribution in pedicle trabecular bone from micro-CT images. *Journal of Physiological Measurements.*, *35*(12), 2415–2428. doi:10.1088/0967-3334/35/12/2415 PMID:25391037

Gorelick, L., & Shechtman, E. (2007, December). Actions as Space-Time Shapes. *IEEE Transactions on Pattern Analysis and Machine Intelligence*, *29*(12), 2247–2253. doi:10.1109/TPAMI.2007.70711 PMID:17934233

Jain, J´egou, & Bouthemy. (2013). Better exploiting motion for better action recognition. *CVPR*.

Jiang, J., Jiang, T., & Zhai, S. (2014). A novel recognition system for human activity based on wavelet packet and support vector machine optimized by improved adaptive genetic algorithm. *Physical Communication*, *13*(Part C), 211–220. doi:10.1016/j.phycom.2014.04.006

Keestra, M. (2015). Understanding human action: Integrating meanings, mechanisms, causes, and contexts. In Interdisciplinary Research: Case Studies of Integrative Understandings of Complex Problems. Sage Publications.

Khare & Jeon. (2017). Towards Discrete Wavelet Transform-based human activity recognition. *Proc. SPIE 10443, Second International Workshop on Pattern Recognition*, 19.

Larson, D. R. (2007). Wavelet Analysis and Applications (See: Unitary systems and wavelet sets). *Appl. Numer. Harmon. Anal. Birkhäuser*, 143–171.

Liu, J., Luo, J., & Shah, M. (2009). Recognizing Realistic Actions from Videos "in the Wild". *IEEE International Conference on Computer Vision and Pattern Recognition (CVPR)*. 10.1109/CVPR.2009.5206744

Lyons, M., Akamatsu, S., Kamachi, M., & Gyoba, J. (1998). *Coding facial expressions with Gabor wavelets*. doi:10.1109/AFGR.1998.670949

Moghaddam, H. A. (2019). Spatiotemporal wavelet correlogram for human action recognition. *International Journal of Multimedia Information Retrieval*, *8*(3), 167–180. doi:10.100713735-018-00167-2

Raju, Sharma, & Devender. (2020). Composite Feature Vector Assisted Human Action Recognition through Supervised Learning. *International Journal of Recent Technology and Engineering, 8*(6).

Raju, S. S., Pati, P. B., & Ramakrishnan, A. G. (2004). Gabor filter based block energy analysis for text extraction from digital document images. *First International Workshop on Document Image Analysis for Libraries, 2004. Proceedings*, 233–243. 10.1109/DIAL.2004.1263252

Sabari Raju, S., Pati, P. B., & Ramakrishnan, A. G. (2005). Text Localization and Extraction from Complex Color Images. *Proc. First International Conference on Advances in Visual Computing (ISVC05), 486-493.*

Schuldt, C., Laptev, I., & Caputo, B. (2004). Recognizing Human Actions: A Local SVM Approach. *Proc. ICPR'04*. 10.1109/ICPR.2004.1334462

Shen, Y., & Miao, Z. (2010). Oriented Gradients for Human Action Recognition. *ICIMCS*, *10*(December), 30–31.

Thanikachalam & Thyagarajan. (2012). Human Action Recognition using Accumulated motion and gradient of motion from video. *ICCCNT-2012*.

Tripathi, V., Gangodkar, D., Mittal, A., & Kanth, V. (2017). Robust Action Recognition framework using Segmented Block and Distance Mean Histogram of Gradients Approach. *Procedia Computer Science*, *115*, 493–500. doi:10.1016/j.procs.2017.09.094

Vishwakarma, D. K., & Dhiman, C. (2019). A unified model for human activity recognition using spatial distribution of gradient and difference of Gaussian kernel. *The Visual Computer*, *35*(11), 1595–1613. doi:10.100700371-018-1560-4

Vishwakarma, D. K., Rawat, P., & Kapoor, R. (2015). Human Activity Recognition Using Gabor Wavelet Transform and Ridgelet Transform. *Procedia Computer Science*, *57*, 630–636. doi:10.1016/j. procs.2015.07.425

Wang, H., Klaaser, A., Schmid, C., & Liu, C. L. (2013). Dense trajectories and motion boundary descriptors for action recognition. IJCV, 103(60-79). doi:10.100711263-012-0594-8

Wang, H., Kl̈aser, A., Schmid, C., & Liu, C. L. (2011). Action Recognition by Dense Trajectories. *IEEE Conference on Computer Vision & Pattern Recognition*, 3169–3176. 10.1109/CVPR.2011.5995407

Wang, H., Oneata, D., Verbeek, J., & Schmid, C. (2015). *A robust and efficient video representation for action recognition*. IJCV.

Wang, H., & Schmid, C. (2013). Action recognition with improved trajectories. In *IEEE International Conference on Computer Vision*, Sydney, Australia.

Weinland, D., Ronfard, R., & Boyer, E. (2006). Free viewpoint action recognition using motion history volumes. *Computer Vision and Image Understanding*, *104*(2-3), 249–257. doi:10.1016/j.cviu.2006.07.013

Chapter 2
Unsupervised Summarization Approach With Computational Statistics of Microblog Data

Abhishek Bhattacharya
Institute of Engineering and Management, Kolkata, India

Arijit Ghosal
St. Thomas' College of Engineering and Technology, India

Ahmed J. Obaid
University of Kufa, Iraq

Salahddine Krit
 https://orcid.org/0000-0003-3868-472X
Ibn Zohr University, Morocco

Vinod Kumar Shukla
Amity University, Dubai, UAE

Krishnasis Mandal
Miles Technologies, India

Sabyasachi Pramanik
 https://orcid.org/0000-0002-9431-8751
Haldia Institute of Technology, India

ABSTRACT

Microblogging, where millions of users exchange messages to share their opinions on different trending and non-trending topics, is one of the popular communication media in recent times. Several researchers are concentrating on these data due to a huge source of information exchanges in online social media. In platforms such as Twitter, dataset-generated lacks coherence, and manually extracting meaning or knowledge from them proves to be painstakingly difficult. It opens up the challenges to the researchers

DOI: 10.4018/978-1-7998-7701-1.ch002

for knowledge extraction driven by a summarization approach. Therefore, automated summary genera-tion tools are recommended to get a meaningful summary out of a given topic becomes crucial in the age of big data. In this work, an unsupervised, extractive summarization model has been proposed. For categorization of data, k-means algorithm has been used, and based on scoring of each document in the corpus, summarization model is designed. The proposed methodology achieves an improved outcome over existing methods, such as lexical rank, sum basic, LSA, etc. evaluated by rouge tool.

INTRODUCTION

In recent time, Twitter is recognised as one of the well accepted Microblogging platform which started in 2006. More than 340 million Tweets are daily posted by more than 500 million of users. The majority of the posts short messages as microblog accept limited text which is known as Tweets which are informal or not particularly significant, about 3.6% (Wikipedia, 2021) of the Tweets represent topics of conven-tional news. During the time of any important event an enormous set of posts are made between a large numbers of users. This huge rate of message exchange causes information overload. Twitter provides a list of popular topics which are known as Trending Topics in chronological order. For researchers, Twitter makes available an API to download data by streaming approach in chronological order that contains a topic (hashtag)(REST API Resources, 2021)(Sandy Hook Elementary School shooting, 2012). All the data are sorted by recency irrespective of relevancy. If anybody has to understand any topic, the user has to go through all the tweets which are almost a tedious job. This is one of the challenging areas in microblog analysis. So, several researchers are currently focusing on Tweet summarization to handle this information retrieval in a more efficient way.

Huge range of users is posting millions of messages in social media during any important events. And not only valid users, spammers are also using the same platforms to spared spam content and fake news as well. There are two different types of summarization, abstractive and extractive. Abstractive summarization is the process of reducing large text content into short text/paragraphs that convey a con-cise meaning of the original content. The large content needs to be made concise to make out meaning from the huge amount of data. Extractive summarization is the process to identify selected portion text which can represent the entire database. In several text summarization research works a large number of researches are carried out, and mostly on microblog dataset. Microblog data is redundant and irrel-evant in nature, therefore, to summarize any huge dataset redundant and similar types of data need to be categorized. So, clustering on this social media that is another correlated challenge comes with sum-mary generation. Similar data are posted multiple times so categorization is another challenging task by grouping syntactical and semantic similar content together. The key issue of microblog summarization is characteristically a multi-document summarization problem. Nonetheless, the cutting edge techniques in summarization of single document are additionally important, by utilizing the microblogs as inputs to make up a single document. It has some discrete constraints, fundamentally because of the tiny size of particular microblogs, and the casual representation of texts in microblogs, that turns it complex to decipher the semantic similarities of microblogs. The earlier state of the art research shows that various researches had been carries out such as – (1) Cluster-rank, (2) COWTS, (3) Frequency Summarizer, (4) LexRank, (5) LSA, (6) LUHN, (7) Mead, (8) SumBasic and (9) SumDSDR. These algorithms mostly select dissimilar sets of tweets in the output summaries. Therefore, different summarization methods typically choose diverse groups of tweets in the summaries for the similar set of input tweets. The distinctive summarization techniques are probably going to assess the general significance of tweets dependent on various components, and thus different rundowns are probably going to reveal features of the input dataset.

In this work, an extractive summarization method is proposed based on unsupervised categorization technique. The search of any non-trending/ trending subject in Twitter, results in thousands of tweets which can be categorized into a few sub-topics using clustering methods. Then from each subtopic most significant tweets are identified to generate the summary which can give a precise overview of the searched topic. The proposed method has been experimented on five recent disaster events. Then to compare the performance evaluation of the proposed method, the experiments are also performed on above mentioned nine different summarization algorithms. The result shows that the quantity of normal tweets is mostly unique summaries from a similar information set of microblogs, with unique tweets in common between the summaries delivered by various methods. In addition, the performance of the different algorithms is evaluated using the standard ROUGE measure for comparison. The result table shows the outcome analysis of different algorithms. The rest of the paper is planned as follows; Section 2 includes related works on clustering. Section 3 illustrates the summarization of microblogging dataset and proposed summarization algorithm. Section 4 elaborates the outcome and the comparison among the different summarization approaches. In Section 5 the work concludes with a few prospective future study aspects.

LITERATURE REVIEW

Several researchers are currently focused on the Summarization problem. Automatically summarizing Micro-Blog topics is a challenge of online social media analysis where the dataset is too huge, noisy and too much. Some summarization algorithms have been used in recent literature. For the identification of summary content, famous algorithms were pro- posed by (Radev et al., 2004) in SumBasic (Lin, 2004) and the centroid method. SumBasic predicts the presence of inclusion of a word in human summary with higher probability which occurs more frequently across the documents. Whereas, the centroid algorithm assessed a centrality measure of each sentence to identify the correlation to the complete subject of the document cluster. Analysing word frequency (Luhn, 1958) proposed a summarization method extracting information from technical articles. Using a hierarchical text graph-based approach TOPIC system (Reimer et al., 1988) generates text on the topics to generate summaries. In recent years, for Microblog summarization various research works have been proposed by several authors (Olariu, 2014),(Shou et al., 2013), (Wang et al., 2015),(Zubiaga et al., 2012). In this work, (Shou et. Al., 2013) discussed a novel approach based on first clustering similar tweets. Using feature selection representative tweets are identified from each cluster which are further ranked based on significance via a graph-based approach (LexRank) (Hassel et al., 2007). An abstractive graph-based summarization has been proposed by (Olariu, 2014) identification of bigrams from the tweets which are mapped as the graph-nodes. A few other authors also used graph based approach for tweet summarization (Dutta et al., 2015) (Xu et al., 2013) .(Kupiec et al., 1995) suggested statistical methods based on supervised learning in summarization using Naive

Bayes classifier on scientific documents. Using Infomap, community detection technique, (Soumi et al., 2015) discussed a graph-based tweet summarization algorithm. The SCISOR system (Rau et sl., 1989) proposed a conceptual summarization method based on a concept graph. Using evolutionary approach (Soumi et al., 2017b) proposed an unsupervised method for twee categorization. SumFocus is proposed by (Lin, 2004) to focus on the issue topic changing during summary generation. To handle the information overloading in Micro-Blog data, (Soumi et al., 2017a) presented a clustering methodology based on feature selection which can improves the efficiency of clustering as well as summarization. Authors in (Brin et al.,1998) proposed TextRank which is another graph-based approach which finds the most highly ranked sentences considering an adjacency matrix. For the positioning of sentences, PageRank technique (Page et al., 1999) has been utilized in this strategy. Assessing the overall significance of sentences and text units in a document, the LexRank method creates a adjacency matrix among the textual units utilizing an IDF-changed cosine distance measure (Hoch et al., 1994). For multi-document summarization problem multilingual summarization method, MEAD(Radev et al., 2004) is proposed. Using a script based extraction of relevant information the FRUMP system (Hoch et al., 1994) proposed a summarized model to analyse UPI news stories in about 50 domains. Document summarization (Hassel et al., 2007) is one of the much explored areas in recent years where most researchers have concerted on the extractive summarization techniques rather than abstractive summarization technique. With the growth of the Internet, Web information mining and summarization is a challenging task, which includes web page summarization (Mahesh, 1997)(Sun et al., 2005), blogs (Tigelaar et al., 2010) and discussion forums (Tao et al., 2013)(Hu et al., 2007). The pattern of entity distribution in a text is identified by Barzilay and Lapata(Barzilay et al., 2008) to form an entity-grid representation of discourse. Then summaries are generated representing the text as a set of entity transition sequences. Alongside broad microblog summarization methods, a couple of studies have additionally concentrated explicitly around synopsis of news stories and tweets posted during crisis occasions (Kedzie et al., 2015),(Nguyen et al., 2015), (Rudra et al., 2015). (Shou et al., 2013) propose a strategy dependent on first clustering of analogous tweets and afterward choosing representative tweets from each bunch, at long last positioning these as per significance by means of a graph based approach (Erkan et al, 2004). (Mackie et al., 2014)looked at techniques for extractive summarization of Twitter texts. Two kinds of methods were discussed that select tweets to create summaries– (I) hybrid tf-idf based method and (ii) a cluster based approach.

Summarization Algorithms

In this part, the extractive summarization techniques that are considered for correlation in the current work have been outlined. It is to be mentioned that some of these strategies were initially proposed for summarization of a one document, where the sentences of the given text document are ranked by some significance scores, and afterward some significant sentences are chosen for the summary. These methods can be effectively applied to summarization of set of text documents of microblogs, where each microblog is practically equivalent to a sentence.

1. **Cluster-rank** $\boxed{CR}$:This technique (Hu et al., 207) is a graph based unsupervised methodology which was initially intended for extractive summarization of meeting records. ClusterRank method is the augmentation of another technique called TextRank that is additionally a graphical strategy

for extricating sentences from different news. It initially fragments the record into clusters that are denoted as nodes in the graph. Similarity among all sets of nearby clusters is then estimated, and the duo with the most elevated likeness is converged into one cluster. Subsequently, a centroid-based methodology is utilized to quantify every sentence inside a significant cluster. Pertinence of the sentences is likewise estimated as well as taking care of poorly constructed sentences with high repetition. At long last, the method chooses the most elevated scoring sentence and puts sentences into the summary until the length restraint is fulfilled.

2. **COWTS** $\boxed{CW}$ **:** This method is explicitly intended for making summarization of microblogs that are made during calamity circumstances. This is a strategy of summarization for identifying significant situational data from microblog texts generated during calamity situations. Utilizing syntactic attributes, the classifier initially recognizes situational/non-situational data. Because of utilization of jargon, the classifier works precisely in cross-domain situations, e.g., when the classifier is prepared over tweets posted during previous disasters and afterward conveyed on tweets generated after catastrophe occasion. At that point the situational tweet stream is summed up by words (numerals,verbs, nouns) utilizing an Integer Linear Programming technique (ILP).

3. **Frequency Summarizer** $[FS]$ **:** It is a basic technique of summarization that endeavours to extricate a split of sentences that cover the primary subjects of a given document. The method deals with the basic idea that sentences which include the most intermittent words in the cont aims to rank the sentences in a given corpus according to the frequency of each word in the bag of words, present in the sentences. The more number of words available in the sentence that are contained in the bag of words, the more their rank. These are then sorted according to their ranks and the summary is formed.

4. **LexRank** $\boxed{LR}$ **:** **LexRank**(Hassel et al., 2007) is a graphical stochastic technique for figuring relative significance of textual units in a report. LexRank is an algorithm identical to the TextRank algorithm that involves estimating the importance of a sentence present in the given corpus using random walks and eigenvector centrality. Here, a graph is produced which is made out of all the sentences in the document. Every sentence is addressed as a vertex, whereas the edges mean resemblance amongst sentences in the document. Cosine similarity is calculated as weight of in that said graph by visualizing each sentence as a bag-of-words model. A similarity matrix is created utilizing the comparability measure that can be utilized as a similarity graph among sentences. Edges whose weights are less than the threshold are eliminated and from the subsequent similarity matrix, the key sentences are identified. A technique dependent on Eigenvector centrality is likewise engaged to rank the sentences. Those sentences then are considered for the summary dependent on their significance scores.

5. **LSA** $\boxed{LS}$ **:** LSA (Yew, 2008) is a nonexclusive text summarization technique to detect semantically significant sentences for creating summary. This unsupervised strategy is used for inferring vector space semantic portrayals from huge records, and needn't bother about training or outside information. It is an unsupervised learning algorithm that does not require any training or outside knowledge. LSA uses the context of the corpus and extracts the sentences that have words that are occurring together in the corpus. A high number of related words or a pattern of words indicate

that the sentences are related in a semantic manner. LSA utilizes setting of the documents and identifying data, for example, set of words are utilized together and set of regular words are found in various sentences. Elevated number of normal words amongst sentences denotes that the sentences are further semantically connected. LSA depends on a numerical method which is termed Singular Value Decomposition (SVD) (Erkan et al., 2014) that is utilized to discover the between relations among sentences and words. It i,e the text documentis initially changed over into a grid where every row addresses a word and every column addresses a sentence. Every cell estimates the significance of the word. Then SVD is used on this matrix to select the sentences to produce the summary.

6.   **LUHN** $\boxed{LH}$: Luhn's method (Sun et al., 2005)deals with the instinct that a few words in a report are elucidating of its content, in addition to the sentences that contain on the whole important information in the text are the ones that hold numerous such enlightening words . The words that happen regularly in a record are probably going to be related with the main subject of the document. Be that as it may, an exemption for this perception is stop-words. Henceforth, Luhn projected the possibility of stop-words like determiners, pronouns and prepositions, which don't have a lot of significant worth in illuminating the subject of the document. So he recommended eliminating these words. LH recognized descriptive words utilizing empirically decided low and high frequency thresholds. The high limits sifting through words happen frequently all through the article and the low edges sifting through words that happen too rarely. The excess words in the record are the distinct words, which demonstrate that content that is significant. A 'importance factor' is figured for each sentence, that could be determined for a sentence by organizing the critical words in the sentence by squaring the quantity of huge words and afterward separating by the no. of words. Sentences are recognized as significant and remembered for the synopsis dependent on the importance factor values. This algorithm involves performing a frequency analysis to find out which words are important in the given corpus and then determining if they are significant in the English language. Those words are then used to find out which sentences are important. Luhn proposed to assign more weight to sentences at the beginning of the document or a paragraph.

7.   **Mead** $\boxed{MD}$: Mead (Mackie et al., 2014) is a multi-lingual summarizer that is based on centroid. This is a platform that involves many summarization algorithms like position-based, centroid-based, etc. To begin with, subjects are identified by agglomerative grouping that works over the tfidf vector portrayals of the reports. Then, a centroid-based technique is utilized to distinguish sentences in each group that are integral to the subject of the whole cluster. For every sentence, three exclusive features are registered: its centroid esteem, positional worth, first-sentence overlap. The Composite score is calculated for each sentence is produced as a mix of the three scores. The score is additionally refined in the wake of thinking about conceivable cross-sentence conditions, e.g., rehashed sentences, sequential requesting, source inclinations.) Sentences are at long last chosen dependent on this score.

8.   **SumBasic** $\boxed{SB}$: SumBasic (Lin, 2004) is addressed as a multi-document summarizer that is based on frequency of words. A multinomial distribution function is used by SumBasic to discover the probability of scattering over the words in a sentence. Considering the mean probabilities of ap-

pearance of the words in the sentence, ranks are allocated to each sentence and those top scores sentences is then picked up. Logically, sentence scores and the word probabilities are invigorated until the ideal summary length is accomplished. The updating of word probabilities addresses a trademark strategy to address the overabundance issue in the multi-document input. The words have higher likelihood of consideration in the human summary than words happening less frequently, which occur frequently in the document cluster. So, Sumbasic gives importance to word frequency.

9. **SumDSDR** $\overline{\left[\overline{SD}\right]}$ **:** It is a document summarization framework based on data reconstruction. Given a sentence by term matrix, each row of the matrix is a weighted term frequency vector of a sentence. The summary that is generated by SumDSDR, can be fragmented up into two portions. Firstly, for every sentence in the text, SumDSDR chooses the connected sentences from the candidate set to recreate the specified sentence by learning a reconstruction function for the specified sentences. Next, for the whole document, SumDSDR tries to generate an optimal set of representative sentences to estimate the complete document, by reducing the reconstruction error with a given function. It chooses the most representative sentences that can best recreate the whole archive. Two kinds of recreation (direct and nonnegative) have been presented and create proficient optimization techniques for them. The linear reconstruction issue is addressed utilizing a greedy technique and the nonnegative reconstruction issue is settled utilizing a multiplicative updating.

10. **Integer Linear Programming [IP]:** ILP Integer Linear Programming Summarization algorithm is used to mainly implement abstractive summarization algorithms. Abstractive algorithms aim to form sentences using words that are deemed important in a given corpus.

PROPOSED METHODOLOGY

In this section an extractive ranking based summarization approach has been discussed here. A set of Tweets has been considered as input and as an outcome a subset of the tweets are extracted identifying significant tweets which can signify the whole dataset. Experimental dataset and the proposed methods have been described in this section.

Data Set

In this section detailed descriptions of the dataset are given which are used not only for experiments but also for comparison purposes with other existing methods. Some of the events are selected which are natural disasters in different areas of this world. Therefore, the language and linguistic style in the tweets can be likely to be various as well. A few sample tweets of the experimental dataset is presented in Table 1. Here are the datasets as follows:

1. NEquake – a massive earthquake in Nepal (Wikipedia (2015)
2. THagupit – a strong cyclone code-named Typhoon Hagupit hit Philippines (Wikipedia (2014).
3. UFlood – massive landslide and floods in the Uttarakhand, India (Wikipedia (2012)
4. SHShoot – an assaulter killed 20 children along with 6 adults at the Sandy Hook elementary school in USA [30].

5.   HDBlast – two explosions in Hyderabad, India (Wikipedia (2013).

Table 1. Some sample tweets of the Uttarakhand Flood of India in 2013

Tweet ID	Tweets
1	@ArvindKejriwal: 50,000 people stranded in Gangotri, Kedarnath and other places in Uttarakhand. They're in quite bad shape. I pray to God
2	@PIBIndia: IAF launches operation 'Rahat' to assist stranded pilgrims tourists inUttarakhand HP.https://t.co/8J8n4OLG1u
3	@DrKumarVishwas: Uttarakhand flood helpline numbers: 0135-2710334,0135-2710335, 0135-2710233. Nainital around: 05946-250138
4	UAir Force choppers carry reinforcement material, food to rain-battered Kedarnath https://t.co/aumyhNVMm
5	@BJPRajnathSingh @narendramodiUttarakhand needs help of whole India to reconstruction of whole Chardham Roads bridges Temples
6	Uttarakhand rains: Harbhajan Singh turns counsellor for trapped pilgrims, touristshttps://t.co/YJ2rxtMO6V

Data Pre-Processing

To work with micro-blog dataset, pre-processing is compulsory due to noisy nature. The pre-processing step comprises elimination of stop words, non-textual symbols like @usernames, smileys, smileys, special characters (apart from"#" symbol, as the hashtag shows an important role in micro-blogging). The com rpus is tokenized into words and tweets. The bag of words is formed after omitting the stop words, i.e., words that do not provide meaning to the corpus are removed. Words and tweets are stored in respective data structures for scoring and evaluation. The sample data set consists of posts that have been made on Twitter. Therefore, unnecessary data like link-backs, and acronyms were present in the dataset. These did not provide any meaning to the posts, therefore redundant. Effective summaries cannot be formed from the corpus, if it is not sanitized. During any important event a enormous no. of messages are posted on Twitter which are collected through the Twitter API (Wang et al., 2015) based on keywords for example 'bomb', 'Hyderabad' and 'blast' to assemble those tweets related to Hyderabad Blast and keywords 'shooting' and 'Sandy Hook' were used to gather the tweets for to the SHShoot event. From all the data set duplicate or near duplicate tweets are removed as it has no such effect on tweet summarization which are re-tweeted by multiple users (Shou et al., 2013). words (excluding URLs and standard English stop words) is considered here for each and every tweets in the dataset. To identify the similarity between the tweets Jaccard similarity is considered and duplicate tweets are removed based on similarity factors.

Clustering of Dataset

After pre-procession are clustered accordingly, using the classical clustering approach, K-means. This is an unsupervised machine learning algorithm that aims to create clusters for a given set of input data.

Clusters will allow the model to reduce redundancy in the summary and form a polarized summary. K-means is a technique for vector quantization which is well known for cluster analysis in information mining. It intends to segment n observations into k number of segments in which every observation has a place with the cluster with the closest mean, which acts as the prototype of the cluster. For experimental purposes we have also collected the ground truth from the experts to have an accurate idea of the number of clusters, K. Apart from K -Means, a few different clustering methods also considered, such as DB-Scan and Hierarchical Clustering. As the experimental dataset is considering social media data, so community detection algorithms are also considered. So, Informal and Louvain are also executed to cluster the dataset.

Scoring/Ranking of Words

According to the significance of words present in the cluster scores/ranks are evaluated for each word. Firstly, their ranks in the clusters are taken into account while assigning the score of the tweet. Secondly, it is checked if a given word is a proper noun with the help of POS tagging. If it is a proper noun, then it is given a better point. This ensures that the tweets are given importance on the basis of if they contain the subject of the corpus, and if they are more important in their respective clusters. A scoring function is determined to provide accurate ranks to the tweets in the corpus, for an effective summary to be formed, shown in equation 1.

$$P = 1+(1-P_{f1}) +(1-P_{f2}) + (1-P_{f3})+..+(1-1) = \sum (1-P_{fn}) \tag{1}$$

P_{fn} denotes the rank factor of the word present in the cluster. The rank factor is increased by a constant c for every increasing rank. In this proposed system, for testing purposes c = 0.05. Therefore, the top word of the cluster has a value of 1, the second word in the cluster has a value of 0.95, and so on, until it reaches 0, i.e., $P_{f1} = 0$, $P_{f2} = 0.05$ and $P_{fn}= 1$, in the test model.

Determining the Summary

In this step, top ranked/scored tweets are selected from each cluster to be extracted from the corpus and be included in the summary. Selection is based on the individual scores of the tweets present in the corpus. These tweets, as mentioned, have been scored according to their importance in their respective clusters and if it is making a statement about the subject in the corpus. For the test data, the top tweets from two clusters are chosen for the final summary. The summary is effective and polarized, resulting from the step of clustering, shown in equation 2.

$$PK = \sum\nolimits_{N*\mu} \tag{2}$$

There, P = Score of individual sentences in the corpus. This is the score that has been assigned to each sentence in the corpus. As aforementioned, this is determined by the rank of the word in the cluster, and if the sentence has a proper noun.

N = Number of sentences in the corpus. This is simply the number of sentences that have been tokenized and stored in a data structure.

μ = Cut off factor. This is used to increase or decrease the cut off of the final summary in the model. If it is more, lesser sentences are deemed important, and if it is less, then more sentences are deemed important. For testing of the system, μ = 2.56. The cut off factor should be increased when using a large dataset and decreased when the corpus is small, to ensure optimum outputs.

The determination factor K is used to decide if a tweet is important enough to be included in the final summary. For the tweet to be considered significant, it has to have a score more than K. If it has a score more than K, then it is more important, therefore, it will be included in the final summary. The core concept of this work, aims to reduce redundancy and include polarization in the summary that will be generated. To achieve this goal, the model design incorporates clustering before the summary is generated. The algorithm for clustering determines a number clusters from a given dataset of tweets. These tweets are tokenized to generate the set of words and raking for each cluster. Based on word ranking, a score is given to each tweet, therefore signifying the importance of the tweet in the corpus. The scoring is based on two factors. Firstly, the rank of the word in the cluster is determined to score the tweet. Score is decreased with decreasing importance of the word in the cluster. Secondly, if the tweet consists of a proper noun, it gets a slightly better score. Therefore, the model has a polarizing summary, which comprises important information, which is actionable and can be used to generate decisions. More importance given to proper nouns ensures that the subject or subjects in the corpus are given more importance.

EXPERIMENTS AND RESULTS

Evaluation of Summarization Algorithms

To assess the effectiveness of a summary, generated by the proposed algorithm, the standard procedure is adopted of creating gold standard summaries with the help from human annotators. The proposed extractive summary is compared with the gold standard ones. Three human expert's opinions are considered in this process. The performance of the established algorithms is discussed as follows. Different summarizer algorithms use dissimilar sets of tweets for summaries to evaluate the relative significance of tweets based on different factors. There are a few tweets common between the sets. Table 2 shows the performance of different algorithms in base summarization. The experimental outcome says that the proposed algorithm Louvain_Rank is performing best and the result is highlighted in boldface.

Evaluation Metrics

Tests are done using the Rouge toolkit (Yew, 2008) against various pre-established algorithms, described in the previous section. The conventional evaluation metrics are considered for comparison, such as- Recall (R), Precision (P) and F-Measure (F). Table 3 shows the outcome, which signifies that the proposed algorithm performs superior than rest of the algorithms except ILP Summary and MEAD performs equally good with the proposed method.

Table 2. Summaries generated by Different Base Algorithms for Sandy Hook Dataset

Algorithms	CR	CW	FS	LR	LS	LH	MD	SB	SM	IP	K-Means Rank	DB-Scan Rank	Hierarchical Rank	Infomap Rank	Louvain Rank
$\overline{CR}$	-	0	0	0	0	0	0	0	0	1	2	0	2	3	3
$\overline{CW}$	0	-	0	0	4	3	1	2	1	0	0	0	1	0	0
$\overline{FS}$	0	0	-	0	3	2	0	0	0	1	0	0	1	0	1
$\overline{LR}$	0	0	0	-	0	0	0	1	2	0	1	1	0	2	1
$\overline{LS}$	0	4	3	0	-	6	0	1	0	2	2	1	1	1	1
$\overline{LH}$	0	3	2	0	6	-	2	1	0	2	1	0	0	3	0
$\overline{MD}$	0	1	0	0	0	2	-	1	0	0	2	1	1	2	0
$\overline{SB}$	0	2	0	1	1	1	1	-	2	4	0	2	0	2	1
$\overline{SM}$	0	1	0	2	0	0	0	2	-	2	2	1	1	1	1
IP	0.	1	0	0	1	0	0	1	2	-	0	1	1	1	2
Proposed Algorithms															
K-Means Rank	0	0	0	0	0	1	0	2	2	1	-	1	2	4	2
DB-Scan Rank	0	1	1	0	1	0	1	1	0	1	2	-	2	1	3
Hierarchical Rank	1	0	1	0	2	0	0	0	1	0	1	1	-	1	2
Infomap Rank	0	1	2	0	2	1	1	1	0	2	1	2	3	-	3
Louvain Rank	1	2	1	0	0	1	0	0	1	0	0	2	2	2	-

For comparative analysis of the performances of the proposed methods standard ROUGE Recall scores are considered. ROUGE-2, and ROUGE-L variants scores (Lin, 2004) gives the performance measurements for evaluation of the quality of the summaries. The normal technique of producing gold standard summaries that is carried out typically by human annotators is used. These values separately measure what portion of the (i) longest matching sequence of words and (ii) bigrams in the gold standard summaries are covered by the summaries created by the proposed methods.

Table 4 shows the performance of ROUGE-2 and ROUGE-L variants for base summarization and proposed algorithms, averaged over all datasets. From the performance analysis it is quite evident that the proposed algorithm **Louvain_Rank**gives better results than the other approaches stated above. As the proposed algorithm is based on a clustering algorithm, community detection algorithms worked well for microblogging Dataset.

Table 3. Performance table for base algorithms and proposed algorithms

Algorithm	Precision (P)	Recall (R)	F-Measure (F)
$\overline{CR}$	0.42329	0.36876	0.39415
$\overline{CW}$	0.51386	0.37116	0.43101
$\overline{FS}$	0.19963	0.57447	0.2963
$\overline{LR}$	0.38632	0.26523	0.31452
$\overline{LS}$	0.41959	0.34135	0.37645
$\overline{LH}$	0.48799	0.36607	0.42478
$\overline{MD}$	0.38078	0.3766	0.37868
$\overline{SB}$	0.24769	0.59292	0.34941
IP	0.41857	0.35133	0.3678
Proposed Algorithm			
K-Means_Rank	0.609	0.539	0.457
DB-Scan_Rank	0.567	0.482	0.461
Hierarchical_Rank	0.439	0.378	0.428
Infomap_Rank	0.626	0.517	0.579
Louvain_Rank	**0.629**	**0.5956**	**0.581**

CONCLUSION

This work proposed an effective methodology for Extractive Text Summarization for micro-blog data based on unsupervised approach. The proposed methodology has been compared with traditional methods, such as- Lexical Rank, Sum Basic, LSA, Frequency Sum and so on. As given in the results table it can be shown that the proposed methodology has performed better than these classic methods in most cases. As the proposed algorithm is based on a clustering algorithm, community detection algorithms worked well for microblogging dataset. As future work, correlation between the intra-document can be considered for better categorization of documents for better summary generation. Ensemble approach also can give better results which will be considering the merits of all other summarizing methods. In Similar way for clustering an ensemble method can be proposed considering a voting approach for the summarization process. Along with that we intend to make the algorithm dynamic so that it can put single tweets in runtime in appropriate clusters and see whether that affects the generated summary or not.

Table 4. Rouge Scores of the base summarization and proposed algorithms, averaged over all datasets. The best score is by the **Louvain_Rank** *algorithm*

Algorithm	Rouge-2 Recall	Rouge-L Recall
$\overline{CR}$	0.08478	0.25938
$\overline{CW}$	0.17493	0.45349
$\overline{FS}$	0.14641	0.36135
$\overline{LR}$	0.04783	0.15376
$\overline{LS}$	0.15782	0.41783
$\overline{LH}$	0.16093	0.4112
$\overline{MD}$	0.11234	0.3790
$\overline{SB}$	0.1156	0.32783
$\overline{SM}$	0.0893	0.2782
IP	0.15169	0.3572
Proposed Algorithm		
K-Means_Rank	0.15034	0.3608
DB-Scan_Rank	0.1756	0.3987
Hierarchical_Rank	0.1478	0.4215
Infomap_Rank	0.1812	0.5871
Louvain_Rank	**0.1896**	**0.6587**

REFERENCES

Barzilay, R., & Lapata, M. (2008). Modeling local coherence: An entity-based approach. *Comput. Linguist., 34*(1), 1–34. doi:10.1162/coli.2008.34.1.1

Brin, S., & Page, L. (1998). The anatomy of a large-scale hypertextual web search engine. *Comput. Netw. ISDN Syst., 30*(1-7), 107–117. doi:10.1016/S0169-7552(98)00110-X

Dutta, S., Ghatak, S., Roy, M., Ghosh, S., & Das, A. K. (2015). A graph based clustering technique for tweet summarization. In *Reliability, Infocom Technologies and Optimization (ICRITO)(Trends and Future Directions), 2015 4th International Conference on* (pp. 1–6). IEEE. 10.1109/ICRITO.2015.7359276

Dutta, S., Ghatak, S., Roy, M., Ghosh, S., & Das, A. K. (2015). A graph-based clustering technique for tweet summarization. In *2015 4th Inter- national Conference on Reliability, Infocom Technologies and Opti- mization (ICRITO) (Trends and Future Directions)* (pp. 1–6). 10.1109/ICRITO.2015.7359276

Dutta & Ghatak. (2017). A.D.M.G.S.D.: Feature selection based cluster- ing on micro-blogging data. *International Conference on Computational Intelligence in Data Mining (ICCIDM-2017).*

Dutta & Ghatak. (2017). S.G.A.K.D.: A genetic algorithm based tweet clustering technique. *2017 International Conference on Computer Communication and Informatics (ICCCI)*, 1–6. 10.1109/ICCCI.2017.8117721

Erkan, G., & Radev, D.R. (2004). *LexRank: Graph-based lexical centrality as salience in text summarization*. Academic Press.

Hassel, M. (2007). *Universitets service, T., Ab, U.: Resource lean and portable automatic text summarization*. Tech. rep.

Hoch, R. (1994). Using ir techniques for text classification in document analysis. In *Proceedings of the 17th Annual International ACM SIGIR Conference on Research and Development in Information Retrieval, SIGIR '94* (pp. 31–40). Springer-Verlag. https://dl.acm.org/citation.cfm?id=188490.188498

Hu, M., Sun, A., & Lim, E. P. (2007). Comments-oriented blog summarization by sentence extraction. In *Proceedings of the Sixteenth ACM Conference on Conference on Information and Knowledge Management, CIKM '07* (pp. 901–904). ACM. doi:10.1145/1321440.1321571

Hyderabad blasts. (2013). https://en.wikipedia.org/wiki/2013Hyderabadblasts

Kedzie, C., McKeown, K., & Diaz, F. (2015). Predicting Salient Updates for Disaster Summarization. *Proc. ACL.*

Kupiec, J., Pedersen, J., & Chen, F. (1995). A trainable document summarizer. In *Proceedings of the 18th Annual International ACM SIGIR Conference on Research and Development in Information Retrieval, SIGIR '95* (pp. 68–73). ACM. doi:10.1145/215206.215333

Lin, C.-Y. (2004). ROUGE: A package for automatic evaluation of summaries. In *Proc. Workshop on Text Summarization Branches Out*. ACL.

Luhn, H.P. (1958). The automatic creation of literature abstracts. *IBM J. Res. Dev., 2*(2), 159–165. doi:10.1147/rd.22.0159

Mackie, S., McCreadie, R., Macdonald, C., & Ounis, I. (2014). Comparing Algorithms for Microblog Summarisation. *Proc. CLEF.*

Mahesh, K.(1997). *Hypertext Summary Extraction for Fast Document Browsing*. Academic Press.

Nepal earthquake. (2015). https://en.wikipedia.org/wiki/2015Nepalearthquake

Nguyen, M. T., Kitamoto, A., & Nguyen, T. T. (2015). Tsum4act: A framework for retrieving and summarizing actionable tweets during a disaster for reaction. *Proc. PAKDD*. 10.1007/978-3-319-18032-8_6

North India floods. (2013). https://en.wikipedia.org/wiki/2013NorthIndiafloods

Olariu, A. (2014). Efficient online summarization of microblogging streams. *Proc. EACL(short paper)*, 236–240. 10.3115/v1/E14-4046

Page, L., Brin, S., Motwani, R., & Winograd, T. (1999). *The pagerank citation ranking: Bringing order to the web*. Academic Press.

Radev, D. R., Allison, T., Blair-Goldensohn, S., Blitzer, J. C., Elebi, A., Dimitrov, S., Dr'abek, E., Hakim, A., Lam, W., Liu, D., Otterbacher, J., Qi, H., Saggion, H., Teufel, S., Topper, M., Winkel, A., & Zhang, Z. (2004). MEAD - A platform for multidocument multilingual text summarization. *Proceedings of the Fourth International Con- ference on Language Resources and Evaluation, LREC 2004.* http://www. lrec-conf. org/proceedings/lrec2004/pdf/757.pdf

Rau, L. F., Jacobs, P. S., & Zernik, U. (1989). Information extraction and text summarization using linguistic knowledge acquisition. *Information Processing & Management, 25*(4), 419–428. doi:10.1016/0306-4573(89)90069-1

Reimer, U., & Hahn, U. (1988). Text condensation as knowledge base abstraction. *Artificial Intelligence Applications, 1988., Proceedings of the Fourth Conference on,* 338–344. 10.1109/CAIA.1988.196128

REST API Resources. (n.d.). *Twitter Developers.* https://dev.twitter.com/docs/api

Rudra, K., Ghosh, S., Goyal, P., Ganguly, N., & Ghosh, S. (2015). Extracting situational information from microblogs during disaster events: A classification-summarization approach. *Proc. ACM CIKM.* 10.1145/2806416.2806485

Sandy Hook Elementary School shooting. (2012). https://en.wikipedia.org/wiki/SandyHookElementarySchoolshooting

Shou, L., Wang, Z., Chen, K., & Chen, G. (2013). Sumblr: Continuous summarization of evolving tweet streams. *Proc. ACM SIGIR.* 10.1145/2484028.2484045

Sun, J. T., Shen, D., Zeng, H. J., Yang, Q., Lu, Y., & Chen, Z. (2005). Web-page summarization using clickthrough data. In *Proceedings of the 28th Annual International ACM SIGIR Conference on Research and Development in Information Retrieval, SIGIR '05* (pp. 194–201). ACM. doi:10.1145/1076034.1076070

Tao, K., Abel, F., Hauff, C., Houben, G. J., & Gadiraju, U. (2013). Groundhog Day: Nearduplicate Detection on Twitter. *Proc. Conference on World Wide Web (WWW).*

Tigelaar, A.S., Opdenakker, R., & Hiemstra, D. (2010). Automatic summarisation of discussion fora. *Nat. Lang. Eng., 16*(2), 161–192. DOI doi:10.1017/S135132491000001X

Typhoon Hagupit. (2014). https://en.wikipedia.org/wiki/TyphoonHagupit

Wang, Z., Shou, L., Chen, K., Chen, G., & Mehrotra, S. (2015). On summarization and timeline generation for evolutionary tweet streams. *IEEE Transactions on Knowledge and Data Engineering, 27*(5), 1301–1314. doi:10.1109/TKDE.2014.2345379

Xu, W., Grishman, R., Meyers, A., & Ritter, A. (2013). A preliminary study of tweet summarization using information extraction. *Proc. NAACL 2013,* 20.

Yew Lin, C. (2004). *Rouge: A package for automatic evaluation of summaries.* Academic Press.

Zubiaga, A., Spina, D., Amigo, E., & Gonzalo, J. (2012). Towards Real-Time Summarization of Scheduled Events from Twitter Streams. Hypertext(Poster). doi:10.1145/2309996.2310053

Chapter 3
Machine Intelligence of Pi From Geometrical Figures With Variable Parameters Using SCILab

Ankita Mandal

Institute of Engineering and Management, Kolkata, India

Soumi Dutta

Institute of Engineering and Management, Kolkata, India

Sabyasachi Pramanik

iD https://orcid.org/0000-0002-9431-8751

Haldia Institute of Technology, India

ABSTRACT

In the present research work, the use of geometrical figures have been made for the calculation of the value of pi. Instead of circle and square, ellipse and rectangle had been used to derive the value of pi. Ellipse can be considered as an extension of a circle where it had been stretched in two dimensions in unequal manner giving rise to the concept of major axis and minor axis. These axes are considered as the length and breadth of the considered rectangle. The ellipse has been considered within the rectangle and some random points are generated to see the position occurrence of the generated points. If the point lies within the ellipse, then the specific counter is incremented; otherwise, the counter for the rectangle is incremented.

INTRODUCTION

The value of pi is of great importance in the calculation of various mathematical formulas, derivations and modeling since ancient times. It is a constant generally denoted by the greek letter π. It is, in general

DOI: 10.4018/978-1-7998-7701-1.ch003

defined as the ratio of the circumference to diameter of a circle. There are many parallel definitions to support the existing and most prevalent definition. It is an irrational number with the value being calculated approximately as 3.14159. The value of pi was first estimated by Greek philosopher Archimedes, for which the value is also referred sometimes as Archimedes' constant. Since pi is an irrational number, it cannot be represented by any common fraction. However the value 22/7 is commonly used to give an approximated value of it. It has certain features to characterize the value such as it is a recurring decimal number where the numbers do not repeat within a fixed interval. This implies that the in the computational value of pi, the numbers occurring after the decimal point are always random in nature, validating the statistical randomness and supposition of the numerical value. It is not the root of any polynomial whose coefficients are rational making the value a transcendental number. This trancedity of pi had been a great challenge for mathematicians in earlier times to manually calculate the value using inscribed polygon within a circle which was circumscribed inside another polygon, where straight edges and compasses had been used to get a mere approximation of the value. Today the value of pi plays a significant role in calculation areas and volumes of various 2-dimensional and 3 dimensional shapes with greater accuracy. In analytical mathematics, the value is defined using the infinite multiple set of integers belonging to the unbounded spectrum of of real number where the value is represented simply as a period or Eigen value without any reference of 2D and 3D structures.

With the advent of technology over various years, numerous new approaches had been developed by mathematicians for the calculations. With the increased computational power and higher efficiency of the implementation systems, the representation of the value had been extended to many numbers. Various efficient algorithms have been developed and implemented using the workability of super computers. The various approaches to calculate the value of pi theoretically include using of polygons, infinite series and calculus for the approximation.

The calculation of pi used the concept of infinite series where a number of terms of a particular sequence or pattern had been used . An infinite series is an infinite sequence of the sum or product of the terms involved. It did not involve any formula but required a large number of terms for the computation purpose. There are many infinite series which have been used. They are

The Viette's Series is one of the first infinite series of the form represented by equation (1).

$$\frac{2}{\pi} = \frac{\sqrt{2}}{2} \cdot \frac{\sqrt{2+\sqrt{2}}}{2} \cdot \frac{\sqrt{2+\sqrt{2+\sqrt{2}}}}{2} \tag{1}$$

The Wallis Series is of the form represented by equation (2)

$$\frac{1}{2} \cdot \frac{2}{3} \cdot \frac{4}{3} \cdot \frac{4}{5} \cdot \frac{6}{5} \ldots \tag{2}$$

Next came the Leibnitz Series which was represented by equation (3) as

$$\pi = \frac{4}{1} - \frac{4}{3} + \frac{4}{5} - \frac{4}{7} + \ldots \tag{3}$$

One of the most popular series is the Nilkantha Series of the form represented by the equation (4) as

$$\pi = 3 + \frac{4}{2*3*4} - \frac{4}{4*5*6} + \frac{4}{6*7*8} - \dots \tag{4}$$

Apart from the above mentioned series, there are many more infinite series used for the calculation of the value. One of the main problems of using infinite series is that a large number of terms are required (generally more than 300) to attain higher accuracy. By using the computational power of supercomputers these calculations have been made much easier.

Since earlier times, geometrical figures had been one of the pioneer instrument for the estimation of one of the significant constants in mathematics. Since it is known that pi is an irrational number, there had been four relationships or approaches in the calculation of the value using closed geometrical figures. The first approach is the relationship between the circumference and diameter of a circle. The second approach is the relationship between the area of a circle and square of the diameter. The third approach relates the area of a sphere to its diameter. The fourth approach deals with the relation between the volume of a sphere and the cube of its diameter. In one of the earliest work by the mathematicians, the value of pi was calculated by the ratio between the circumference and diameter of a circle. It was a tedious task to consider the numerous points on the circumference in the absence of computational methods. Gradually it was deduced that a circle is a regular polygon with infinite sides. This lead to the consideration of 3 sided,4 sided polygons where the value was given by the ration between the perimeter and the diameter of the circumscribed circle. The introduction of calculus gave the modern mathematicians as well as computer scientists an advantage in developing new approaches. One of the most popular approaches to calculate the value of pi used by present day researchers is the ratio between the area of the circle to the area of the square where the circle is inscribed within the square.

The present work deals with the use of ellipse and rectangle for the calculation of the value using Monte Carlo simulation where Scilab has been the platform for the mathematical modeling. There are several software available in the market for mathematical calculations ad modeling such as Mathematica, Maple, Scilab, Matlab, etc. Amongst all the software's available, Scilab has been chosen as the platform in the present research work due to its free, open source, platform independent features. The availability of the unlicensed software with various packages makes it one of the most popular engineering and scientific calculation platform. In the process of calculation of the value, numerous approaches have been taken, one of them being the use of polygons and circle, relating to the ratio of the area of the circle and the polygon. In the present research work, the ratio between the areas of an ellipse to a rectangle has been made for the calculation. Among various statistical simulation models, Monte Carlo Simulation model has been used. It is a probabilistic technique used to predict the outcome of uncertain events with the intervention of random variables. It is one of the most reliable techniques in forecasting and prediction analysis. Central tendency parameters like variance and standard deviation are the main basis of formulation of Monte Carlo Simulation. The primary logic behind the implementation of this simulation is to allocate multiple values random in nature to an uncertain variable denoting an uncertain event at any particular instance, obtain multiple outcomes with respect to the values given. This process is repeated over multiple times taking different values of the uncertain event and finally estimate the final output or value by taking a mean of the derived outcome. Monte Carlo Simulations generates graphical results of the different outcomes which makes it easier to visualize the problem realistically and understand the improvement of the solutions when repeated over and over again. Moreover it gives an overall sensitive analysis with respect to the optimum input value in terms of the random variables allocated to the uncertain events in order to determine the most appropriate input for a specific result. It also provides an overall idea about the scenario or real life environment in which the various inputs are grouped together and the variance of the corresponding outputs. The interdependency between the input variables in the accuracy of the output which is an estimation of related input parameters.

BACKGROUND

The evolution of pi has been described in the work by Allen G.D.(1999) which hints at the earliest being done in Egypt where the value of the circle and square considered were much higher. Work of Archimedes has also been described where regular polygons were inscribed and circumscribed on a circle to compute the value of pi to be 3.15 approximately. Srimani S.(2015) gave various methods for geometrically deriving the value.The work of Janorkar, D.S(2018) segments the circumference of a circle into 6 equal parts upon which mathematical calculations have been implemented to get the exact computed value of pi as 3.141592653 which is also termed as Goba constant by the authors to symbolise the equations. Barauh N.D et al(2016) conducted a survey work related to the application of the inverse of pi and its evolution. The ideas during evolution ranged from singular modular to complementary modular quantity. The transition of pi into multiple series formulation and conversions has also been worked with. Various books by the authors discuss about various numerical methods used by scientists which provide a base for computative applications.(Rao S.,2002 ; Steven C. et al.,2010; Salleh. Z, 2011).The generation of random numbers had been an important criteria for the simulation of real life situations in order to get the probability of uncertainity. Extensive work had been done in the field of generation of polar variables random in nature (Bailey R.,1994),generating normal deviates random in nature(Box G.E.P. et al.,1958),quasi random sequence generation(Bratley P.,1988 ; Joe S., Kuo F., 2003),random number variation as well as applications(Howes L.,2007 ; Langdon B.,2009 ; Sriram V.,2007 ; David B.,2009 ; Park S. K. , 1988 ; Marsaglia G.,2003 ; Marsaglia G.,1991 ; Panneton F.,2005 ; Brent R.P.,2004 ; L'ecuyer P. et al.,2007 ; L'ecuyer P.,1999),generating random numbers using gaussian model under efficient hardware(Thomas D.B.,2008), generation of simple random numbers(L' Ecuyer P.,1990 ; L' Ecuyer P.,2003 ; Matsumoto M.,1997 ; Srinivasan A.,1999 ; Coddington P.D.,1996 ; Wu P., Huang 2006) . Hissoiny, S. et al. (2011) and Anderson J.A. et al. (2008) used units based on graphics for random number generation. Shetty, M. (2004) gave various methods about the geometrical estimation of the value.

Metropolis N.(1949) brought in the concept of Monte Carlo methods in real life applications (Metropolis, N.,1987). Garland M. et al,2008 uses Monte Carlo Simulation for extensive parallelism where multiple events have been simulated which are random in nature. Some pseudo random numbers are generated using the Random Number Generator(RNG) and graphics processing unit has been taken as the base for the implementation of the collection. Reuillon R. et al.(2008) used Monte Carlo Simulation in the field of Grid Computing where the simulation has been distributed over localized clusters with the usage of the GATE software. Generic object oriented natured toolbox has been used in the referred work along with the enabling grid for e-science thereby decreasing the otherwise three year long computation to one week process time. Matrix analysis of different structures had also been studied and worked upon extensively(Kassimali, A.,1999 ; Weaver Jr., W.,1986).Chong J.et al(2010) used Monte Carlo methods extensively for parallel computing to use suitable random generators for efficient implementation of

Monte Carlo Simulation based applications for different structural, computational, parallel algorithm, implementation strategy and parallel execution patterns. It has been widely used in analyzing estimation of market risk associated with financial values. Dixon M.F. et al (2012) helped developers working on quantitative applications in the domain of finance by developing a highly efficient risk analysis system using Monte Carlo techniques. Hubbard, D. (2007) implemented it to find out risk related to any tangible quantity. It also aided in giving a new accuracy in estimating market risks(Dixon M.F. et al.,2009 ; Risk Management Systems in the Aftermath of the Financial Crisis Flaws, Fixes and Future Plans, a GARP report prepared in association with SYBASE, 2010 ; Giles M.B.,2008 ; Glasserman P., 2003 ; Jondeau E.,2007 ; Jorion P., 2007 ; Moro B.,1995 ; Singla N.,2008 ; Srinivasan A.,2002 ; Youngman p.2009 ;). Singh U.(2013) made a relative study in his work and evaluated pi using Monte Carlo method where C++ had been used for coding.It has been used for various probabilistic simulations(Karmshu, 2012) . Kim S. et al. (2020) used Monte Carlo Simulation process to generate wind power for electro voltaic charging stations. Robotics along with image processing has also used Scilab in Drowsy driver detection system(Yauri-Machaca M.,et al.2008),tracking of object in real time (Murshed Sk.Z.,2016 ; Chelva M. S.,2016 ; Ojha S.,2015 ; Dr.Vijayalakshmi M.N,2016 ; Ikhankar R.,2015 ; Kalyani A.,2015 ; Janowc-zyk A.,2008 ; Fatema A.K. et al,2013 ; Khutubuddin S. et al.,2014 ; Samanta D. et al.,2013 ; Pradhan K.N. et al,2019 ; Karthikeyan M.P. et al.,2021 ;).Various types of number generators such as random, parallel, fibonacci and sequential have used this simulation in varied fields for application(Mascagni M.,2004 ; Mascagni M.,2000 ; Srinivasan A., 1999 ; James F.,1990 ; Marsaglia G.,1987 ; Matsumoto M. et al.,1998 ; Marsaglia G.,1997 ; L'ecuyer P.1988 ; Marsaglia G.,1995 ; Podlozhnyuk V. 2007). Lazaro D., et al.,2004 & Lazaro D., 2005 used monte carlo Simulation for 3- dimensional reconstruction of a single photo point emitted in tomography. Large ecosystems have also been implemented by simulation through Monte Carlo method (Hill D.R.C.,1997), where the stochastic and faster quantitative simulations have been achieved(Pawlikowski K.,2003 ; Garland M.,2008 ; Samanta D. et al.,2020 ; Biswal A.K. et al.,2021 ; Maheswari M. et al.,n.d. ; Gomathy, V et al.,2020 ;) making use of the high end parallel grid storage functionality(Anderson D.P.,2004 ; Maigne L.,et al.,2004 ; Matsumoto M.et al.,2000 ; Traore M., 2001) .Performance of large scale computational grides have been improved by distribution of large scale simulation tool kits available(Li Y.,et al., 2003 ; Mascagni M. et al.,2004 ; Reuillon R. et al.,2008 ; Keutzer K. et al.,2009 ; Althar R.R. et al.,2021).One of the exciting application of Monte Carlo Simula-tion is in the field of astronomy, astrophysics and galaxial interpretations(Macgillivray H. T.,2004 ; Jan S. et al., 2004 ; Keutzer K. et al,2009 ; Andieu C., n.d. ; Baeurle S.,n.d ; Lönnblad L.,1994 ; Intel. Intel Math Kernel Library.).

The use of Scilab has been a general choice amongst scholars and academicians in recent years for its extensive use in numerous kinds of statistical modeling and simulation of different real time environments(Stephen L. C et al.,2006 ; Gilberto E.,2001).Virtual laboratories and architecture imple-mentation of various automation processes is one of the major application areas of Scilab(Huba M. et al.,2006 ; Jakab F. et al.,2006 ; Leão C. P. et al.,2004 ; Liguš J. et al.,2005 ; Restivo M. T. et al.,2009). Salleh, Z. et al. (2012) used Scilab for the computation of basic numerical computational methods.It has been used in challenges related to object detection and recognition such as tracking vehicles, surveillance systems and tracking systems. Chopparapu S. T., & Seventline J, D. B. (Eds.). (2020) implemented Scilab in object recognition from images and videos using the operations of the robots. The authors have used the technique of blob analysis in order to evaluate the the attributes related to the shape of the object being recognized and have attached Infrared sensors to the Arduino for recording the robotic operations. Blob analysis and Arduino interfacing both have been implemented using Scilab. It has been used by researchers in the area of signal processing variedly. Simulation of the Wigner Villie Distribution has been done using Scilab which produces the visualization of information in terms of frequency and time dimensions(Chioncel C.P.et al 2011).It has further been used as simulation platform for speed control digitally and control circuit calculations in a specific time domain.(Chioncel C.P.et al,2016 7; Chioncel C.P.et al, 2014).Various models have been simulated by the Scilab environment such as Simulation of stochastic model for calculating the effect of an epidemic(Ang K.C.,2009),Linear and non linear behavioral dynamic simulation models(Gillich G.R.,2005 ; Sivakumar, P. et al.,2020 ; Khamparia, A. et al.,2020 ; Althar, R.R. et al.,2021 ; Guha, A. et al.,2021 ; Samanta, D. et al.,2020 ; Nyland L.,n.d.),time dependent equation based hybrid power system models(Panciatici P.,2011). Simulation models have been implemented to generate higher dimensional sequences with minimum deviations(Wang X. et al., 2008).Wenjiang L.,(2009) used Scilab for automatic control.Model formulation for generating computing sequences keeping in mind different financial and corporate issues have also been exercised (Antonov I.,n.d. ; Black F.,n.d.).Annigeri S.(2009) discussed about the wide usage of Scilab in visualization of various mathematical problems. In the work by the author analysis of plane frames have been done by the Direct Stiffness Method which leads to final element analysis method. The implementation of Scilab has been extensive in the field of medical science and research activity. Scilab had also been used in the linux environment for the analysis of processes having same structure(Badal A.,2006 ; Fractal Analysis of geomorphologic processes in the Linux environment using SCILAB.,2011 ; NVIDIA Corporation Inc. NVIDIA CUDA Compute Unified Device Architecture Programming Guide version 2.3, 2009.).Siddique S. et al (2016) used Scilab to develop a system for faster and efficient counting and segmentation of the RBCs present in the image of a blood smear sample. Scilab has been used for image enhancement after acquisition, segmenting and displaying the RBC count. It can also be used for classification and counting of all types of blood cells(Vargese J.T. et al,2017; Thiruvinal V. J. et al.,2017; Sharif J. M.,2012 ; Mahmood N.H.,2012 ; Maitra M.,2012 ; Nguyen N.T.,2011 ; Praveen B. et al.,2020 ; Guha A. et al.,2020 ; Chakrabarti S et al.,2013 ; Samanta D.et al.,2012 ; Mukherjee M.et al.,2014 ; Kuchy S.A. et al.,2017 ; Kureethara V.et al.,n.d.).Moulick R. (2019) has effectively used Scilab to calculate the value of pi from the conventional idea of finding the ratio of area of circle to area of square under random number generation system. Chopparapu, S. T.(2020) used Scilab for object detection under comparative environment. Various aspects of image processing has also used Scilab for different experimentations(Hall, A., 1977 ; Gonzalez R. G.2002 ; Fabbri R.,n.d. ; Maini R., n.d.).

PROBLEM STATEMENT

The inclusion of infinite series for calculating the value of pi requires many numbers to be generated for the calculation to proceed. This leads to the increase in complexity at the input level. Moreover the infinite generation of numbers based on a specific pattern of the series makes the implementation complex where any system needs to be iterated or repeated fixed number of times for getting the output from the system. The present methodologies dealing with the calculation of the value of pi from geometrical figures includes most commonly the usage of figures with single parameter of variation. Circles and squares are examples of such figures. The radius of a circle is constant which implies that the distance of any point on the circumference of the circle will always be same. Such methodologies implementing figures with single variable parameter fail to function for any situation where the radius of the geometrical figure varies according to the occurance of the point which is a very real life problem.

SOLUTION

The above mentioned problem is a common problem which has been addressed previously with geometrical figures with minimum variations in their parameters. The present research work uses figures like ellipse and rectangle which has variable parameters based on the angle of inclination.Here an ellipse is considered within a rectangle where the major axis and the minor axis of the ellipse are considered as lines parallel to the length and breadth of the considered rectangle respectively.

Figure 1. Pictorial idea about the system considered

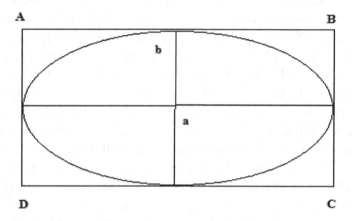

The Figure 1 gives a pictorial idea about the system taken into consideration. Here ABCD is the rectangle within which an ellipse has been inscribed. The major axis is 'a' and the minor axis is 'b'. These are also the length and breadth of the considered rectangle. The general equation of an ellipse is given by equation (5) as

$$\frac{x^2}{a^2} + \frac{y^2}{b^2} = 1 \tag{5}$$

Where (x,y) are any arbitrary point on it.
   The parametric equation of the same is given by equations (6) and (7)

$$x = a\cos\theta \tag{6}$$

$$y = a\sin\theta \tag{7}$$

Where θ is the inclination of the point to the horizontal axis.
   On the above set of figures Monte Carlo simulation is applied. The main idea is to generate random numbers rx and ry in the 2-dimensional plane which in the ordered pair (rx, ry) denote any point within

the domain of the rectangle. Since the points are generated randomly, some points will lie within the ellipse while some will lie outside the ellipse but within the rectangle.

Figure 2. Randomly generated points in the given system

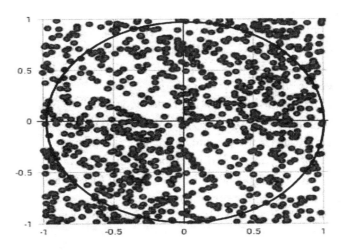

Figure 2 gives a pictorial view of the points randomly generated within a specified domain where some of the occurances may be within the ellipse while some of the points may occur outside the ellipse but within the considered rectangle.Then the ratio between the points that lie inside the ellipse and total number of points generated lying outside the ellipse but inside the defined rectangle is calculated .This is implemented by finding the ratio between the area of the ellipse and the area of the rectangle.

The area of the ellipse is given by the formula given in equation (8)

$$E = \frac{\pi ab}{4} \tag{8}$$

The area of the rectangle is given by by equation (9)

$$R = ab \tag{9}$$

This ratio gives the probability P of the random points lying inside the ellipse which gives the formula for calculation of the value of pi.

$$\text{Therefore P} = \frac{Area\ of\ the\ ellipse}{Area\ of\ the\ rectangle} = \frac{\pi}{4}$$

So,

$$\pi = 4*P \tag{10}$$

Equation (10). gives the formula for the calculation of the value of pi using the ratio of the area of ellipse and rectangle.

The following algorithm has been used for the implementation of the proposed methodology.

Step 1: Initialize area of ellipse elp to 0.

Step 2: Initialize the area of the rectangle rec to 0.

Step 3: Initialize the major axis(a) and minor axis(b) of the ellipse.

Step 4: Consider any random numbers rx and ry which in ordered pair denote the co-ordinates of any point on the ellipse.

Step 5: Find the radius of the ellipse .

Step 6: Compute the distance of the random point generated in Step 4. From the centre of the ellipse.

Step 7: If the distance computed in step 6 is less than or equal to the radius of the ellipse,then the point lies inside the ellipse and elp is incremented, else rec is incremented.

Step 8: The Step 4 to Step 8. is repeated till the number of iterations given by the user is satisfied.

Step 9: The ratio between the area of the ellipse and the area of the rectangle gives the probability P of the random number generated lying inside the ellipse.

Step 10: Finally the Equation 5 given above is used to compute the value of pi from the probability P.

The above algorithm has been implemented using Scilab. The number of random points generated had been plotted where the points lying within the ellipse have been marked by blue dots and the points lying outside the ellipse but inside the rectangle have been marked by red dots. All the points have been generated in the continuous interval [0,1] to give a concise result.

The number of random points generated denoted by the number of iterations have also been varied to see the change in the computed value.

The following section gives the Scilab code along with the computed value of pi and the corresponding graph about the orientation of the random points generated.

Number of random points generated: 5000

```
--> r=1;
        --> a=2;
--> b=1;
--> elp=0;
--> rec=0;
--> for i=1:5000
  > rx=r*rand();
  > ry=r*rand();
  > r2=rx^2+ry^2;
  > t=linspace(0,2*%pi,100);
  > z=(a*cos(t))^2+(b*sin(t))^2;
  > if(r2<=z) then
  > elp=elp+1;
  > plot(rx,ry,'b.')
  > else
  > plot(rx,ry,'r.')
```

```
   > end
   > rec=rec+1;
             > end

--> pi=4*(elp/rec);
--> disp(pi)
Output:
   3.12
```

Figure 3. Representation of 5000 random points lying within the ellipse and rectangle

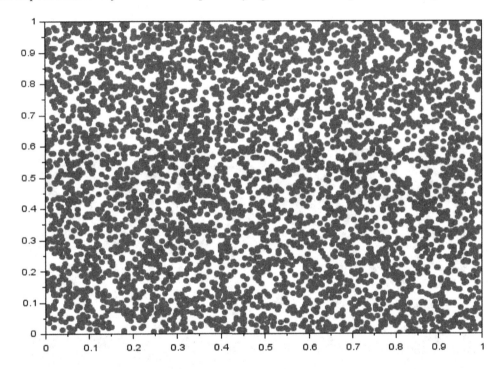

Number of random points generated: 7000

```
--> r=1;
         --> a=2;
--> b=1;
--> elp=0;
--> rec=0;
-->for i=1:7000
   >   rx=r*rand();
   >   ry=r*rand();
   >   r2=rx^2+ry^2;
   >   t=linspace(0,2*%pi,100);
```

```
>    z=(a*cos(t))^2+(b*sin(t))^2;
>    if(r2<=z) then
>    elp=elp+1;
>    plot(rx,ry,'b.')
>    else
>    plot(rx,ry,'r.')
>    end
>    rec=rec+1;
>    end

-->  pi=4*(elp/rec);
-->  disp(pi)
```

```
Output:
  3.1371429
```

Figure 4. Representation of 7000 random points lying within the ellipse and rectangle

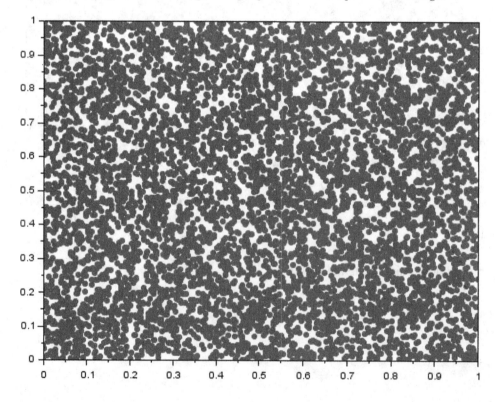

Number of random points generated: 8000

```
--> r=1;
--> a=2;
```

```
--> b=1;
--> elp=0;
--> rec=0;
-->for i=1:8000
>    rx=r*rand();
>    ry=r*rand();
>    r2=rx^2+ry^2;
>    t=linspace(0,2*%pi,100);
>    z=(a*cos(t))^2+(b*sin(t))^2;
>    if(r2<=z) then
>    elp=elp+1;
>    plot(rx,ry,'b.')
>    else
>    plot(rx,ry,'r.')
>    end
>    rec=rec+1;
>    end

-->   pi=4*(elp/rec);
-->   disp(pi)
```

Figure 5. Representation of 8000 random points lying within the ellipse and rectangle

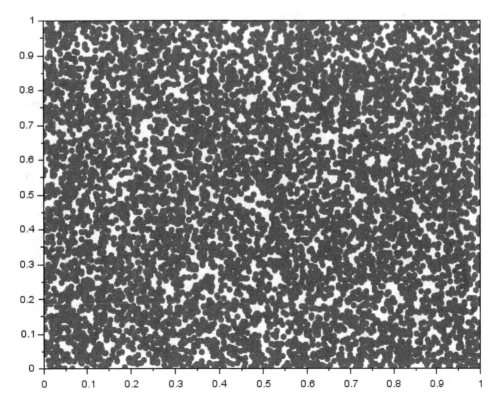

```
Output:
  3.1395
```

Number of random points generated: 8500

```
--> r=1;
--> a=2;
--> b=1;
--> elp=0;
--> rec=0;
-->  for i=1:8500
  >    rx=r*rand();
  >    ry=r*rand();
  >    r2=rx^2+ry^2;
  >    t=linspace(0,2*%pi,100);
  >    z=(a*cos(t))^2+(b*sin(t))^2;
  >    if(r2<=z) then
  >    elp=elp+1;
  >    plot(rx,ry,'b.')
  >    else
  >    plot(rx,ry,'r.')
```

Figure 6. Representation of 8500 random points lying within the ellipse and rectangle

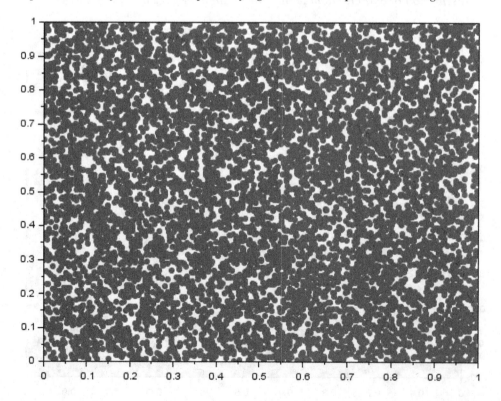

```
>    end
>    rec=rec+1;
>    end

-->   pi=4*(elp/rec);

-->   disp(pi)
Output
   3.1487059
```

Number of random points generated: 9000

```
--> r=1;
         --> a=2;
--> b=1;
--> elp=0;
--> rec=0;
-->for i=1:9000
>    rx=r*rand();
>    ry=r*rand();
>    r2=rx^2+ry^2;
```

Figure 7. Representation of 9000 random points lying within the ellipse and rectangle

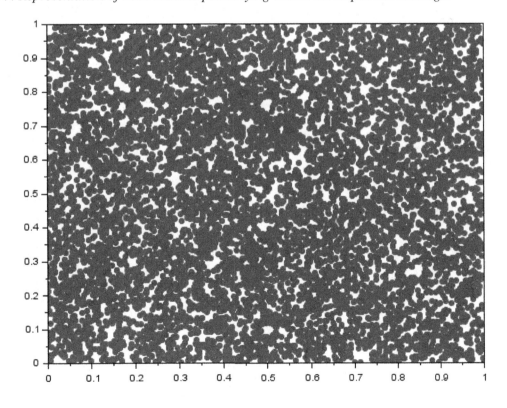

```
>    t=linspace(0,2*%pi,100);
>    z=(a*cos(t))^2+(b*sin(t))^2;
>    if(r2<=z) then
>    elp=elp+1;
>    plot(rx,ry,'b.')
>    else
>    plot(rx,ry,'r.')
>    end
>    rec=rec+1;
>    end

-->   pi=4*(elp/rec);
-->   disp(pi)
```

Output:
   3.176

Number of random points generated: 10000

```
--> r=1;
           --> a=2;
--> b=1;
--> elp=0;
--> rec=0;
-->for i=1:10000
>    rx=r*rand();
>    ry=r*rand();
>    r2=rx^2+ry^2;
>    t=linspace(0,2*%pi,100);
>    z=(a*cos(t))^2+(b*sin(t))^2;
>    if(r2<=z) then
>    elp=elp+1;
>    plot(rx,ry,'b.')
>    else
>    plot(rx,ry,'r.')
>    end
>    rec=rec+1;
>    end

-->   pi=4*(elp/rec);
-->   disp(pi)
```

Output:
   3.1592

Figure 8. Representation of 10000 random points lying within the ellipse and rectangle

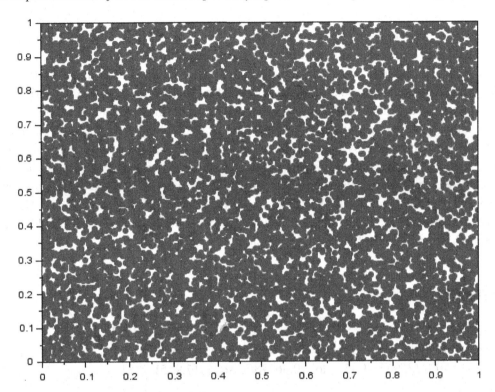

Figure 3, Figure 4, Figure 5, Figure 6, Figure 7 and Figure 8 are the graphical representation of the points which are randomly generated. The blue points denote the generated random numbers which lie within the ellipse and the red dots denote the points lying within the rectangle but outside the ellipse. By making a comparative study it can been deduced that the density of the points becomes more concentrated at each of the respective areas making the boundary between the two distinct areas more prominent. It has been observed by repeating the simulation for large number for iterations that the increase in the number of iterations increases the accuracy of the system. The use of Monte Carlo Simulation specifically gives an advantage of choosing random instances of numbers which gives an effective computational value considering the uncertainty of the occurrence of the events. The value of pi changes every time the same iterations are repeated as random numbers are generated at every instance of the simulation.

FUTURE RESEARCH DIRECTIONS

The methodology can be modified further by using normally distributed random numbers over a specified spectrum which will help to study the variation in the results upon change in the input value spectrum. Furthermore the usage of three dimensional structures for the calculation of the value of pi can also be considered in future thereby increasing the usage of geometrical figures in scientific and mathematical calculations which is presently more concentrated towards two one or two dimensional structures.

CONCLUSION

This chapter intends to extend the use of geometrical figures from calculation of scientific constants to figures where there are more than one parameter of variation. Ellipse and rectangle have been used for the initial system consideration as an alternative to the popular convention of using circle and square as the initial system. Parametric equations have been used to consider any point on the perimeter of the ellipse in order to find the radius of the ellipse which varies according to the angle of inclination of the point on the major axis. The implementation of Monte Carlo Simulation has made the instantiation of random variables into the present system thereby calculating the probability of the uncertain events which occur into terms of the random numbers generated. The random number generator used in the Monte Carlo method chooses uniformly distributed random numbers at every instance of the occurance with the least chance of redundancy. Using Scilab has made the implementation much easier. The complex mathematical calculations and generation of random numbers have been easily done within the Scilab environment which is most suitable for complex scientific calculations.

REFERENCES

Ahmad Kuchy, S., Ahmed Khadri, S. K., Mukherjee, M., Samanta, D., & Le, D.-N. (2017). An Aggregation Approach Based on Elasticsearch. *Journal of Engineering and Applied Sciences (Asian Research Publishing Network)*, *12*, 9451–9454. doi:10.36478/jeasci.2017.9451.9454

Aleya, K. F., & Samanta, D. (2013). *Automated damaged flower detection using image processing.* https://www.semanticscholar.org/paper/AUTOMATED-DAMAGED-FLOWER-DETECTION-USING-IMAGE-Aleya-Samanta/11f8ebd4082acef98b7329cecc81601b6ec20bc8

Allen, G. D. (1999). *A Brief History.* https://www.math.tamu.edu/~dallen/masters/alg_numtheory/pi.pdf

Althar, R. R., & Samanta, D. (2021). Building Intelligent Integrated Development Environment for IoT in the Context of Statistical Modeling for Software Source Code. In R. Kumar, R. Sharma, & P. K. Pattnaik (Eds.), Multimedia Technologies in the Internet of Things Environment. Studies in Big Data (Vol. 79). Springer. https://doi.org/10.1007/978-981-15-7965-3_7.

Althar, R. R., & Samanta, D. (2021). The realist approach for evaluation of computational intelligence in software engineering. *Innovations Syst Softw Eng.* doi:10.100711334-020-00383-2

Anderson, D. P. (2004). BOINC: A System for Public-Resource Computing and Storage. Grid, 4-10.

Anderson, J. A., Lorenz, C. D., & Travesset, A. (2008). General purpose molecular dynamics simulations fully implemented on graphics processing units. *Journal of Computational Physics, 227*(10), 5342–5359.

Andieu, C., de Freitas, N., Doucet, A., & Jordan, M. (n.d.). *An introduction to MCMC for machine learning.* http://people.cs.ubc.ca/ nando/papers/mlintro.pdf

Ang, K. C. (2009). A simple stochastic model for an epidemic {numerical experiments with matlab. *The Electronic Journal of Mathematics & Technology, 1*(2), 117–128.

Annigeri, S. (2009). *Matrix Structural Analysis of Plane Frames using Scilab*. https://www.researchgate.net/publication/242759801

Antonov, I., & Saleev, V. (n.d.). *An economic method of computing lpt-sequences*. Academic Press.

Badal, A., & Sempau, J. (2006). A package of Linux scripts for the parallelization of Monte Carlo simulations. *Computer Physics Communications, 175*(6), 440 – 450.

Bailey, R. (1994). Polar generation of random variates with the t-distribution. *Mathematics of Computation, 62*, 779–781.

Baruah, N. D., Berndt, B. C., & Chan, H. H. (2016). Ramanujan's Series for $1/\pi$: A Survey. *The American Mathematical Monthly, 116*(7), 566–587. doi:10.4169/193009709X458555

Biswal, A. K., Singh, D., Pattanayak, B. K., Samanta, D., & Yang, M.-H. (2021). IoT-Based Smart Alert System for Drowsy Driver Detection. *Wireless Communications and Mobile Computing.* doi:10.1155/2021/6627217

Black, F., & Scholes, M. S. (n.d.). The pricing of options and corporate liabilities. Journal of Political Economy, 637-654.

Box, G. E. P., & Muller, M. E. (1958). A note on the generation of random normal deviates. *Annals of Mathematical Statistics, 29*(2), 610–611.

Bratley, P., & Fox, B. L. (1988). Implementing Sobol's quasirandom sequence generator. *ACM Transactions on Mathematical Software, 14*(1), 88–100.

Brent, R. P. (2004). Note on Marsaglias xorshift random number generators. *Journal of Statistical Software, 11*(5), 1–4.

Chakrabarti, S., & Samanta, D. (2016). Image Steganography Using Priority-Based Neural Network and Pyramid. In N. Shetty, N. Prasad, & N. Nalini (Eds.), Emerging Research in Computing, Information, Communication and Applications. Springer. https://doi.org/10.1007/978-981-10-0287-8_15.

Chapra, S. C., & Canale, R. P. (2010). *Numerical Methods for Engineers* (6th ed.). McGraw Hill.

Chelva, M. S., Halse, S.V., & Ratha, B.K. (2016). Object Tracking In Real Time Embedded System Using Image Processing. *International conference on Signal Processing, Communication, Power and Embedded System (SCOPES).*

Chioncel, C. P., Chioncel, P., Gillich, N., & Tirian, O. G. (2011). Wigner Ville Distribution in Signal Processing, using Scilab Environment. *Analele Universităţii "Eftimie Murgu" Reşiţa: Fascicola I, Inginerie, 18*(2), 101–106. http://anale-ing.uem.ro/

Chioncel, P., Gillich, N., Chioncel, C. P., & Elizabeta, S. (2016). *Digital Speed Cascade Control, using Scilab / Xcos Environment.* https://www.researchgate.net/publication/310124174

Chioncel, P., Silviu, D., & Chioncel, C. P. (2014). Calculation of Control Circuits in Time Domain using Scilab/ Xcos Environment. *Analele Universităţii "Eftimie Murgu" Reşiţa: Fascicola I, Inginerie, 21*(3). https://doaj.org/toc/1453-7397

Chong, J., Gonina, E., & Keutzer, K. (2010). Monte Carlo methods: a computational pattern for our pattern language. *Workshop on Parallel Programming Patterns.* 10.1145/1953611.1953626

Chopparapu, S., & Seventline Dr, B. J. (2020). GUI for Object Detection using Voila Method in MAT-LAB. *International Journal of Electrical Engineering and Technology*, *11*(4), 169–174.

Chopparapu, S. T., & Seventline, J. D. B. (Eds.). (2020). Object detection using Matlab, Scilab and Python. IAEME Publication. doi:10.34218/IJEET.11.6.2020.010

Coddington, P. D. (1996). *Random number generator for parallel computers. NHSE Review.*

David, B. T., Lee, H., & Wayne, L. (2009). A comparison of CPUs, GPUs, FPGAs, and massively parallel processor arrays for random number generation. FPGA, 63–72.

Dixon, M., Chong, J., & Keutzer, K. (2009). Acceleration of market value-at-risk estimation. *Workshop on High Performance Computing in Finance at Super Computing 2009.*

Dixon, M. F. (2012). Monte Carlo Based Financial Market Value-at-Risk Estimation on GPUs. In T. Bradley, J. Chong, & K. Keutzer (Eds.), *GPU Computing Gems Jade Edition* (pp. 337–353). Morgan Kaufmann. doi:10.1016/B978-0-12-385963-1.00025-3

Eleftheriou, M., Moreira, J., & Ryu, K. (Eds.). (2009). *WHPCF 2009: Proceedings of the 2nd Workshop on High Performance Computational Finance.* ACM.

Fabbri, R. (n.d.). *Scilab & SIP for Image Processing.* Institute of Mathematical and Computer Sciences, University of Sao Paulo, Brazil.

Fractal Analysis of geomorphologic processes in the Linux environment using SCILAB. (2011). https://www.academia.edu/12089182/

Garland, M., Grand, S. L., Nickolls, J., Anderson, J. A., Hardwick, J., Morton, S., Phillips, E., Zhang, Y., & Volkov, V. (2008) Parallel Computing Experiences with CUDA. *Micro, IEEE, 28*(4), 13–27.

Gilberto, E. (2001). Probability Distributions with SCILAB. Academic Press.

Giles, M. B., Kuo, F. Y., Sloan, I. H., & Waterhouse, B. J. (2008). Quasi-Monte Carlo for finance applications. *The ANZIAM Journal*, *50*, 308–323.

Gillich, G. R., & Chioncel, C. P. (2005). Simulation of dynamical systems with linear and non-linear behavior in SCICOS environment. *Annals of „Dunărea de Jos" University of Galati, Fascicle XIV. Mechanical Engineering*, 55–60.

Glasserman, P. (2003). *Monte Carlo Methods in Financial Engineering. Appl. of Math., 53.*

Gomathy, V., Padhy, N., & Samanta, D. (2020). Malicious node detection using heterogeneous cluster based secure routing protocol (HCBS) in wireless adhoc sensor networks. *Journal of Ambient Intelligence and Humanized Computing*, *11*, 4995–5001. https://doi.org/10.1007/s12652-020-01797-3

Gonzalez, R. G. (2002). *Digital Image Processing* (2nd ed.). Prentice Hall.

Guha, A., & Samanta, D. (2020). Real-Time Application of Document Classification Based on Machine Learning. In L. Jain, S. L. Peng, B. Alhadidi, & S. Pal (Eds.), Intelligent Computing Paradigm and Cutting-edge Technologies. ICICCT 2019. Learning and Analytics in Intelligent Systems (Vol. 9). Springer. https://doi.org/10.1007/978-3-030-38501-9_37.

Guha, A., & Samanta, D. (2021). Hybrid Approach to Document Anomaly Detection: An Application to Facilitate RPA in Title Insurance. *Int. J. Autom. Comput., 18*, 55–72. doi:10.100711633-020-1247-y

Hall, A., & Kabaila, A. P. (1977). *Basic Concepts of Structural Analysis*. Pitman Publishing.

Hill, D. R. C. (1997). Object-oriented pattern for distributed simulation of large scale ecosystems. *SCS Summer Computer Simulation Conference*, 945-950.

Hissoiny, S., Després, P., & Ozell, B. (2011). *Using graphics processing units to generate random numbers*. Academic Press.

Howes, L., & Thomas, D. (2007). Efficient random number generation and application using CUDA. In H. Nguyen (Ed.), *GPU Gems 3, NVIDIA*. Addison Wesley.

Huba, M., Bisták, P., Fikar, M., & Kamenský, M. (2006). Blended Learning Course 'Constrained PID Control'. *7th IFAC Symposium on Advances in Control Education ACE'06*, Madrid, Spain.

Hubbard, D. (2007). *How to Measure Anything: Finding the Value of Intangibles in Business*. John Wiley & Sons.

Ikhankar, R., Kuthe, V., Ulabhaje, S., Balpande, S., & Dhadwe, M. (2015). Pibot:The Raspberry Pi Controlled MultiEnvironment Robot For Surveillance & Live Streaming. In *2015 International Conference on Industrial Instrumentation and Control (ICIC)*. College of Engineering Pune,.

Jakab, F., Andoga, V., Kapova, L., & Nagy, M. (2006). Virtual Laboratory:Component Based Architecture Implementation Experience. Electronic Computer and Informatics.

James, F. (1990). A review of pseudorandom number generators. *Computer Physics Communications, 60*, 329–344.

Jan, S. (2004). GATE: A simulation toolkit for PET and SPECT. *Physics in Medicine and Biology, 49*, 4543–4561.

Janorkar, D. S. (2018). True Value of Pi (π) Now is 3.141592653 we Call This as Goba Constant we Symbolic it as This Goba, This Letter. *International Journal of Mathematics Trends and Technology, 59*(1), 27–34. doi:10.14445/22315373/IJMTT-V59P505

Janowczyk, A., Chandran, S., & Aluru, S. (2008). Fast, Processor-Cardinality Agnostic PRNG with a Tracking Application. In *Computer Vision, Graphics and Image Processing, 2008. ICVGIP08. Sixth Indian Conference on*, (pp. 171–178). Academic Press.

Joe, S., & Kuo, F. (2003). Remark on algorithm 659: Implementing Sobol's quasi-random sequence generator. *ACM Transactions on Mathematical Software, 29*(1), 49–57.

Jondeau, E., Poon, S., & Rockinger, M. (2007). *Financial Modeling Under Non-Gaussian Distributions*. Springer Finance.

Jorion, P. (2007). *Value-at-Risk: The New Benchmark for Managing Financial Risk* (3rd ed.). McGraw-Hill.

Kalyani, A., Premalatha, B., & Ravi Kiran, K. (2018). Real Time Emotion Recognition from Facial Images using Raspberry Pi. IJATIR, 10(1), 13-16.

Karmshu. (2012). *Probabilistic Simulation and Monte Carlo Method*. INSPIRE Science Camp, ISM Dhanbad.

Karthikeyan, M. P., Samanta, D., Banerjee, A., Roy, A., & Inokawa, H. (2021). Design and Development of Terahertz Medical Screening Devices. In M. Chakraborty, R. K. Jha, V. E. Balas, S. N. Sur, & D. Kandar (Eds.), Trends in Wireless Communication and Information Security. Lecture Notes in Electrical Engineering (Vol. 740). Springer. https://doi.org/10.1007/978-981-33-6393-9_40.

Kassimali, A. (1999). *Matrix Analysis of Structures*. Brooks/Cole Publishing Company.

Keutzer, K., & Mattson, T. (2009). *Our pattern language (opl)*. Academic Press.

Khadri, S. K. A. (2014). Approach of Message Communication Using Fibonacci Series. Cryptology. Lecture Notes on Information Theory. doi:10.12720/lnit.2.2.168-171

Khamparia, A., Singh, P. K., Rani, P., Samanta, D., Khanna, A., & Bhushan, B. (2020). An internet of health things-driven deep learning framework for detection and classification of skin cancer using transfer learning. *Trans Emerging Tel Tech*. doi:10.1002/ett.3963

Kim, S., & Hur, J. (2020). A Probabilistic Modeling Based on Monte Carlo Simulation of Wind Powered EV Charging Stations for Steady-States Security Analysis. *MDPI*. doi:10.3390/en13205260

Kureethara, V., Biswas, J., & Debabrata Samanta, N. G. (n.d.). Balanced Constrained Partitioning of Distinct Objects. *International Journal of Innovative Technology and Exploring Engineering*. Doi:10.35940/ijitee.K1023.09811S19

L'ecuyer, P. (1988). Efficient and portable combined random number generators. *Commun. ACM, 31*(6), 742–751.

L'Ecuyer, P. (1990). Random numbers for simulation. Communications of the ACM, 85-98.

L'ecuyer, P. (1999). Tables of linear congruential generators of different sizes and good lattice structure. *Math. Comput., 68*(225), 249–260.

L'Ecuyer, P., & Simard, R. (2003). TESTU01: a software library in ANSI C for empirical testing of random number generators. Department d'Informatique et de Recherche Operationnelle, University of Montreal.

L'ecuyer, P., & Simard, R. (2007). TestU01: A C library for empirical testing of random number generators. *ACM Trans. Math. Softw., 33*(4), 22.

Langdon, B. (2009). A Fast High Quality Pseudo Random Number Generator for nVidia CUDA. *GECCO 2009 Workshop, Tutorial and Competition on Computational Intelligence on Consumer Games and Graphics Hardware CIGPU*.

Lazaro, D., Breton, V., & Buvat, I. (2004). Feasibility and value of fully 3D Monte-Carlo reconstruction in single photon emission computed tomography. *Nuclear Instruments & Methods in Physics Research. Section A, Accelerators, Spectrometers, Detectors and Associated Equipment, 527*, 195–200.

Lazaro, D., El Bitar, Z., Breton, V., Hill, D. R. C., & Buvat, I. (2005). Fully 3D Monte Carlo reconstruction in SPECT: A feasibility study. *Physics in Medicine and Biology, 50*, 3739–3754.

Leão, C. P., & Rodrigues, A. E. (2004). Transient and steady-state models for simulated moving bed processes: Numerical solutions. *Computers & Chemical Engineering, 28*(9), 1725–1741.

Li, Y., & Mascagni, M. (2003). Improving Performance via Computational Replication on a Large-Scale Computational Grid. *CCGRID, 3rd International Symposium on Cluster Computing and the Grid*, 442-446.

Liguš, J., Ligušová, J., & Zolotová, I. (2005). *Distributed Remote Laboratories in Automation Education.* 16th EAEEIE Annual Conf. on Innovation in Education for Electr. and Information Eng., Lappeenranta, Finland.

Lönnblad, L. (1994). CLHEP – a project for designing a C++ class library for high energy physics. *Computer Physics Communications, 84*, 307–316.

Macgillivray, H. T., & Dodd, R. J. (2004). Monte-Carlo simulations of galaxy systems. Academic Press.

Maheswari, & Geetha, Kumar, Karuppiah, Samanta, & Park. (n.d.). PEVRM: Probabilistic Evolution based Version Recommendation Model for Mobile Applications. *IEEE Access: Practical Innovations, Open Solutions.* Advance online publication. doi:10.1109/ACCESS.2021.3053583

Mahmood, N.H., & Mansor, M.A. (2012). Red Blood Cells Estimation Using Hough Transform Technique. *SIPIJ, 3*(2).

Maigne, L., Hill, D. R. C., Calvat, P., Breton, V., Reuillon, R., Lazaro, D., Legre, Y., & Donnarieix, D. (2004). Parallelization of Monte Carlo simulations and submission to a grid environment. *Parallel Processing Letters, 14*, 177–196.

Maini, R., & Aggarwal, H. (n.d.). Study and Computational of Various Image Edge Detection Techniques. *International Journal of Image Processing, 3*(1).

Maitra, M., Gupta, R.K., & Mukherjee, M. (2012). Detection and Counting of Red Blood Cells in Blood Cell Images using Hough Transform. *International Journal of Computer Application, 53*(16).

Marsaglia, G., & Zaman, A. (1987). *Toward a Universal Random Number Generator.* Florida State University.

Marsaglia, G. (1995). *Diehard, a battery of tests for random number generators.* Academic Press.

Marsaglia, G. (1997). *A random number generator for C.* Sci. Math. Num-analysis news group.

Marsaglia, G. (2003). Xorshift RNGs. *Journal of Statistical Software, 8*(14), 2003.

Marsaglia, G. (2003). Random number generation. In *Encyclopedia of Computer Science* (pp. 1499–1503). John Wiley and Sons Ltd.

Marsaglia, G., & Zaman, A. (1991). A New Class of Random Number Generators. *The Annals of Applied Probability, 1*(3), 462-480.

Mascagni, M., Ceperley, D., & Srinivasan, A. (2000). SPRNG: A scalable library for pseudorandom number generation. *ACM Transactions on Mathematical Software, 26*, 618–619.

Mascagni, M., & Chi, H. (2004). Parallel linear congruential generators with Sophie-Germain moduli. *Parallel Computing, 30*, 1217–1231.

Mascagni, M., & Srinivasan, A. (2004). Parameterizing parallel multiplicative lagged-Fibonacci generators. *Parallel Computing, 30*, 899–916.

Matsumoto, M., & Nishimura, T. (1997). Mersenne Twister: A 623-dimensionally equidistributed uniform pseudorandom number generator. *Proceedings of the 29th conference on Winter simulation*, 127-134.

Matsumoto, M., & Nishimura, T. (1998). Mersenne twister: a 623-dimensionally equidistributed uniform pseudo-random number generator. *ACM Trans. Model. Comput. Simul., 8*(1), 3–30.

Matsumoto, M., & Nishimura, T. (2000). *Dynamic creation of pseudorandom number generators* (Vol. 1998). Monte Carlo and Quasi- Monte Carlo Methods.

Metropolis, N. (1987). The Beginning of the Monte Carlo Method. *Los Alamos Science*, 125–130.

Moro, B. (1995). The full monte. *Risk Mag., 8*(2), 57–58.

Moulick, R. (2019). Calculating the value of Pi (π): A Monte Carlo Scheme in Scilab. *International Journal of Emerging Technologies and Innovative Research, 6*(1), 600-603. www.jetir.org

Mukherjee, M., & Samanta, D. (2014, June). Fibonacci Based Text Hiding Using Image Cryptography. *Lecture Notes on Information Theory, 2*(2), 172–176. doi:10.12720/lnit.2.2.172-176

Murshed, S. Z., Dutta, A., Chatterjee, S., Mondal, I., Saha, A., Saha, S., Kundu, D., Ghosh, S., & Das Gupta, S. (2016). Controlling an Embedded Robot through Image Processing based Object Tracking using MATLAB. *10th International Conference on Intelligent Systems and Control (ISCO)*. 10.1109/ISCO.2016.7726922

Nguyen, N.T., Duong, A.D., & Vu, H.Q. (2011). Cell Splitting with High Degree of Overlapping in Peripheral Blood Smear. *International Journal of Computer Theory and Engineering, 3*(3).

NVIDIA Corporation Inc. (2009). *NVIDIA CUDA Compute Unified Device Architecture Programming Guide version 2.3*. Author.

Nyland, L., Harris, M., & Prins, J. (2007). Fast N-Body Simulation with CUDA. In H. Nguyen (Ed.), *GPU Gems 3*. Addison Wesley Professional.

Ojha, S., & Sakhare, S. (2015). Image Processing Techniques for Object Tracking in Video Surveillance-A Survey. *2015 International Conference on Pervasive Computing (ICPC)*.

Panciatici, P., & Chieh, A. S. (2011). Equation-based hybrid modeling of power systems for time-domain simulation. *Power and Energy Society General Meeting. IEEE*, 1-9. 10.1109/PES.2011.6039155

Panneton, F., & L'ecuyer, P. (2005). On the xorshift random number generators. *ACM Trans. Model. Comput. Simul., 15*(4), 346–361.

Park, S. K., & Miller, K. W. (1988). Random number generators: good ones are hard to find. *Commun. ACM, 31*(10), 1192–1201.

Pawlikowski, K. (2003). Towards credible and fast quantitative stochastic simulation. *Proceedings of International SCS Conference on Design, Analysis and Simulation of Distributed Systems, DASD'03.*

Podlozhnyuk, V. (2007). *Parallel Mersenne Twister.* NVIDIA Corporation Inc.

Pradhan, Siddappa, Kavitha, & Samanta. (2019). Analysis & Improvement of Wireless Network Security Based on Biometrics. In *Proceedings of International Conference on Sustainable Computing in Science, Technology and Management (SUSCOM).* Amity University Rajasthan. https://ssrn.com/abstract=3356360

Praveen, B., Samanta, D., Prasad, G., Ranjith Kumar, C., & Prasad, M. L. M. (2020). Protecting Medical Research Data Using Next Gen Steganography Approach. In L. Jain, S. L. Peng, B. Alhadidi, & S. Pal (Eds.), Intelligent Computing Paradigm and Cutting-edge Technologies. ICICCT 2019. Learning and Analytics in Intelligent Systems (Vol. 9). Springer. https://doi.org/10.1007/978-3-030-38501-9_34.

Rao, S. (2002). *Applied Numerical Methods for Engineers and Scientist* (3rd ed.). Pearson Prentice Hall Education.

Restivo, M. T., Mendes, J., Lopes, A. M., Silva, C. M., & Chouzal, F. (2009). A Remote Lab in Engineering Measurement. *IEEE Transactions on Industrial Electronics, 56.*

Reuillon, R., Hill, D.R.C, & Bitar, Z. (2008). Rigorous Distribution of Stochastic Simulations Using the Dist Me Toolkit. *IEEE Transactions on Nuclear Science.*

Reuillon, R., Hill, D. R. C., Gouinaud, C., Bitar, Z. E., Breton, V., & Buvat, I. (2008). Monte Carlo simulation with the GATE software using grid computing. *8th International Conference on New Technologies in Distributed Systems.*

Risk Management Systems in the Aftermath of the Financial Crisis Flaws, Fixes and Future Plans. (2010). A GARP report prepared in association with SYBASE.

Salleh, Z. (2011). *Fundamental of Numerical Methods for Scientists and Engineers.* Lambert Academic Publishing.

Salleh, Z., & Yusop, M. Y. M. (2012). Basic of numerical computational using Scilab programming. *2nd International Conference on Mathematical Applications in Engineering (ICMAE2012).*

Samanta, Mousumi, Khutubuddin, & Khadri. (2013). Message Communication Using Phase Shifting Method (PSM). *International Journal of Advanced Research in Computer Science, 4*(11), 9–11. doi:10.26483/ijarcs.v4i11.1936

Samanta, D., Sivaram, M., Rashed, A., Boopathi, C. S., Sadegh Amiri, I., & Yupapin, P. (2020). Distributed Feedback Laser (DFB) for Signal Power Amplitude Level Improvement in Long Spectral Band. *Journal of Optical Communications.* . doi:10.1515/joc-2019-0252

Samanta, D. (2020). Distributed Feedback Laser (DFB) for Signal Power Amplitude Level Improvement in Long Spectral Band. *Journal of Optical Communications.* www.degruyter.com

Samanta, D., & Sanyal, G. (2012). Novel Shannon's Entropy Based Segmentation Technique for SAR Images. In K. R. Venugopal & L. M. Patnaik (Eds.), Wireless Networks and Computational Intelligence. ICIP 2012. Communications in Computer and Information Science (Vol. 292). Springer. https://doi.org/10.1007/978-3-642-31686-9_22

Sharif, J. M., Miswan, M. F., Ngadi, M. A., & Salam Md, S. H. (2012). Red Blood Cell Segmentation Using Masking And Watershed Algorithm: A Preliminary Study. *International Conference On Biomedical Engineering*, 27-28.

Shetty, M. (2004). *Geometric Estimation of Value of Pi.* http://www.ijoart.org/docs/Geometric-Estimation-of-Value-of-Pi.pdf

Siddique, S., & Sayyed, R. (2016, March). Automated RBCs Segmentation & Counting using SCILAB. *International Journal of Engineering Research & Technology*, 5(3). Advance online publication. doi:10.17577/IJERTV5IS030872

Singh, U. (2013). *Estimation of the value of using Monte-Carlo Method and Related Study of Errors.* https://www.academia.edu/1887423/

Singla, N., Hall, M., Shands, B., & Chamberlain, R. D. (2008). Financial Monte Carlo simulation on architecturally diverse systems. *Workshop on High Performance Computational Finance, Supercomputing 08*, 1–7.

Sivakumar, P., Nagaraju, R., & Samanta, D. (2020). A novel free space communication system using nonlinear InGaAsP microsystem resonators for enabling power-control toward smart cities. *Wireless Networks*, 26, 2317–2328. https://doi.org/10.1007/s11276-019-02075-7

Srimani, S. (2015). A Geometrical Derivation of π (Pi). *IOSR Journal of Mathematics (IOSR-JM), 11*(6), 19–22. doi:10.9790/5728-11611922

Srinivasan, A. (2002). Parallel and distributed computing issues in pricing financial derivatives through Quasi Monte Carlo. *Proceedings of the 16th International Parallel and Distributed Processing Symposium*, 14–19.

Srinivasan, A., Ceperley, D. M., & Mascagni, M. (1999). Random number generators for parallel applications. In D. M. Ferguson, J. I. Siepmann, & D. G. Truhlar (Eds.), Advances in Chemical Physics Series: Vol. 105. *Monte Carlo Methods in Chemical Physics* (pp. 13–36). John Wiley and Sons.

Sriram, V., & Kearney, D. (2007). High Throughput Multi-port MT19937 Uniform Random Number Generator. *Parallel and Distributed Computing Applications and Technologies, International Conference on*, 157–158.

Stephen, L. C., Chancelier, J. P., & Nikoukhah, R. (2006). *Modeling and Simulation in Scilab/Scicos.* Springer.

Syed, K. A. K., Samanta, D., & Paul, M. (2014). Approach of Message Communication Using Fibonacci Series: In Cryptology. Lecture Notes on Information Theory, 2(2), 168-171. doi:10.12720/lnit.2.2.168-171

Thiruvinal, V. J., & Ram, S. P. (2017). Automated Blood Cell Counting and Classification Using Image Processing. *International Journal of Advanced Research in Electrical, Electronics and Instrumentation Engineering, 6*(1). doi:10.15662/IJAREEIE.2017.0601010

Thomas, D. B., & Luk, W. (2008). Multivariate gaussian random number generation targeting reconfigurable hardware. *ACM Trans. Reconfigurable Technol. Syst., 1*(2), 1–29.

Traore, M., & Hill, D. R. C. (2001). The use of random number generation for stochastic distributed simulation: application to ecological modeling. *Proceedings of the 13th European Simulation Symposium*, 555-559.

Vijayalakshmi, M. N., & Senthilvadivu, M. (2016). Performance Evaluation of Object Detection Techniques for Object Detection. *2016 International Conference on Inventive Computation Technologies (ICICT)*. 10.1109/Inventive.2016.7830065

Wang, X., & Sloan, I. H. (2008). Low discrepancy sequences in high dimensions: How well are their projections distributed? *Journal of Computational and Applied Mathematics, 213*(2), 366–386.

Weaver, W., Jr., & Gere, J. M. (1986). Matrix Analysis of Framed Structures (2nd ed.). CBS Publishers and Distributors.

Wenjiang, L., Nanping, D., & TongShun, F. (2009). The application of Scilab / Scicos in the lecture of automatic control. *Open-source Software for Scientiifc Computation (OSSC), IEEE International Workshop*, 85–87.

Wu,, P., & Huang, K. (2006). Parallel use of multiplicative congruential random number generators. *Computer Physics Communications, 175*, 25–29.

Yauri-Machaca, M., Meneses-Claudio, B., & Vargas-Cuentas, N. (2018). Design of a Vehicle Driver Drowsiness Detection System through Image Processing using Matlab. IEEE.

Youngman, P. (2009). Procyclicality and Value-at-Risk. Bank of Canada Financial System Review Report.

Chapter 4
Computational Statistics–Based Prediction Algorithms Using Machine Learning

Venkat Narayana Rao T.
Sreenidhi Institute of Science and Technology, Hyderabad, India

Manogna Thumukunta
Sreenidhi Institute of Science and Technology, Hyderabad, India

Muralidhar Kurni
https://orcid.org/0000-0002-3324-893X
Anantha Lakshmi Institute of Technology and Sciences, Ananthapuram, India

Saritha K.
Sri Venkateswara Degree and PG College, Ananthapuram, India

ABSTRACT

Artificial intelligence and automation are believed by many to be the new age of industrial revolution. Machine learning is an artificial intelligence section that recognizes patterns from vast amounts of data and projects useful information. Prediction, as an application of machine learning, has been sought after by all kinds of industries. Predictive models with higher efficiencies have proven effective in reducing market risks, predicting natural disasters, indicating health risks, and predicting stock values. The quality of decision making through these algorithms has left a lasting impression on several businesses and is bound to alter how the world looks at analytics. This chapter includes an introduction to machine learning and prediction using machine learning. It also sheds light on its approach and its applications.

INTRODUCTION

Evolution has been an inherent part of human beings since the advent of time. Humans have always been in search of more effective and convenient solutions to livelihood. The concept of using a mechani-

DOI: 10.4018/978-1-7998-7701-1.ch004

cal device to simplify human tasks has also been a result of the very need for simplification. Initially functioning at reducing human effort in mundane tasks, a machine can now mimic human behavior and thought ability and do so with great accuracy.

Artificial Intelligence has proven an effective solution to many problems that were initially deemed impossible to decode. Artificial Intelligence has set the stakes higher in all the fields it has penetrated. In this day and age, machines can comprehend high-dimensional data and produce results of immense value. The main goal of Artificial Intelligence is to produce results with as little human intervention as possible (Sodhi et al., 2019). Artificial Intelligence aims at perfecting machine independence and replication of human processing ability.

The current success of Artificial Intelligence can be accounted to the development of new and modern systems that make use of pools of information along with human expertise to perfect their performance. The edge that Artificial Intelligence has over humans is that machines never tire out. The machine will go through multiple iterations to learn from each of them and improve. The added advantage of computational speed and accuracy makes them invaluable, effective, and quick solutions.

WHAT IS MACHINE LEARNING?

The term Machine Learning was coined in the year 1959 by Arthur Samuel. Machine Learning is an Artificial Intelligence application wherein the machine learns and improves through experiential learning(Cioffi et al., 2020). A Machine Learning algorithm can access heaps of data and learn by examining it (Sodhi et al., 2019). Learning begins with the machine accessing the data, observing patterns in the data, and learning through those examples and improving, thus making better decisions. Machine Learning algorithms are generally classified into supervised and unsupervised (Salian, 2018). In supervised learning, the machine is trained using a known training dataset and uses the knowledge gained to predict unknown datasets' output (Salian, 2018). In unsupervised learning, the machine trains with the data that is not classified or labeled (Salian, 2018)(Sodhi et al., 2019).

The process of Machine Learning starts with identifying the problem statement. This step involves understanding the problem at hand, determining what has to be predicted. The next step in the process is to gather all relevant data from suitable and trusted sources. The data collected can be inconsistent and have missing values, redundant or duplicate values in it. Such data can hinder the algorithm's performance and accuracy; thus, the data must be cleaned to eliminate them. The cleaned data can then be studied for patterns and correlations among the variables. The insights gained from the previous step can now be used to build a model. This involves splitting the data into training and testing samples. The training data will be used to analyze and structure the model. The testing data helps check the model's accuracy and performance. Based on these results, the model can be improved and optimized. The machine, at this point, is finally ready to make predictions.

PREDICTIVE MODELING

Predictive modeling is a concept that employs a Machine Learning algorithm that is capable of making predictions by learning from a training dataset. Predictive modeling is a process of predicting outcomes in the future by analyzing the results of the past (Kumar & Garg, 2018). A predictive model's function is to identify that a data unit from a different source with similarity to existing patterns exhibits a specific behavior(Sodhi et al., 2019). The available data with known values and labels are used to train the training dataset model. The dataset with unknown values and labels that the model has to predict is known as the test dataset(Xin et al., 2018).

Finding the best predictive model involves creating the model, testing, and validating it with respect to outcomes predicted based on given input data. A model can make use of multiple algorithms to best analyze the data. There are multiple algorithms, each more suitable to specific kinds of data or output (Kalyanam, 2017). Employing models that best suit the data at hand generates more accurate results. Inculcating this technology acts as an added advantage to companies. For instance, using predictive modeling, the marketing sector can accurately target customer segments based on various factors. To sum up, predictive modeling is a data-mining technology that creates a model to help predict future results by analyzing historical and current data.

There are two types of predictive modeling: pattern classification and regression.

1. **Classification:** It is a supervised learning approach where the algorithm learns through training data to appropriately classify new data (Salian, 2018)(F.Y. et al., 2017). Classification is the process of categorizing information into discrete classes with specific labels (Sodhi et al., 2019). The classification can either be a bi-class or a multi-class. The output can take one of two bi-class options, like classifying a person as major or not (minor). Whereas in multi-class, the output can take a label from multiple options, for example, the nationality of a person being German, Indian, Australian, etc.
2. **Regression:** It is a supervised machine learning algorithm. It is a predictive modeling technique that establishes a relationship between dependent and independent variables. The dependent variable or target is the predicted value using independent variables or predictors, where the target quantity is continuous in nature (Sodhi et al., 2019). In this technique, a curve or line fits into the data points so that the difference between the distances of the data points from that curve is minimized. There are different regression techniques and their applications. Some of the instances for regression usage are weather forecasting, determining the cost of a place based on area, etc.

Classification Models

There are multiple classification models, each suitable for various applications and data. Some of the most popular classification models are discussed below.

Decision Trees: A decision tree classification model is a supervised machine learning approach that is tree-structured. It disintegrates data into smaller subsets and simultaneously builds a decision tree (F.Y. et al., 2017). A decision tree has three parts- decision nodes, edges or branches, and leaf nodes. A decision node branches out into two or more branches and checks an attribute for a specific condition. The edges or branches are connections of a decision node to other decision nodes or leaf nodes. The leaf nodes are the terminal nodes that represent an outcome or a decision. A root node is the topmost decision node, which represents the best condition or predictor. The advantage of decision trees is that they can take both categorical and numeric data(Sodhi et al., 2019). Figure 1a represents a company's data, that his age and salary as the attributes and the product purchased as the outcome. The decision tree for this is represented in Figure 1b.

Figure 1a. Company data

S.NO.	AGE	SALARY	PURCHASED
1.	21	50,000	yes
2.	35	70,000	yes
3.	47	72,000	no
4.	62	90,000	no
5.	22	30,000	yes
6.	33	80,000	yes
7.	50	1,56,000	no
8.	33	85,000	no
9.	22	1,00,000	no
10.	32	1,00,000	no
11.	42	60,000	no
12.	20	25,000	yes
13.	60	58,000	no
14.	41	1,15,000	no
15.	23	95,000	no
16.	17	20,000	yes

Figure 1b. Decision Tree

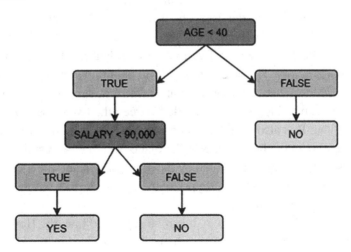

The root node in the above example is a decision node that validates the condition of age being less than 40. If true, it leads us to another decision node for checking if the salary is less than 90,000. If these conditions are met, the decision tree's result is that the customer will likely purchase the product. If not met, then according to the decision tree, the customer is not likely to purchase the product.

Note: To understand the concept, small data set with fewer variables and a small decision tree is represented. Actual data is most likely highly complex with multiple variables and a large decision tree.

Random Forest: As the name suggests, a random forest is a supervised learning classification algorithm with several separate decision trees that work collectively and predict the outcome(F.Y. et al., 2017). The working of a random forest is similar to that of a committee deciding an outcome based on votes. In a random forest classifier, each of the individual decision trees predicts an outcome or a class. The class with the most number of predictions from the decision trees is decided as the entire model's outcome. The random forest model's high success rate can be attributed to its functionality that establishes a larger randomness scale(F.Y. et al., 2017). The low correlation between the individual decision trees also allows for a more comprehensive analysis of the information, thus providing a better prediction. The higher success rate can also be accounted for because a random forest model has a shallow scope for error. Even with some decision trees being wrong, collectively, they are less likely to arrive at the wrong outcome.

A random forest can be used to perform both regression and classification tasks. It is simple and easy to understand and implement; hence it is widely used. The larger the number of decision trees in the random forest, the greater the model's accuracy (F.Y. et al., 2017). The lesser the correlation between the trees, the better the model (F.Y. et al., 2017). Also, for the model to predict accurate results, the dataset needs to be as accurate as possible. Lesser computational time and higher accuracy, even with larger datasets, are some of the reasons why the random forest is one of the most used algorithms (F.Y. et al., 2017). For a better understanding of the concept behind the algorithm, let us consider an example. A person consulting a single friend before buying a mobile phone is an example of a decision tree. When that person consults several of his friends and buys the mobile phone that most of his friends suggested, it is a random forest.

Support Vector Machine: Support Vector Machine (SVM) is a supervised Machine Learning algorithm that uses a hyperplane to classify the data. The data is plotted as data points on an n-dimensional space, 'n' corresponding to the data's number of features (Kumar & Garg, 2018). The hyper-plane divides the n-dimensional space to best suit the classification of these data points. Each of these individual coordinates is called support vectors. The question may arise about how the model will identify the right hyperplane (F.Y. et al., 2017). This can simply be answered by identifying the hyper-plane that best divides the classes. Plane C best identifies the divide between the two classes among the three planes in the example below. Plane B is also observed to be correctly dividing the two classes, in which case the model calculates the perpendicular distance of the plane from the nearest data points in each of the classes. This distance is called Margin (indicated by the orange line), and the hyperplane with the highest Margin is considered optimal. In Figure 2, hyper-plane C has a higher margin compared to plane B. In consequence, Plane C is the best frontier.

In the example above, the data is linear; i.e., it can be separated linearly. However, in real life, data is hardly linear, which makes it difficult for linear separation. This is when the SVM algorithm's kernel trick comes into the picture. The kernel is a function that transforms data into higher dimensions in a less expensive way. Thus, it separates data that is not linearly separable in the same dimensional space.

Figure 2. Support Vectors

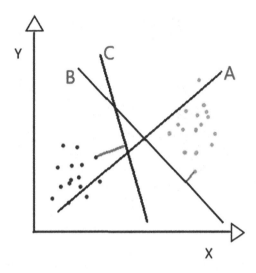

The advantage of the Support Vector Machine algorithm is its high accuracy with low computational power. SVM is also said to act effectively with smaller datasets.

Naive Bayes: Naive Bayes classification is a machine learning algorithm based on the Bayes Theorem. To understand the working of the algorithm, let us study Bayes' theorem. Bayes' theorem is based on conditional probability. Conditional probability can determine the happening of an event based on events that have occurred previously. The mathematical representation of Bayes' theorem is shown in equation (1) below:

$$P\left(A / B\right) = \frac{P\left(B / A\right)P\left(A\right)}{P\left(B\right)}$$

(1)

P(A) = Probability of A happening
P(B) = Probability of B happening
P(A/B) = Probability of A happening, given B, is true
P(B/A) = Probability of B happening, given A, is true

Here, A is called the hypothesis, i.e., the prediction factor. B is called the evidence or the known data. Using this theorem, we can predict the probability of a particular event happening, having defined evidence at hand (F.Y. et al., 2017). This theorem makes an assumption that the predictors are independent and not related, thus the word naive. The assumption is that features are independent of each other, rarely true with datasets in real life. Naive Bayesian classifier also assumes equal weightage or importance to each of its features. Certain features might affect the outcome with actual data more than the other, which naive Bayes fails to consider. Nevertheless, this classifier's advantage is that it only requires a small training dataset to estimate output values. The classifier is mainly used for text classification problems because of its independence and multi-class factors.

Figure 3. Neural Networks

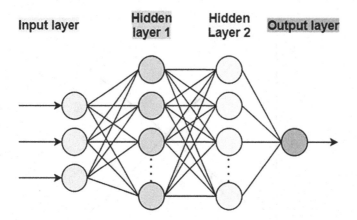

Let us consider a model that classifies a patient as either person with diabetes or not. Some of the features used to predict the outcome are age, gender, body mass index (BMI), insulin level, etc. The model identifies the probability of a person having diabetes based on these factors but does not consider interdependence. Sentiment Analysis and Spam mail filtering are the major applications of this classifier model.

Neural Networks: Neural networks, as the name explicitly suggests, are modeled on the brain's neural structure. Inspired by the neurons' architecture and working, a neural network classification model can process complex data by mimicking the natural brain activity (Kumar & Garg, 2018). The artificial neuron or node receives an input, processes it, and passes the next neuron's output. The nodes' connections are called edges, which are generally weighted and modified as the algorithm learns. The weights indicate the importance of that particular input. The weights are adjusted as the learning progresses. A neural network's basic structure has an input layer, single or multiple hidden layers, and an output layer (Mei et al., 2018), (F.Y. et al., 2017). The input layer takes in the features for which the output is predicted through the output layer. One of the many applications of neural networks is image processing. The input layer takes in pixel values broken down from the image. The multiple functional layers then process this image by converting it into data points that the machine can understand (Mei et al., 2018), (F.Y. et al., 2017). The model starts to understand the trends by examining multiple images and identifying similarities. Figure 3 shows a simplified version of a neural network with two hidden layers, an input layer, and the output layer.

Neural networks are generally not programmed with any specific rules and learn as they proceed. For instance, a neural network algorithm used in image recognition learns to identify the image, say boat, by analyzing examples of other images that are labeled boat or not. The model will identify the image through these examples by generating features without any prior knowledge of said features in those images. The model's learning rate is the number of steps it takes to correct or adjust its errors while processing each observation. Optimization is the process of minimizing errors, thus increasing the Artificial neural network's accuracy and reliability.

Regression Models

Linear regression: Linear regression is used to analyze and find the relationship between two variables. In predictive modeling, linear regression is used to predict the value of a variable. It is a supervised learning approach. The linear regression task determines the value of a dependent variable (y) based on the independent variable's known value (x). This regression aims to determine the linear relationship between the two sets of variables(Silhavy et al., 2017). Equation(2) can represent linear regression:

$$y = ax + b \tag{2}$$

Where y represents a dependent variable,

x represents known data or independent variables,
b is the intercept and
a is the x-coefficient.

The best fit is determined by a straight line that most correctly fits the data points. To find the best fit, we determine the sum of squares of vertical distances of each line's data points. These distances are termed as deviations or sometimes called errors. The goal is the minimize the sum as much as possible. The sum can be denoted as $\Sigma\ d_i^2$, where di is the distance of each data point from the regression line. The lesser the sum, the better the linear regression fit(Silhavy et al., 2017). The model fits the best line by considering the best values of a and b. Shown below is a graph (Figure 4) that establishes a linear relationship between an employee's salary and experience.Here, salary is the dependent variable, and experience is the independent variable. So, to predict an employee's salary whose experience is known, from the particular value of experience on the x-axis, draw a vertical line that meets the regression line. Then draw a horizontal line from the point of contact to the y-axis. The value on the point of intersection is the predicted value of the regression model.

Figure 4. Linear Regression

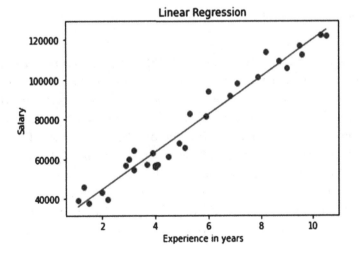

Multiple regression: Multiple regression is an extension of linear regression, with multiple regression having the capability to handle multiple independent variables instead of a single one as that of linear regression. The model's overall fit can be established in multiple regression, predicting each dependent variable(Silhavy et al., 2017). The simple regression extension is reflected in the addition of coefficients for each predictor to the existing equation. The general representation is depicted by equation (3) as below:

$$y = a_0 x_0 + a_1 x_1 + ... + a_n x_n + b \qquad (3)$$

where y is the dependent variable and $x_0, x_1 ... x_n$ are the independent variables.

Since there are multiple independent variables, the multiple regression model tries to fit the regression line through a multidimensional space where the data points are plotted. The target variable(y) has to be continuous in nature, while the factors can either be continuous or discrete in nature. Outliers in the dataset can have a significant impact on the outcome of the multiple regression model. The algorithm is also prone to over-fitting, but using dimensionality reduction, the problem can be overcome.

Let us take an example of the university admission data. Using multiple regression, we can predict the admission probability taking into account several factors like CGPA, English proficiency score, GRE score, work experience, publications, etc. Multiple regression yields a graph with multiple dimensions. All of these factors act as the independent variables, and the outcome or the probability of admission is the factor that depends on them. The more the number of attributes, the more the dimensions. Thus, a graph cannot be plotted with more than three dimensions.

Polynomial regression: The need for polynomial regression arises when a straight line cannot capture the patterns projected by the data (Haussmann, 2020). An equation with higher complexity is needed to fit the data effectively. A higher-order equation serves the purpose. Powers can be added to existing features to get the higher order. The fit is a quadratic curve in nature, thus better describing and capturing the data patterns. Polynomial regression is still a linear model since the coefficients are linear. Equations (4) and (5) shown below represent the linear form and the transformed higher-order equations:

Linear model: $y = ax + b$ \qquad (4)

Polynomial model: $y = a_1 x + a_2 x^2 + ... + a_n x^n + b$ \qquad (5)

The key to polynomial regression is the isolation of the relationship between each of the independent variables and the dependent variable. If these variables are related, then there arises the problem with the fit of the model. This problem is termed multicollinearity. Multicollinearity is when the variables are highly dependent on each other. If this degree of correlation is high enough, then it can disrupt the fit of the model. Thus, minimizing multicollinearity will improve the performance of the model. A wide range of curved data can be fit through polynomial regression. It also perfectly captures the relationship between the dependent and independent variables. However, outliers can vastly affect the outcome of the model.

The graphs shown below depict the linear and polynomial regression fit, respectively. In Figure 5a, the linear model fails to capture the patterns and distribution of data. In contrast, there is an accurate fit in the second graph (Figure 5b), where a polynomial regression model is used.

Figure 5a. Linear Regression

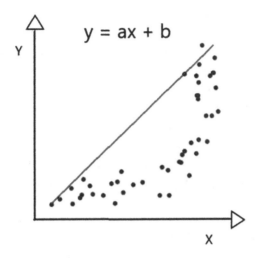

Figure 5b. Polynomial Regression

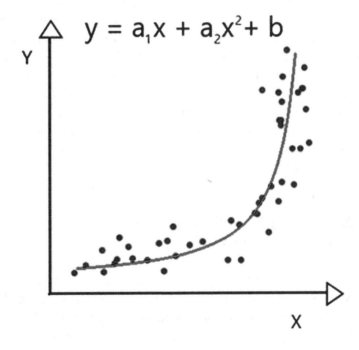

WHAT IS INDUSTRY 4.0

Industry 4.0 is the term given for the fourth industrial revolution. Below are each of the industrial revolutions described briefly.

- **Industry 1.0** - The 18th century saw an advance in production and mechanics. The introduction of steam used to power engines, the establishment of railway lines, power mills, etc., allowed for a multiplied production rate. The increased production rate allowed for individual-owned singular entities to businesses that serve a broader customer base.

- **Industry 2.0 -** The second industrial revolution occurred with the rise of the automotive industry, electricity, more efficient machines, assembly-line car manufacturing. This era saw the development and birth of several manufacturing firms and management jobs. The segregation of labor into various levels of management was a common practice. Electricity and automation allowed for an exponential increase in production, and mass production plants were popular. With production automation, companies focused on creating consistent and quality goods.

- **Industry 3.0 -** The third industrial revolution began with the advent of early computers. These were bulky and performed simple tasks but laid a foundation for the modern age computing we see today. This age gave rise to large-scale automation and the interference of computers to help aid mundane tasks. This era saw a shift from manual management to digital management. Computers helped manage simple tasks but created numerous fields of employment. This lead to a cascading of new and creative products in every field imaginable.

- **Industry 4.0 -** Industry 4.0 describes a large-scale application of Artificial Intelligence and predicting machines' ability to automate further and revolutionize various industry sectors (Cioffi et al., 2020)(Sodhi et al., 2019). This era of smart computing is made possible by technologies like Cloud computing, the Internet of Things (IoT), Human-machine interaction, smart cities, automated cars, among many. Industry 4.0 combines advanced manufacturing techniques with advanced technologies like the Internet of Things to create systems that drive the future towards an interconnected and artificially advanced world. It allows business owners to have higher control and understanding of all aspects of their businesses by integrating physical aspects of manufacturing with the new age digital technological solutions. It is not just about increasing efficiency in production, but on a greater level is about transforming the way the industry operates and functions (Alcácer & Cruz-Machado, 2019). To thrive as a business in the market today, reforming business practices to better adopt new technologies is the key.

APPLICATIONS OF ML (PREDICTIVE MODELING) IN VARIOUS FIELDS OF INDUSTRY 4.0

Every possible sector has received a technological makeover that has drastically altered or impacted those fields' workings. The very concept of Machine Learning and Predictive modeling has opened doors to an array of possibilities and advancements. The accuracy, convenience, cost-effectiveness, and abundant computational ability have much appealed to many businesses and have created innovation opportunities on a large scale. The number of applications employing predictive modeling has only increased ever since its conception. With the exponential growth in Artificial Intelligence, more sophisticated and accurate models are taking shape (Hallak & Azar, 2020)(Sodhi et al., 2019). There are numerous applications of predictive modeling. A few applications in some of the major sectors of Industry 4.0 are mentioned below.

1. **Business:** Predictive modeling has proven invaluable to the business sector (Indriasari et al., 2019). By allowing management to make educated decisions, the market risk has drastically reduced, thus marginally improving the company's performance. Predictive analysis is being utilized in all levels and departments. Some of the uses include identifying target customers, anticipating losses, thus

preventing them, helping build and developing a financial model, and estimating sales, among many (Kumar & Garg, 2018)(Vinay & Rahul, 2020). Calculation of customer lifetime value, customer segmentation, and dynamic pricing are among many Machine Learning applications in the business sector. With predictions from these models in hand, companies are focusing more on expansion in creative jobs (Sodhi et al., 2019). Not just large-scale businesses, but businesses on a smaller scale are also employing machine learning due to its robustness and cost-efficiency (Sodhi et al., 2019).

2. **Financial:** The financial sector has been a play of intuition and market experience. With predictive modeling in the game, it has evened the playing field for a wider audience. Financial leaders use predictive analysis in many ways, including predicting revenue, analyzing patterns in the market, deciding investment, predicting returns, forecasting stock values, etc. (Sodhi et al., 2019), (Indriasari et al., 2019). The models can account for a wide range of factors that affect the sector. Past trends and historical data allow the model to analyze relationships between various factors and predict the output. The financial sector is considered a number game, and who better than a computer to handle these numbers accurately (Sodhi et al., 2019).

3. **Medical:** The healthcare sector has seen an incredible advancement following the adoption of Artificial Intelligence (Kumar & Garg, 2018)(Sodhi et al., 2019). With a high success rate, there have been endless applications of Machine Learning in this domain. From patient diagnosis to robotic surgeries, Artificial Intelligence has been applied everywhere. Predictive analysis has especially proven vital in helping doctors make decisions by providing all the data and information needed. Various diseases can be predicted to high accuracy based on a person's symptoms by identifying a disease's risk in an individual (Hallak & Azar, 2020)(Sodhi et al., 2019). Diabetes, mental illness, and chronic diseases are numerous ailments that the existing models can identify(Sodhi et al., 2019). Artificial Intelligence is helping reach a wider audience through its means, thus spreading awareness on a large scale.

4. **Social Media:** The recent explosion of social media platforms and their usage has only added to the amount of information worldwide, thus accounting for better identifying trends (Kalyanam, 2017). Social media has become a leading sector, and predictive analysis plays a significant role in these platforms' inherent workings. For example, targeted advertisements, marketing, and sales result from the concept of predictive analysis (Carol, 2015) (Magesh &Kirishnan, 2016). Targeted marketing allows companies to streamline their advertisements to the sector of customers that most relate to it. This, in turn, fetches the company a greater exposure with lesser marketing costs. If not already have, all the major industries are shifting towards digital marketing, which is the new era of marketing (Michael& Kang, 2016) (Magesh &Kirishnan, 2016).

EXISTING SYSTEM AND ITS LIMITATIONS

It is not up to debate that Artificial Intelligence is one of the most sought-after technologies today. Perfection in prediction is what Artificial Intelligence seeks to achieve through better and efficient approaches(Cioffi et al., 2020). There are numerous algorithms and methods, each more suited to specific data and requisites currently being used extensively (Indriasari et al., 2019). Some of the many applications of predictive modeling have been discussed in the previous section. Artificial Intelligence is slowly building towards perfecting human behavior in all aspects (Hallak & Azar, 2020)(Wuestet al., 2016). Vision, speech, intuition, and hearing are sections of human senses that the machine is already trying to mimic by performing image recognition, speech synthesis, predictive analysis, and voice recognition, respectively. Existing machine learning models can perform these functions with high accuracy and precision, but with more data and efficient methods, these Machine Learning algorithms' performance is expected to rise drastically(Sodhi et al., 2019).

Even with the ever-growing craze for adaption of these methods, Machine Learning and Artificial Intelligence, in general, have their fair share of limitations. Some of the significant drawbacks of Machine Learning are listed below.

1. The model needs to be trained with vast amounts of data to increase accuracy and precision, proving to be a tedious task.
2. Incomplete or incorrect data can have an enormous impact on the outcome, leading to anincorrect prediction.
3. The process of machine learning still requires some form of human involvement, which defeats the purpose of machine independence.
4. With large amounts of data being used by these models, there could be a potential security threat.
5. Relying blindly on these algorithms' outcomes can lead to serious problems, especially with models that predict health conditions. A wrong diagnosis can have massive repercussions. Thus, it is always advisable not to completely rely on such a diagnosis.
6. When considering the larger picture, Machine Learning helps save valuable time with its massive computational ability, but it also requires investing much time to study data, train the model, and perfect it.
7. The models are not capable of high-level planning or logical reasoning.

Artificial Intelligence is relentlessly advancing on improving and overcoming said limitations. The end goal of Artificial Intelligence is to reduce human effort and open new doors for opportunities, not to create new threats to mankind. However, it should be kept in mind that Artificial Intelligence aims not to replace human intuition but rather help sharpen it (Wuestet. al., 2016).

MACHINE LEARNING USECASES

With the advent of Machine Learning, it became one of the main catalysts for promoting innovation in various sectors. The following are some of the use cases that help improve multiple organizations' customer experience and streamline their operations in Industry 4.0.

• **Product Recommendation Systems:** Numerous e-commerce websites often list product recommendations to each of their customers individually (Guo et. al., 2018). Product Recommendation Systems are a way of upselling an e-commerce business' customers and create additional value. The recommendation system processes information such as search and browsing history, previous orders, and other customer behavioral traits on the internet (Kaur et. al., 2018). A machine learning model then analyzes this information to generate a list of product recommendations personalized to that individual. It enables businesses to have better reach and guarantees more purchases, and the product is targeted at only suitable customers.

- **Personal Assistants:** Personal Assistants are reinventing customer support in several fields, either in the form of a voice assistant or a chatbot (Bohouta et. al., 2018). They are trained using machine learning to interpret natural language and make intelligent decisions to respond to a customer's query(Adamopoulou & Moussiades, 2020). As a result, many issues can be resolved without any human intervention. The efficiency of these personal assistants is evident from their wide usage in the online sector. Personal assistants help resolve issues that do not require human effort and do so quicker than humans (Tulshan et. al., 2019). These can thus be used to resolve queries that do not necessarily require human effort. It allows for quick and effective implementation of customer support. This, in turn, reduces customer frustration, thereby increasing customer satisfaction.

- **Dynamic Pricing:** This machine learning-based pricing is mostly prevalent in the travel industry. The price of tickets is determined by the date of travel and date of booking, availability, and compare prices. These attributes are not always constant (Qiang&Bayati, 2016). A machine learning algorithm is trained to generate prices accordingly and adapt to these continually changing factors. A similar strategy is also observed in the pricing of products on e-commerce websites (Qiang&Bayati, 2016).

- **Self-driving cars:** Self-driving cars use a variety of algorithms under the umbrella of machine learning. They detect objects around the car, traffic signals, and the driver's state using image processing algorithms(Weaver, 2020). They use location and path planning to navigate. They can eliminate traffic collisions and delays and be used in many industries such as shipping, emergency, and public transportation. Already in implementation, these cars are only imagined to get more refined and take over human driving in the future (Weaver, 2020). In a way, they allow individuals to make better use of their time by taking over the driving wheel. Experimentation on self-driving technology dates back to the 1920s. However, since the last few years, the technology has seen wide-scale implementation with regulations being set and permits being issued in many countries.

- **Fraud Detection:** The banking sector uses fraud detection algorithms to notify their customers or avoid any suspicious transactions. It is used for the flagging activity, which appears to be fraudulent based on time, location, transaction method, and several other characteristics(Randhawa et al., 2018). Anomaly detection is the concept behind detecting fraudulent transactions. The algorithm identifies unconventional factors during a transaction and notifies such a suspicion. Large-scale implementation of security measures is made possible by these algorithms. Fraud Detection can help protect customers' money from theft and build customer loyalty (Favila&Shivam, 2018) (Raghavan et. al., 2019).

- **Email Filtering:** Email services have dedicated folders for junk, spam, and primary mails(Sodhi et al., 2019). Every email received is analyzed by a machine learning algorithm and categorized by being assigned one of the labels. These emails are then displayed in their respective folders based on their label. This automatic filtering allows users to navigate away from unwanted spam or promotional emails, which otherwise would have proven to be very frustrating to the user (Bhowmick& Hazarika, 2018).

CONCLUSION

Artificial Intelligence has been the most influential of the technologies, gaining popularity, thereby being implemented on a large scale. The abundance of data being sourced from multiple applications like email,

Facebook, Instagram, Twitter, etc., has led to a data-driven world. Machine Learning has escorted to be one of the top problem-solving approaches when paired with the systems' tremendous computational ability. The Global machine learning market is estimated to be worth 30 Billion dollars by2024, with a compound annual growth rate of 48% from2018 to 2024. This technology's craze is evident from the vast number of people showing interest in learning it and starting their career in this field. In the last few years, Machine learning has seen a massive shift in popularity towards it. Machine Learning has been the tipping point as a revolutionary technology for the new age Industry 4.0. Organizations have adapted to this technology with ease, which has guaranteed them a competitive advantage in their respective fields. With Industry 5.0 already building its roots into the market, it has become more critical than ever for manufacturing firms and organizations to inculcate smart techniques for manufacturing and management. Industry 5.0 is aiming at integrating the culture of working alongside robots and smart machines. Productivity, efficiency, and manufacturing quality are set to soar to new heights regarding adopting technology. Furthermore, predictive modeling and Machine Learning are at the center of such development and growth. There is immense scope for machine learning as better and creative models emerge every day. With improved and advanced models entering the market, new branches and solutions emerge.Machine Learning and Artificial Intelligence have impacted daily lives on a massive scale and have induced a creativeapproach to finding solutions. The future entails high scope concerning more efficient solutions through Artificial Intelligence.

REFERENCES

Adamopoulou, E., & Moussiades, L. (2020). Chatbots: History, technology, and applications. *Machine Learning with Applications, 2*(November). doi:10.1016/j.mlwa.2020.100006

Alcácer, V., & Cruz-Machado, V. (2019). Scanning the Industry 4.0: A Literature Review on Technologies for Manufacturing Systems. *Engineering Science and Technology, an International Journal, 22*(3), 899–919. doi:10.1016/j.jestch.2019.01.006

Bhowmick, A., & Hazarika, S. (2018). E-Mail Spam Filtering: A Review of Techniques and Trends. *Advances in Electronics, Communication and Computing, Lecture Notes in Electrical Engineering,* 583-590. doi:10.1007/978-981-10-4765-7_61

Bohouta, G., & Këpuska, V. (2018). Next-Generation of Virtual Personal Assistants (Microsoft Cortana, Apple Siri, Amazon Alexa and Google Home). *IEEE 8th Annual Computing and Communication Workshop and Conference (CCWC).* 10.1109/CCWC.2018.8301638

Cioffi, R., Travaglioni, M., Piscitelli, G., Petrillo, A., & De Felice, F. (2020). Artificial intelligence and machine learning applications in smart production: Progress, trends, and directions. *Sustainability (Switzerland), 12*(2), 492. Advance online publication. doi:10.3390u12020492

Favila, A., & Shivam, P. (2018). *Systems and methods for online fraud detection.* U.S. Patent Application No. 15/236,077.

F.Y., O., J.E.T., A., O., A., O., H. J., O., O., & J., A. (2017). Supervised Machine Learning Algorithms: Classification and Comparison. *International Journal of Computer Trends and Technology, 48*(3), 128–138. doi:10.14445/22312803/IJCTT-V48P126

Gujre, V. S., & Anand, R. (2020). Machine learning algorithms for failure prediction and yield improvement during electric resistance welded tube manufacturing. *Journal of Experimental & Theoretical Artificial Intelligence*, *32*(4), 601–622. doi:10.1080/0952813X.2019.1653995

Guo, Y., Yin, C., Li, M., Ren, X., & Liu, P. (2018). Mobile e-Commerce Recommendation System Based on Multi-Source Information Fusion for Sustainable e-Business. *Sustainability*, *10*(2), 147. Advance online publication. doi:10.3390u10010147

Hallak, J. A., & Azar, D. T. (2020). The AI revolution and how to prepare for it. *Translational Vision Science & Technology*, *9*(2), 1–3. doi:10.1167/tvst.9.2.16 PMID:32818078

Haussmann, A. (2020). *Polynomial Regression: The Only Introduction You'll Need*. Towardsdatascience. https://towardsdatascience.com/polynomial-regression-the-only-introduction-youll-need-49a6fb2b86de

Indriasari, E., Soeparno, H., Gaol, F. L., & Matsuo, T. (2019). Application of Predictive Analytics at Financial Institutions: A Systematic Literature Review. *Proceedings - 2019 8th International Congress on Advanced Applied Informatics, IIAI-AAI 2019, February 2020*, 877–883. 10.1109/IIAI-AAI.2019.00178

Kalyanam, J. (2017). *Machine Learning and Applications on Social Media Data*. UC San Diego. Retrieved from https://escholarship.org/uc/item/6545w71z

Kaur, J., Bedi, R., & Gupta, S. K. (2018). Product Recommendation Systems a Comprehensive Review. *International Journal on Computer Science and Engineering*, *6*(6), 1192–1195. doi:10.26438/ijcse/v6i6.11921195

Kumar, V., & Garg, M. L. (2018). Predictive Analytics: A Review of Trends and Techniques. *International Journal of Computers and Applications*, *182*(1), 31–37. doi:10.5120/ijca2018917434

Lash & Zhao. (2016). Early Predictions of Movie Success: the Who, What, and When of Profitability Artificial Intelligence (cs.AI). *Social and Information Networks*.

Magesh, S., & Krishnan, N. (2016). *A survey on machine learning approaches to social media analytics*. Academic Press.

McDonald. (2015). *Parallel and Iterative Processing for Machine Learning Recommendations with Spark*. https://www.mapr.com/blog/parallel-and-iterative-processing-machine-learning-recommendations-spark

Mei, S., Montanari, A., & Nguyen, P. M. (2018). A mean field view of the landscape of two-layer neural networks. *Proceedings of the National Academy of Sciences of the United States of America*, *115*(33), E7665–E7671. doi:10.1073/pnas.1806579115 PMID:30054315

QiangS.BayatiM. (2016). Dynamic Pricing with Demand Covariates. doi:10.2139srn.2765257

Raghavan, P., & Gayar, N. (2019). Fraud Detection using Machine Learning and Deep Learning. *International Conference on Computational Intelligence and Knowledge Economy (ICCIKE)*, 334-339. 10.1109/ICCIKE47802.2019.9004231

Randhawa, K., Loo, C. K., Seera, M., Lim, C. P., & Nandi, A. K. (2018). Credit Card Fraud Detection Using AdaBoost and Majority Voting. *IEEE Access: Practical Innovations, Open Solutions*, *6*, 14277–14284. doi:10.1109/ACCESS.2018.2806420

Salian, I. (2018). *SuperVize Me: What's the Difference Between Supervised, Unsupervised, Semi- Supervised and Reinforcement Learning?* NVIDIA. https://blogs.nvidia.com/blog/2018/08/02/supervised-unsupervised-learning/

Silhavy, R., Silhavy, P., & Prokopova, Z. (2017). Analysis and selection of a regression model for the Use Case Points method using a stepwise approach. *Journal of Systems and Software, 125*, 1–14. doi:10.1016/j.jss.2016.11.029

Sodhi, P., Awasthi, N., & Sharma, V. (2019). Introduction to Machine Learning and Its Basic Application in Python. SSRN *Electronic Journal*, 1354–1375. doi:10.2139srn.3323796

Tulshan, A., & Dhage, S. (2019). Survey on Virtual Assistant: Google Assistant, Siri, Cortana, Alexa. *4th International Symposium SIRS 2018*. 10.1007/978-981-13-5758-9_17

Weaver, B. C. (2020). Self-Driving Cars Learn to Read the Body Language of People on the Street. *IEEE Spectrum*, 1–5.

Wuest, T., Weimer, D., Irgens, C., & Thoben, K.-D. (2016). Machine learning in manufacturing: Advantages, challenges, and applications. *Production & Manufacturing Research, 4*(1), 23–45. doi:10.1080/21693277.2016.1192517

Xin, Y., Kong, L., Liu, Z., Chen, Y., Li, Y., Zhu, H., Gao, M., Hou, H., & Wang, C. (2018). Machine Learning and Deep Learning Methods for Cybersecurity. *IEEE Access: Practical Innovations, Open Solutions, 6*, 35365–35381. doi:10.1109/ACCESS.2018.2836950

Chapter 5
Computational Statistics of Data Science for Secured Software Engineering

Raghavendra Rao Althar

First American India, India & Christ University (Deemed), India

Debabrata Samanta

iD https://orcid.org/0000-0003-4118-2480

Christ University (Deemed), India

ABSTRACT

The chapter focuses on exploring the work done for applying data science for software engineering, focusing on secured software systems development. With requirements management being the first stage of the life cycle, all the approaches that can help security mindset right at the beginning are explored. By exploring the work done in this area, various key themes of security and its data sources are explored, which will mark the setup of base for advanced exploration of the better approaches to make software systems mature. Based on the assessments of some of the work done in this area, possible prospects are explored. This exploration also helps to emphasize the key challenges that are causing trouble for the software development community. The work also explores the possible collaboration across machine learning, deep learning, and natural language processing approaches. The work helps to throw light on critical dimensions of software development where security plays a key role.

INTRODUCTION

With the explosion of data globally and particularly in software engineering, there is a heavy focus needed on the analyst who can leverage from the knowledge hidden in this data. Data in the real world begs for automatic and semi-automatic methods of data exploration. Efforts put in over decades in artificial intelligence have provided excellent base work to develop upon. With the increasingly growing power of computation, it has reached the point of complexity involved to select areas to explore and

DOI: 10.4018/978-1-7998-7701-1.ch005

focus our energy upon. Analytics in software development projects is an active area of exploration, with the knowledge hidden across the data like emails, source code, testing related reports, to name a few. All software engineers' activities leave back the information that can provide useful insights into the processes. All these have also resulted in collaboration among the professionals extending from building on to existing tools, learning best practices, and other areas. Analytics with the software data can help in real-time analytics such as event monitoring and reporting. It would take a significant amount of time and effort without the analytical capability being utilized. Also, in the case of application log monitoring, analytics's powers will provide the necessary efficiency and effectiveness for the process and create value. People, process, and technology being the key focus areas of the software development processes, exploration of the data need to revolve around these aspects. Productivity and quality of the software hold the key from a consumer point of view, which will decide on the user experience. With a large amount of data being processed, there is a need for more robust analytical approaches that can leverage these data. Understanding the stake of all the personnel involved in software engineering will be critical as well. Software analytics revolve around the computational capability to handle large data sets, analysis capabilities like data mining, pattern recognition, and visualization capabilities to derive insight. The software development area has many challenges to tackle. Issues related to the quality of the product, time to market, deployment associated challenges, and traceability capabilities across software development lifecycle are some of them that can be highlighted. Some of the information security breaches that have haunted the software development companies have put this area into top focus.

There is a struggle to bring all the stakeholders up to understand the importance of security in development. Though there is a large amount of knowledge all over, making sense of that information and putting them to use has not been an easy task. Data science has been explored to see if there can be useful learning and improvement to automate some of these areas. Since there is a need to ensure the security-related requirements are understood right at the beginning of the software development life cycle, the requirements management phase must be focused upon. All the stakeholder's involvement in this area needs to be understood and executed well. It is seen that there is a need for creating framework that can integrate all sources of knowledge in the industry and within the companies to help build the best practices across the industry. Understanding the domain of information security will play a key role in devising these systems. It cannot continue to be an area for experts to own, but everyone is needed to understand the need for security in software engineering. Apart from building solutions for the specific scenarios, there is a need to generalize the solutions as much as possible, leading to the institutionalization of the solutions effectively.

FOCUS OF THE CHAPTER

The chapter focuses on exploring the work done in software security management in the software development lifecycle. The job is primarily done to handle software requirements, focusing on shifting left in the software development life cycle. Later the security issues are figured out; it becomes costlier and difficult to manage them. With the need to utilize the knowledge captured in various security knowledge repositories, work that focuses on synthesizing this knowledge and making it available for practitioners is explored. A combination of automated, semi-automated, and other hybrid approaches for data assimilation with data analytics approaches are explored. From essential machine learning to advanced techniques of deep understanding are scoped for review. Security requirements engineering and software engineering are studied to understand the key aspects that drive software development. With a fair amount of text data that gets created in software engineering, the capabilities of NLP (Natural Language Processing) are explored. Many of the works have explored the combination of machine learning, deep learning, and NLP approaches for their experimentation. Information retrieval methods covering both structured and unstructured data sources are used as bases in many of the works. Various other activities related to the software systems' security have been explored to take away crucial knowledge and build it into models built for secured software engineering.

EXPLORATION OF SECURED SOFTWARE ENGINEERING

In work (Jindal et al., 2016) author focuses on automated classification of security requirements to categorize security requirements with algorithms like the J48 decision tree algorithm. The info-gain feature selection method is one of the prominent techniques used. Security requirements from the promise data repository are used as a base for the experiments. SRS (Software Requirements Specification) document has been used as a reference to pick the requirements as input for classification. Models are built across various security categories like authentication, access control, authorization, cryptography, and data integrity. Identifying the security needs scattered across the SRS document is a vital part of securing the software applications. The authors use text mining approaches to extract the description and use it as input for J48 decision tree methods. The work's importance is in providing the approach for the software development practitioners for them to use it in the early phase of software development processes. AUC (Area Under Curve) from ROC (Receiver Operating Characteristics) has been used for measuring the accuracy of the work. The authors also plan to extend their work towards the classification of the non-functional requirements, which has a good part in building a software application in addition to functional requirements. Authors in their work (Amoroso et al., 1994;Moffett et al., 2004;Moffett et al., 2003)highlight that there is a mindset of considering security as an after-development concern for the stakeholders.

As shown in figure 1, the system's architecture comprises security requirements gathering, applying text mining techniques, and prediction models. Text mining includes pre-processing, feature selection, and Tf-IDF (Term Frequency and Inverse Document Frequency) weighting. Prediction modeling is performed in multiple versions utilizing the J48 model; each version of the prediction model is for a different category of the security specified earlier. The Info-gain method is one of the main approaches for feature engineering. In this approach, a set of words from the input documents is used to derive the target's inference. J48 algorithm is a tree-based algorithm that is used to classify. Decision trees can be created with pruned trees and unpruned trees. J48 algorithm provided by WEKA software works with a pruned tree. Pruning of the tree is conducted to remove extra information without impacting the performance of the prediction. Also, this method makes sure that there is no over-fitting of the data. J48 also is specialized to adopt tree-based classification or work based on rules set, results being shown in table 1.

In work (Riaz et al., 2014), the author proposes methods to identify natural language artifacts' security requirements in an automated way. In the approach, authors create security requirements objectives using which security requirements templates are created. These templates are mapped to the security requirements objective. These templates will facilitate the derivation of the security objectives. K-NN classifier, modified Levenshtein distance, multinomial naïve Bayes, sequential minimal optimization classifier are the approaches experimented with. Information gain parameter in WEKA software is

*Figure 1. Security requirements architecture proposed in the work
(Jindal et al., 2016)*

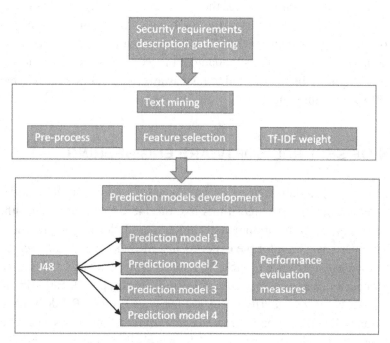

Table 1. Results of Decision tree application with J48 in the work

	AUC	Sensitivity %	Cut-off
Model 1	0.72	65	0.12
Model 2	0.77	75	0.41
Model 3	0.69	69	0.07
Model 4	0.83	80	0.05

(Jindal et al., 2016)

used for a classifier that is applied. Establishing security objectives based on domain and then deriving keywords against those objectives is the work's key theme. This work focus on security requirement for health care domain. Keyword identification for the security objectives is the key theme of the work. This work can establish security objectives for various other domains, then deriving critical words from those objectives to ensure generalized learning and extending the approach to other domains. In the case of the software development operations that follow the agile model of development, this approach can be leveraged to create user story templates for documenting requirements covering the customer's security requirements. The work's approach focuses on bringing out the implicit security requirements hidden in the natural language conversations. Work here focus on deriving the security requirements based on the context and provision to translate them into functional security. Machine learning approaches are used in work to derive security requirements. Confidentiality, Integrity, and Availability are the security objectives that are focused on classification. Research in this work tends to explore what all critical categories of security need to be covered. Also, to study the extent of implicit and explicit specification of the security requirements. Figuring out the standard terms across security objectives and their content is also the focus.

K-NN (K-Nearest Neighbor) algorithm applied in work operates by taking a majority vote of the existing classifications. Sentence with the closest association is figured out with Levenshtein distance, which is customized. Levenshtein distance is customized in this work by changing its format from looking up for the total count of edits to transform one string to another for arriving at distance value to looking up for several transformations needed in words to change one sentence to another. N-fold cross-validation technique is employed for modeling purposes. Key themes used for security requirements template extraction are deriving common patterns and themes in the sentence structure, clustering the sentence with appropriate approaches like k-medoids. In work (Firesmith, 2004), Firesmith states that security requirements can be reused across various systems and has proposed developing templates reusable for security requirements. In work (Mellado et al., 2007), Mellado et al. has argued against the effectiveness of these reuse of the security requirements related to each other. Studies in (Salini et al., 2012) show that though there is the availability of various Software Requirements Engineering methods, only a handful of them have been utilized that calls for more extensive exploration of making some of the effective ways popular.

In work (Li, 2017), the author proposes identifying security requirements based on machine learning and linguistic analysis. The author offers a combined approach of linguistic analysis and machine learning to identify security requirements. Bayes, tree, function, and rule-based algorithms are explored. Logistic model tree, J48, Sequential Minimal Optimization, and Logistic Decision Table approaches are experimented. Precision recall and F-measure are the metrics utilized for assessing the goodness of the approaches implemented. LDA (Latent Dirichlet Allocation) based topic modeling approaches can be combined with linguistic analysis format as an experiment to see if the model's performance can be further enhanced. Cutting edge techniques of the NLP (Natural Language Processing) like that of Word2Vec can also be experimented with for improving the classification of security requirements classification. Identifying the security requirements early in the life cycle can go a long way to ensure cost-effective software development. But since the security requirements are mixed up with explicit requirements, this

Figure 2. Overview of the proposal provided in the work (Li, 2017)

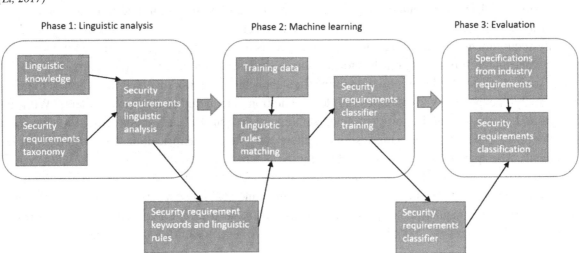

makes it a challenging area. Earlier approaches have been dependent on the probability approach, which has not been fully precise and domain specific. In this work tool-based method is proposed to combine linguistic analysis with machine learning to make it generic. The linguistic approach uses the linguistic knowledge in the data based on the security requirement-based ontology established. These inputs are later used on machine learning approaches.

As per figure 2, the approach proposed in this work includes the linguistic analysis stage, machine learning phase, and evaluation phase. As part of training security requirements classifiers, parse tree generation is involved, followed by parse tree stemming, keyword matching, and linguistic rule matching. This work focus on three research areas as follows. Exploring machine learning that works well for this problem area, the extent to which the classifiers used in this work can classify security requirements. Do these approaches have the capability to beat the benchmark performance? The ability of the classifiers used in this work to be generalized for other domains. Variety of the machine learning approaches provided by the WEKA tool is experimented with as part of the first phase. Based on the best algorithms, prepared datasets are modeled with these. Performance on these datasets is assessed against the benchmark. The third part of the experiment included the experiment being done on the datasets from different domains to explore their generalizability. Linguistic analysis can further be optimized with advanced NLP approaches like word2vec. As per the survey presented in work (Souag et al., 2012), there are twenty security-related ontologies. Authors here confirm their exploration of security taxonomies extending to the position in (Firesmith, 2005; Firesmith, 2004).

In work (Knauss et al., 2011), the author supports requirements engineers with the ability to recognize security issues. Automatic security requirements identification with the Bayesian classifier is experimented in this work. Precision, recall and F1 scores are used as measurements. Approach proposed in this work needs to be extended to industrial applications. The approach looks at incorporating the security identification pattern into the requirements management pipeline, which can help build the reusable capabilities and information base. Approach banks on the previously classified security-related requirements. Bayesian classifiers have provided the ability to identify domain-specific security needs, but they are essential in providing the practitioners' knowledge base. Bayesian approaches are the statistical methods of classification that bank on the probability of the events. Naïve Bayes classification, which works on the Bayesian approach's backdrop, is utilized in this work. Naïve Bayes is one of the popular approaches for classification, but they are only as good as the training data.

Systematic selection of training data is critical for these experiments in addition to avoiding the overfitting, k-fold cross validation helps ensure these. There has been extensive work being done on security requirements engineering, but NLP applications are not being explored. In this work author put together the NLP-based approaches for the same. The author proposes that the approach provided in this work that operates in a tool will facilitate the labor-intensive task involved in the initial requirements analysis expected. Various industrial documentation is explored in these experiments. With the background of non-availability of a text understanding approach that ends in all sense, this approach provides an excellent platform to make sure that there is no wrong identification of security-related requirements. The approach offers to add support for the security experts. It helps build a learning system where the process is traceable and documentable if needed for audit purposes. As a further exploration, it would be interesting to explore similar areas like safety and usability aspects. Some of the work done to identify the good requirements specification documentation from the requirements can be seen in work (Kof, 2005; Lee et al., 2006; Kiyavitskaya et al., 2008). In some of these works, there is a focus on semi-automatic extraction of the requirements documentation's ambiguities. The author feels that their work can be validated with these approaches to make a comparative study.

In work (Elahi et. al., 2009), the author focuses on modeling ontology to integrate vulnerabilities into security requirements as a conceptual foundation. The approach used here is the vulnerability centric modeling ontology, where the integration of empirical knowledge related to the vulnerabilities is used in system development. The authors have explored modeling and analyzing the vulnerabilities and what kind of effect does the same have on the system. Based on these concepts, criteria can be devised to compare and evaluate security frameworks that can address these vulnerabilities. Authors provide examples to explore modeling framework concepts through modeling ontology. Authors have studied the literature covering the security requirements and security engineering to figure the fundamentals required for elicitation of the requirements. Ontology is the meta-model in the abstract; in this study, it associates concept framework components to that of vulnerabilities. Examples illustrated by the authors show that the conceptual structure of various frameworks has its differences. So, looking at the conceptual framework's ontological components from the variety of the results of the analysis can be done. The study highlights that many vulnerable elements are not expressed due to a lack of definite structures in some of the concepts' modeling frameworks. Ontology is adopted for risk models like CORAS and miss-use case diagrams. To make comprehensive ontology, the proposed ontology needs to be expanded to a variety of modeling concepts frameworks, which will also provide usefulness and better expression. Work has focused on empirical case studies on the human subjects to evaluate these ontologies, which has used modeling framework concepts that are extended. Some of the security analysis types that cannot be expressed in ontology or by humans are aimed to be covered in case studies. Work also intends to critically analyze the models by an interview of the subjects and derive inference around the proposed conceptual model and their ability to express. Scalability issues associated with graphics of the visual model that are complex are not covered in this work. The purpose of handling the complexity of the model due to the security aspects being built-in can be taken by filtering out some of the views. The work's flow in the paper covers understanding vulnerability analysis and its conceptual foundation, approaches in the vulnerabilities modeling and its analysis, computer network security vulnerabilities analysis, security requirements engineering vulnerabilities models, and comparison of concepts models. Further exploration is around vulnerabilities ontology modeling, adoption of the same, some examples for use case diagrams and vulnerability models integration and modification of the models of vulnerabilities in case of some of the approaches. Some of the references for attack modeling and analysis where graphs and Bayesian Network are leveraged for vulnerabilities assessment are in work (Phillips et al., 1998; Liu et al., 2005; Frigault et al., 2008).

In work (Hayrapetian et al., 2018), the author focusses on empirical analysis and evaluation of security features in software requirements. The complex project usually has security and other non-functional requirements. Ensuring compliance in this context makes it task extensive. NLP and Machine learning will find their prominence in these scenarios with requirements information being transacted in the free flow format. There is a proposal of system that is semi-automatic to explore the security requirements for their completeness in work. This work ensures there is a collaboration between requirements management and adherence to the compliance needs. Association of the ISO (International Organization for Standardization) standards information and OWASP (Open Web Application Security Project) information with project documentation needs are established in this work.

The outcome of these studies and the details annotated in the documents are used to train Neural Network Model to predict the existence of the security standards in the document. The approach experimented here looks at bringing out the structures hidden in the requirements documents of the software. The exploration of these structures provides the software organization's base to arrive at requirements specifications that address security issues. Also, guidelines needed for the same can be established. The overall thought process of arriving at the precise requirements with security focus ensures that the security issues are focused right at the beginning of the software development lifecycle. This focus enables to address them effectively without much cost and effort. This early focus on security also helps establish guidelines that can help maintain the consistency of the process execution. Another key benefit is to enable the stakeholder with the right visibility on the security features. All these calls for establishing a formal structure for the security requirements elicitation. Work here adds value to bring in the empirical information about the security requirement's incompleteness and help maintain the compliance need before the software development kicks in. NLP and machine learning-based approaches will help bring in textual entailment and neural network-based modeling to evaluate the compliance need and draft the security requirements. F-Scores were assessed across various configurations of the projects to bring out the best configuration to ensure the prediction's completeness. Extension of the work to the larger dataset is one of the future works, bringing in the flexibility to adapt to the various project needs. Extending the work to other functionality requirements is another area to look at, followed by building capability to modify the requirements based on the input and making it ready for the next phase of the software development project. There are efforts to use machine learning and NLP approaches to establish manual annotation of the project requirements concerning industry best practices like ISO and OWASP (ISO, 2009; OWASP, 2015; Fisher, 2007; ISO, 2015).

In work (Malhotra et al., 2016), Analysis and evaluation of security features are taken up for software requirements. Due to the complex projects being dominated by non-functional requirements, there is a large amount of effort needed to identify them, and it's an uphill task. With the natural language-based inputs being in their free flow, ML (Machine Learning) and NLP will find their prominence to ensure comprehensive coverage. In this work, the authors introduce a semi-automated approach to assessing the completeness of security requirements in software development. Requirements specified by ISO

Figure 3. Approach used in work
(Malhotra et al., 2016)

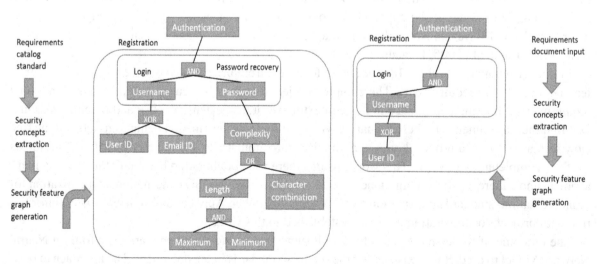

standards are used as a reference to construct models to classify security requirements with ML and NLP. Also, graph analysis is employed in work.

Furthermore, publicly available software documents are used as input in addition to the industry-based standards. In addition to mentioned approaches, other text analysis approaches and relationship processing formats will have to be experimented with in the future. The work depicted in Figure. 3 focus on understanding the underlying expectation in the industry standards abstractly. The creation of the tool suite based on this approach will also be the objective of this work.

Some of the other aspects of targeted requirements are operational, maintainability, performance, look and feel, availability, scalability, and security. Architecture intends to support text engineering where extraction of the security-related requirements can be done by information extraction method. Various speech features are also extracted, Named Entity Recognition identification would be covered in this approach. The pipeline will include the sequence processing components like tokenizer, sentence splitter, POS (Parts Of Speech) tagger, analysis for morphology, and chunker for noun phrases. Work also uses the standard reference tool Protégé to create the structures that help establish relations between phrases and words. The hierarchy created by these relations is called ontology. Concepts associated with the security are built into the Protégé tool; these concepts cover aspects like, 'user must contain the password,' 'user must register,' and so on. Processing of the information targets to gather critical features from the document, which can be utilized later in the development to figure out the relationship among the features, which is called co-reference, and identify if these references are available in the master standard. This approach helps to assess the extent to which the organization's standards comply with these master standards. Features gathered in the first phase can be input into ML and NLP processor to conduct further processing. The features leveraged in the first phase and the second additional advanced pipeline can be built into the project. Work (Wilson et al., 1997)has created an automated tool for the search of documents based on quality indicators. Work (Lami et al., 2004) focus on lexical and syntax-based tool for analysis of the requirements documents. In work (Cleland-Huang et al., 2007), automatic classification of non-functional requirements like security, scalability, and performance are targeted.

In work (Mellado et al., 2010), the author does a systematic review of the security requirements engineering. Security requirements engineering is the core phase of developing secure software systems. With the non-availability of this area's systematic study, the author intends to fulfill the same with their work. Their work intends to provide the background information that will be handy to pursue future research in this area. Information security gets its proper technical attention during the implementation stage. Still, for adequate attention to be given to the security in information systems, there must be an ongoing focus throughout all development stages. This focus becomes more critical when the security requirements are firmed up, as that lays the foundation for the robust information system. With the importance held by the initial steps of software development, the same will be the case for security to build a robust information security system. Work here focuses on providing all the required context to the researchers summarizing the critical information from the security perspective with thorough coverage. This work scores high compared to other previous work regarding its reliability and precision as claimed by authors. This work's main limitation is seen due to work in systematic literature review of software engineering being done only in recent past. Software engineering has more issues and makes it difficult to get a comprehensive compilation of the information. It is anticipated that this work may have some limitations in terms of the bias due to limited knowledge and missing central source of information that can adequately integrate all the aspects. Adding to these is the non-availability of the standards for systematic reviews and missing guidelines that are generally accepted for conducting experiments.

Searching for the right information and evaluating them for software engineering is a bottleneck in a systematic review and leads to bias. Work focuses on information extraction and synthesis of the information. A systematic review conducted in work is based on the formats proposed in work (Kitchenham, 2004; Brereton et al., 2007; Kitchenham et al., 2007). Some of the noteworthy aspects highlighted are an increase in flexibility of the information systems' connectivity, and growing complexities call for increased chances of security breaches. Information systems that we have today have more enormous possibilities of being vulnerable from cyber-attacks, malicious codes and cyber terrorism, and a host of other concerns. The dependency of the computer network applications will create the possibility of extensive financial breaches and danger to human life. With all these concerns, security needs require a fair amount of focus and security experts for the same. Stakeholders do not have this on top of their list except to comply with legal or basic standard requirements. This lack of focus puts security into an initial position while the requirements manager performs requirements elicitation.

In work (Casamayor et al., 2010), the author deal with NFR (Non-Functional Requirements) for textual specifications with a semi-supervised learning approach. It is essential to figure out non-functional requirements early in the life cycle of software development to ensure there is enough focus on the alternatives of the architectural possibilities and design decisions. Leveraging the text data and categorizing the security requirements have been taken up in multiple works to help the professionals explore the NFR, which covers various security needs. In the background, there is a need to have a significant number of labeled data that will be input for supervised learning; this calls for the high effort needed for manual labeling of data points, so that supervised learning is possible for new instances. In this work, the author provides a semi-automatic approach for the text categorization where the categories of NFR can be explored. This exploration would need few requirements identified in the elicitation process to be used as a base for the classifier's initial learning. This learning will be used for further elicitation of requirements in iterations. Work intends to integrate the work into a recommender system that can support practitioners in designing the system's architecture. The approach involves the detection as well as classification of NFRs. Classification considers the knowledge from few unlabeled data points and the learning from the text data.

The classification process gets enhanced with the feedback received from the users. The semi-supervised approach used in this work demonstrates better performance than supervised methods that work with standard data input. Work highlights the fact that the semi-supervised learning needs less intervention from the human than supervised approaches. Continual improvement of the approach leveraging the feedback received from the analyst gives it the further advantage. There is potential to further enhance the approach by further building on the inputs received from the analyst. The possibility of any noise in the data is also limited since previous projects' information is not a primary reference. If the previous project's data were also a need, it would include noise due to the different patterns used by additional requirements elicitation that is done by different team members. Work leverages the text characteristics in the data like the words' co-occurrence, which indicates they belong to the same class of data. The author intends to introduce the active learning concept further into the approach to reduce the need for manual labeling of the data. In work (Fantechi et al., 2003), there is a discussion on applying linguistic techniques for the collection of metrics related to quality, including spot defects. In work (Kitapci et al., 2006; Fernandes et al., 2004), NLP has been leveraged as a bridge between formal and informal requirements specifications.

In work (Cleland-Huang et al., 2007), automated classification of NFR (Non-Functional Requirements) is dealt with. This paper focuses on detecting and classification of NFR in automated methods concentrating on performance, usability, and security. To avoid refactoring of the NFRs later in work, early identification helps to use it in the design process. Requirements will have scattered information of stakeholder concerns. The approach focuses on detecting the same from non-categorized requirements, including the unstructured data available in free form texts like meeting minutes and emails. Work evaluated various classification algorithms on the hand-built data sets of requirements specifications from the DePaul University MS students. A new approach, which is also iterative, is experimented with for further re-training of the classifier, which is to unearth the NFRs, from the training set that is different from the earlier training set. The approach devised is then evaluated with the free-form requirements document from Siemens Logistic and Automotive organization.

Though there is a limitation in extent to which NFRs are discovered, this will facilitate as a vital tool for analysts as part of their high effort work. This work also enables a quick search of NFRs on complex documents. In work (Cysneiros et al., 2011), the author has explored the language extended lexicon method to elicit, track, and analyze NFRs. Work (Dörr et al., 2003) focused on identifying functional use cases then a checklist was devised to identify NFRs, in association with the identified use cases. Work (Kaiya et al., 2004) interestingly utilized GQM (Global Question Metric) to explore NFRs and their dependencies. Information retrieval methods have been studied in this work to identify NFR identification from both structured and free form text. Though there is a need for effort from analysts, the same is reduced to a considerable level with this new approach. Though perfect recall is not possible with this method, there are other key benefits of this approach. Analyst's error-prone manual activity can now be augmented with the automatic system. This approach also brings in the capability of skimming through a large set of complex stakeholder information related to requirements. All the unstructured information that gets created during the requirements elicitation process can be leveraged. This data collation will enable the security analyst to query for the information related to the stakeholders' specific security needs, or the UI (User Interface) designer may query for the look and feel related aspects.

In work (Bettaieb et al., 2019), the author focusses on security control identification for decision support using machine learning. Healthcare and banking domains have a variety of security related needs to be complied with. Security requirements identification is influenced by the specifications provided in the security standards and the best practices that are prevalent in the industry. The biggest challenge faced by the analyst is the need to scan through all the expectations of the security and figuring out the needs of the project. This becomes difficult when there is a lack of expertise of the security experts in the organization. In this work, there is an automated system to identify the security needs of the project based on its context. Machine learning-based approach used in this work leverage the past data to identify the security needs of the project and recommend improvements. Banking domain data has been used for evaluating the performance of the experiment; in the results its seen that there is a good performance when the less applicable controls are removed from historical data. High recall demonstrated in the experiment confirms that amount of security control missed is limited; the effort required for confirming the recommendations is not high as the precision shows good results. This demonstrates that the work helps to support analysts with the approach that helps identify relevant security controls in an effective way with less possibility of missing key controls. In work (Sindre et al., 2005), a systematic approach is devised for elicitation of security requirements taking inputs from the use cases with prioritization of guidelines. Work focuses on providing machine learning-based solutions to support analysts with activities to confirm what security controls do matter for the group. These proposed activities focus on bringing in a security mindset right at the start of the software development processes. Banking domain data provides the real-time data from the security assessment. If the security controls are existing quite regularly, then the solution provided by this work is effective. This work highlights that it's the first attempt to use machine learning for security controls modeling. For cases of few occurrences, the author is proposing to make use of case-based reasoning in their future work. The author also highlights the possibility of building solutions around the decision-making associated with threat and vulnerability identification. These areas that intend to broaden the approach needs attention beyond structural assessment towards information that is stored in an unstructured format. There is also a need for extracting and leveraging the data that sit in the system and asset description.

In work (El-Hadary et al., 2004), the focus is on capturing security requirements in software systems. As highlighted by the authors of all the works, this work also emphasizes the importance of the early identification of the security requirements and putting up a proper mechanism to tackle the same, which will assist in avoiding the rework possibility. Approaches needed for security requirements engineering should consider the fact that elicitation needs a structured thought process. Problem outlines are accounted for when methodologies of security requirements are established. The focus of the methodology in this work is also towards the early adoption of security in software development. Work focus on streamlining the requirements elicitation process for developers in the requirements engineering area with a methodology that assist the developers. Making use of the past security knowledge security catalog is constructed focusing on the frames arrived for the problem. Threats are modeled based on the abuse frames, and security requirements are modeled based on security problems. Evaluation criteria are established to assess security requirements where the conflicts hidden within the requirements can be brought out. Work shows that a comprehensive requirements coverage can be ensured during the elicitation in addition to the support that can be provided for the developers to elicit security requirements in a systematic approach. Abuse frames and security problem frames are referred to in this work from the work of (Lin et al., 2003; Hatebur et al., 2007) respectively. Work (Haley et al., 2004; Haley et al., 2008) is also referred that make use of problem frames for identification of vulnerabilities and assist in the elicitation of security requirements. Problem frames are used as the basis for the integration of methodology of security and requirements engineering. Threat identification and proposing the security controls for remediation is ensured in this work. Security knowledge reuse happens with the problem frames devised from the security catalog, which associates the threat with the security controls. Abuse frames utilization also has assisted in deriving general learning from the security catalog to make sure all other aspects are also addressed. Crosscutting threats are exposed in the approach with a specific focus on the related problem frames. Work also is compared with various other works in various domains like human resources. Accountability and authenticity are the key themes that originated from the same. As a future focus author highlight that there is a need for an empirical study on large-scale software systems for the methodology established. Further exploration with the development community that has varied knowledge of the security can bring more value to the validation of the methodology.

FUTURE RESEARCH DIRECTIONS

Though there has been extensive work in the secured software engineering area, there is a need for exploration of a wide variety of ensemble machine learning approaches that create hybrid approaches for the problem areas. It is also observed that the works have majorly focused only on a few critical metrics for validation of their experiments. Exploring other metrics will help figure out the appropriateness of these metrics to a variety of scenarios. It will bring out the key insights from the data and help devise practical approaches. Availability of the data is another concern where the focus is needed across industries to create a data sharing culture for research and development. Publicly available information has been the main area for the experiment; more industry-based data must be leveraged to explore the approaches' effectiveness in real-world settings. With increasing threat of security concerns across globe, there is a need for an efficient system that can learn from the knowledge that is being generated and provide the required experience to the practitioners in the necessary format and scope. Dependency on the security experts must be cut down so that the security mindset and knowledge become everyone's responsibility. In software development teams, there is a need for collaboration between customers, the requirements management team, and the development team. There is a smooth transition of the business knowledge with enough focus on security towards the project's technical requirements. The mindset of security becoming the focus area only if things go wrong should be done away with. There needs to be a revised security approach becoming part of the software development strategy for all the people involved. Transfer learning is making good marks in the data science world and brings a lot of promise to tackle data insufficiency and optimization of machine learning problems. Since the experiments have been localized to project-specific data, it is essential to develop generalized learning approaches to cover a larger area of operations and globalize the experiments. Global solutions can be proposed. The benefit of global solutions would be to fast-pace the target research and revelation of the new global best practices.Exploring the data science across the software development lifecycle, with a specific focus on each of the phases of requirements, design, construction, and testing is required. Understanding the characteristics of each of the stages will help to customize the data science approaches appropriately. As it is evident that the security issues become costlier as they are identified later in the software development lifecycle. The influence of time on the severity of the vulnerabilities can be understood better. This knowledge also will be essential to design long-lasting approaches. Issues hid in software systems weaken the system over the period, and they create technical debt for the system. This nature makes it very important for timely identification and remediation of security issues. Devising the right remediation approach can also be backed with the proper insight derived from the data. Data visualization is a critical part of data science; there is a need to tackle security issues in software development processes effectively.

CONCLUSION

With the careful analysis, it's evident that the software industry is striving towards software systems that are of high quality and that can strike a balance between cost and time. Practitioners are struggling to handle these attributes under pressure. It will bring a lot of value to the practitioners if they can build their confidence and customer confidence in terms of the team capabilities, product features, and process capabilities. There is a demand for backing decisions with the data so that the practitioners can feel comfortable and focus on improvement rather than firefighting. This improvement focus can happen only if the data eco-system is strengthened. Building these decision enabling systems will need to have two key components: to provide insights during the operations to make course corrections and to help learning-based strategies that can leverage the pattern of the processes. Education is needed to formulate a business problem and transform the same into a statistical problem and see the statistical solution that can be transformed into a business solution. Amid tremendous effort to build the data analytics solution, not much is focused on this critical backbone of problem-solving. There is a greater need to align the organizational measurement systems with the vision to rectify all the loopholes in this system. With this solid background work that has been done in the area of software development process optimization with data science will find better expression and faster growth. For complex areas like security in software development, this organic growth of the foundational aspects holds the key to success. Without concentrating on the base system, it won't be easy to make breakthroughs. Capturing the tribal knowledge for the organization's long-term benefit and industry is a vital area that needs attention. It is essential for security experts and software engineering practitioners to collaborate well with data science experts to build a potent combination to fulfill the aspirations of the practical application of data science practices in secured software engineering.

REFERENCES

Amoroso, E. G. (1994). *Fundamentals of Computer Security Technology.* Prentice-Hall.

Brereton, P., Kitchenham, B. A., Budgen, D., Turner, M., & Khalil, M. (2007). Lessons from applying the systematic literature review process within the software engineering domain. *Journal of Systems and Software, 80*(4), 571-583.

Casamayor, A., Godoy, D., & Campo, M. (2010). Identification of non-functional requirements in textual specifications: A semi-supervised learning approach. *Information and Software Technology, 52*(4), 436-445.

Cleland-Huang, J., Settimi, R., Zou, X., & Solc, P. (2007). Automated classification of non-functional requirements. *Requirements Engineering, 12*(2), 103-120.

Dörr, J., Kerkow, D., Von Knethen, A., & Paech, B. (2003, June). Eliciting efficiency requirements with use cases. In Ninth international workshop on requirements engineering: foundation for software quality. In *Conjunction with CAiSE* (Vol. 3, pp. 22-23). Academic Press.

El-Hadary, H., & El-Kassas, S. (2014). Capturing security requirements for software systems. *Journal of Advanced Research, 5*(4), 463-472.

Firesmith, D. G. (2005, August). A taxonomy of security-related requirements. In *International Workshop on High Assurance Systems (RHAS'05)* (pp. 29-30).Firesmith.

Fisher, J. (2007). *Owasp Application Security Requirements.* Available: https://www.owasp.org/index.php/File:OWASP_Application_Security_Requirements_-_Identification_and_Authorisation_v0.1_(DRAFT).doc

Fernandes, R., & Cowie, A. (2004). *Capturing informal requirements as formal models* (Doctoral dissertation). Deakin University.

Hatebur, D., Heisel, M., & Schmidt, H. (2007, April). A pattern system for security requirements engineering. In *The Second International Conference on Availability, Reliability and Security (ARES'07)* (pp. 356-365). IEEE. 10.1109/ARES.2007.12

Haley, C., Laney, R., Moffett, J., & Nuseibeh, B. (2008). Security requirements engineering: A framework for representation and analysis. *IEEE Transactions on Software Engineering, 34*(1), 133-153.

ISO. (2009). *Evaluation, ISO/IEC 15408: Information technology - Security techniques.* Available: https://www.iso.org/iso/home/store/catalogue_ics/catalogue_detail_ics.htm

ISO. (2015). *ISO/IEC 27001 - Information security management.* Available: https://www.iso.org/iso/home/standards/management-standards/iso27001.htm

Knauss, E., Houmb, S., Schneider, K., Islam, S., & Jürjens, J. (2011, March). Supporting requirements engineers in recognising security issues. In *International Working Conference on Requirements Engineering: Foundation for Software Quality* (pp. 4-18). Springer. 10.1007/978-3-642-19858-8_2

Kof, L. (2005). *Text analysis for requirements engineering* (Doctoral dissertation). Technische Universität München.

Brereton, P., Kitchenham, B. A., Budgen, D., Turner, M., & Khalil, M. (2007). Lessons from applying the systematic literature review process within the software engineering domain. *Journal of Systems and Software, 80*(4), 571–583. doi:10.1016/j.jss.2006.07.009

Kitapci, H., & Boehm, B. W. (2006, September). Using a hybrid method for formalizing informal stakeholder requirements inputs. In *Fourth International Workshop on Comparative Evaluation in Requirements Engineering (CERE'06-RE'06 Workshop)* (pp. 48-59). IEEE. 10.1109/CERE.2006.8

Kiyavitskaya, N., Zeni, N., Breaux, T. D., Antón, A. I., Cordy, J. R., Mich, L., & Mylopoulos, J. (2008, October). Automating the extraction of rights and obligations for regulatory compliance. In *International Conference on Conceptual Modeling* (pp. 154-168). Springer. 10.1007/978-3-540-87877-3_13

Kitchenham, B. (2004). Procedures for perfoming systematic review, software engineering group. Department of Computer Science, Keele University, United Kingdom and Empirical Software Engineering, National ICT Australia Ltd., TR/SE-0401.

Kaiya, H., Osada, A., & Kaijiri, K. (2004, September). Identifying stakeholders and their preferences about NFR by comparing use case diagrams of several existing systems. In *Proceedings. 12th IEEE International Requirements Engineering Conference*, 2004 (pp. 112-121). IEEE. 10.1109/ICRE.2004.1335669

Li, T. (2017, December). Identifying security requirements based on linguistic analysis and machine learning. In *2017 24th Asia-Pacific Software Engineering Conference (APSEC)* (pp. 388-397). IEEE. 10.1109/APSEC.2017.45

Liu, Y., & Man, H. (2005, March). Network vulnerability assessment using Bayesian networks. In *Data mining, intrusion detection, information assurance, and data networks security 2005* (Vol. 5812, pp. 61–71). International Society for Optics and Photonics. doi:10.1117/12.604240

Lami, G., Gnesi, S., Fabbrini, F., Fusani, M., & Trentanni, G. (2004). *An automatic tool for the analysis of natural language requirements. Informe técnico, CNR Information Science and Technology Institute.*

Lee, S. W., Muthurajan, D., Gandhi, R. A., Yavagal, D., & Ahn, G. J. (2006). Building decision support problem domain ontology from natural language requirements for software assurance. *International Journal of Software Engineering and Knowledge Engineering, 16*(06), 851–884. doi:10.1142/S0218194006003051

Lin, L., Nuseibeh, B., Ince, D., Jackson, M., & Moffett, J. (2003). Introducing abuse frames for analyzing security requirements. *Proceedings of the 11th IEEE international requirements engineering conference (RE'03)*, 371–2. 10.1109/ICRE.2003.1232791

Malhotra, R., Chug, A., Hayrapetian, A., & Raje, R. (2016, February). Analyzing and evaluating security features in software requirements. In *2016 International Conference on Innovation and Challenges in Cyber Security (ICICCS-INBUSH)* (pp. 26-30). IEEE. 10.1109/ICICCS.2016.7542334

Moffett, J. D., Haley, C. B., & Nuseibeh, B. (2004). Core security requirements artefacts. Department of Computing, The Open University, Milton Keynes, UK, Technical Report, 23.

Moffett, J. D., & Nuseibeh, B. A. (2003). *A framework for security requirements engineering.* Report-University of York Department of Computer Science YCS.

Mellado, D., Fernández-Medina, E., & Piattini, M. (2007). A common criteria based security requirements engineering process for the development of secure information systems. *Computer Standards & Interfaces*, 29(2), 244–253. doi:10.1016/j.csi.2006.04.002

Mellado, D., Blanco, C., Sánchez, L. E., & Fernández-Medina, E. (2010). A systematic review of security requirements engineering. *Computer Standards & Interfaces*, 32(4), 153–165. doi:10.1016/j.csi.2010.01.006

OWASP. (2015). *Category: OWASP Application Security Verification Standard Project.* Available: https://www.owasp.org/index.php/Main_Page

Phillips, C., & Swiler, L. P. (1998, January). A graph-based system for network-vulnerability analysis. In *Proceedings of the 1998 workshop on New security paradigms* (pp. 71-79). 10.1145/310889.310919

Riaz, M., King, J., Slankas, J., & Williams, L. (2014, August). Hidden in plain sight: Automatically identifying security requirements from natural language artifacts. In *2014 IEEE 22nd international requirements engineering conference (RE)* (pp. 183-192). IEEE.

Sindre, G., & Opdahl, A. L. (2005). Eliciting security requirements with misuse cases. *Requirements Engineering*, 10(1), 34–44. doi:10.100700766-004-0194-4

Salini, P., & Kanmani, S. (2012). Survey and analysis on security requirements engineering. *Computers & Electrical Engineering*, 38(6), 1785–1797. doi:10.1016/j.compeleceng.2012.08.008

Souag, A., Salinesi, C., & Comyn-Wattiau, I. (2012, June). Ontologies for security requirements: A literature survey and classification. In *International conference on advanced information systems engineering* (pp. 61-69). Springer. 10.1007/978-3-642-31069-0_5

Wilson, W., Rosenberg, L., & Hyatt, L. (1997). Automated analysis of requirement specifications, *Proceedings of the 19th ACM international conference on Software engineering*, 161-171. 10.1145/253228.253258

Chapter 6
Quantitative and Visual Exploratory Data Analysis for Machine Intelligence

Dharmendra Trikamlal Patel

https://orcid.org/0000-0002-4769-1289

Smt. Chandaben Mohanbhai Patel Institute of Computer Applications, CHARUSAT, India

ABSTRACT

Exploratory data analysis is a technique to analyze data sets in order to summarize the main characteristics of them using quantitative and visual aspects. The chapter starts with the introduction of exploratory data analysis. It discusses the conventional view of it and describes the main limitations of it. It explores the features of quantitative and visual exploratory data analysis in detail. It deals with the statistical techniques relevant to EDA. It also emphasizes the main visual techniques to represent the data in an efficient way. R has extraordinary capabilities to deal with quantitative and visual aspects to summarize the main characteristics of the data set. The chapter provides the practical exposure of various plotting systems using R. Finally, the chapter deals with current research and future trends of the EDA.

INTRODUCTION

In 1961 John W. Turkey in his research work at Princeton University discussed some aspects about the future of data analysis (JhonW.Turkey,1961). He published that "Data analysis is a larger and more verified field than inference or inclusive procedures or allocation. Statistics has contributed much to data analysis". In 1970 John W. Turkey, in his book published that Exploratory Data Analysis is one kind of philosophy and no fixed rules for how you approach it. In 1984, the author (Mulaik, S. A.,1984) discussed the exploratory statistics for better analysis. Exploratory Data Analysis can be defined by numerous ways based on the concepts given by above literatures:

DOI: 10.4018/978-1-7998-7701-1.ch006

Definition 1: Exploratory Data Analysis does the preliminary exploration of data to discover meaningful insights by using statistics and visual representation techniques.

Definition 2: Exploratory Data Analysis focuses on data in order to gather novel, meaningful and hidden patterns for the effective model generation based on capable statistical and graphical representation techniques.

Definition 3: Exploratory Data Analysis proposes the acceptable model based on the data using data science techniques with little or no assumptions.

Definition 4: Exploratory Data Analysis uses a significant process to generate efficient model based on the data analysis using quantitative and visual representation techniques of data science in order to achieve purposeful insights.

In 1992, author presented the concept of Confirmatory Data Analysis (CDA) and provided a balanced presentation of both EDA and CDA. Confirmatory Data Analysis (CDA) is also known as hypothesis driven data analysis. Exploratory and Confirmatory Data Analysis are complements of each other. EDA helps the data analyst to find clues or trends in order to get the conclusion, whereas CDA assesses confirmation by challenging their hypothesis about the data Table-1 describes the differences between hypothesis driven and exploratory data analysis methodologies.

Table 1. Difference between Hypothesis and Exploratory Data Analysis

Parameter	Hypothesis Driven	Exploratory Data Analysis
Motivating Question	Can I reject the null hypothesis?	How I can describe variation in my dataset?
Nature of Work	Involve subjectively located plots	Collection of fresh data from objectively located plots
Analysis	All the analysis must be planned in advance	NumerousDeviations on analysis.
Statistical Inference	p-values are significant in this context	p-values give only jagged idea.
Automated Data Dividing	Not valid without cross validation	Applicable and very constructive means.
Model Context	Model Validation	Model Building

It is feasible to use hybrid data analysis by combining both approaches for a larger study. Two-phase analysis is required for the same whereas first phase is of Exploratory Data Analysis and the second phase is Confirmatory Data Analysis. In 1993, the author (Joreskog, K et.al,1993, Bradley E.S.et al.,2010, Bonomi,F et al.,2012, Haddara, M et al.,2015, Yoo, C. et al,2014, Krekhov A. et al.,2020) presented discussion of data analysis techniques based on several situations. Graphical representation of data is another important aspect of exploratory data analysis (SAWILOWSKYet.al,2003, Natrella M,2010,Yan L.et al.,2011, Khan M. et al.,2011, High R . et al.,2012,Borkin MA et. al,2016, Godfrey P. et al.,2016, Di Blas et al.,2017, Gibson, D. et al., 2017,Lee PS et.al.,2018, Konopka BM, et al.,2018) but there are several research challenges associated with it(Moorhead R, et al,2006, Johnstone I.M. et al.,2009, Javed W.et al.,2013, Kaisler S. et al.,2013, Kamat, N. et al., 2014,Iyer G.et al.,2017). Designing effectual visualization is a very multifarious process that requires a complicated indulgent of human information-processing capabilities, both visual and cognitive manner. In the modern era, several open source based tools are available which have excellent capabilities of both statistics and visualizations. The R programming language provides an astonishing functions and packages about statistics and visualizations. The R programming language considers the best for exploratory data analysis as it includes both graphical and non-graphical EDA methodologies in the simple sense (Yang W. et al.,n.d., Duffy, F.H. et al.,1990, Thompson J.M., (1992),Jianchang Mao et al.,1995, Erbacher, R. F., 2007, Lewis J. M. et al.,2008, Piringer H. et al.,2008, Kelly J.E.,2015,Matthieu Komorowski, et al.,2016, Roger D. Peng,2016, Ronald K.Pearson,2018).

Classical View and Its Limitations

The two main popular data analysis approaches are: Classical Approach and Exploratory Data Analysis Approach.Figure-1 describe the difference between the two approaches.

Figure 1. Classical vs. EDA Approach

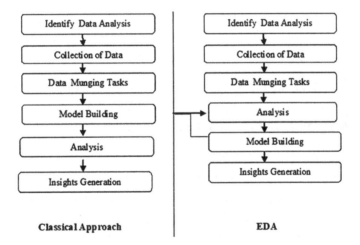

Two approaches are similar in the initial steps. However, the difference arises in the intermediate steps in terms of sequence and focus. In the classical approach, both deterministic (for an example regression model) and probabilistic models (example Markov Chain Models) are derived based on the data directly. On the divergent, EDA permit the data to propose acceptable models that best apposite for the data. In the classical approach, the focus is on the model while EDA focuses on the data. Classical approach uses the quantitative means to represent the data, whereas EDA approach uses visualization techniques for the representation of the data. The classical approach is scrupulous whereas EDA approach is perceptive about the best fit model selection. In the classical approach, there is a possibility of loss of information while in EDA approach; there is no possibility of data loss. The classical approach is based on the assumption while EDA approach uses little or no assumptions. Classical view finds trends in data through some statistical techniques. It helps in sorting real patterns from outliers or noisy data. The classical view studies problem statement first. After that it does data munging kind of tasks. After data munging, data is too fed to model for insights generations and then conclusions derived. There is no emphasis on the quality of the data. EDA on the other hand, emphasizes on the quality of the data. It uses several techniques to analyze the data before it is fed to the model. It carries out important steps such as extracting important variables, detect outliers and anomalies, test underline assumptions etc. Data Analysis of preprocessed data is the key trait of EDA. After proper analysis the data is fed to model by EDA approach.

QUANTITATIVE AND GRAPHICAL EXPLORATORY DATA ANALYSIS

Inferential statistics entails the use of statistical methods in the testing of hypotheses and depicting conclusions from the result of the study(Kolawole, E.B.,2001, A. Meyer-Baese et al.,2004, Huang S. et al.,2005, Perin C. et al.,2013, Furmanova, K. et al.,2017,M. Rodríguez-Ibáñez et al.,2019). There are two familiar types of statistical inferences: (a) Estimation and (b) Null Hypothesis Test of Significance Estimation is vital for the exploratory data analysis.

Estimation

Estimation in statistics has two main forms: Point Estimation and Interval Estimation. The point estimate is a single value that is maximally likely value ofparameter. For example:

- The point estimation of population mean μ = The sample mean of observations $\bar{X}$

  Equation (1) represents Point Estimation of population mean

$$= \left(\sum_{i=1}^{N} xi \right) / N \tag{1}$$

Where **x1, x2,.., xi**are the different observations, **N** is the total number of observations.

- The point estimate of the population variance σ^2 = The sample variance of observations S^2

  Equation (2) represents Point Estimation of population variance

$$= \left(\sum_{i=1}^{N} (xi - {}^-x)^2 \right) / N - 1 \tag{2}$$

where x1, x2,.., xi are the different observations, $^-$x is the mean of observations, **N** is the total number of observations .

- The point estimate of the population standard deviation σ = The sample standard deviation of observations S

  Equation (3) represents Point Estimation of population standard deviation

$$= \left(\sqrt[2]{\sum_{i=1}^{N}(xi - \bar{x})} \right) / N - 1 \tag{3}$$

where x1, x2,.., xi are the different observations, ⁻x is the mean of observations, **N** is the total number of observations .

Exploratory Data Analysis uses interval estimation quantitative technique to construct a range of values within which a variable is likely to fall (Alan Anderson et al., 2015).

Interval Estimation

Interval estimation aims to estimate the interval of likely values of an unfamiliar population parameter. The confidence interval(given by Equation. 4) is one kind of interval estimation used extensively for the exploratory data analysis.

Confidence Interval = Point Estimate +/- Margin of error $\tag{4}$

Margin of error depicts the deviation of the point estimate from one sample to the next. Margin of error is based on the standard deviation and the number of samples being used. The desired level of the confidence is any level between 1-100%, however, most common value is 95%. The confidence interval is based on the Z- Score represented by equation (5).

$$Z = \left(x - \mu \right) / \sigma \tag{5}$$

Where μ is mean, σis standard deviation
The standard error of the mean(equation.(6)) is:

Standard error of Mean $= \sigma / \sqrt{N}$ $\tag{6}$

Where σ is standard deviation, **N** is total number of observations
The critical value of Z is associated with a central area under the standard curve and is represented by equation (7).

Central area $= 1 - 2\alpha$ $\tag{7}$

Where α is the tail area of the curve
In general, confidence interval for the mean with certain probability is calculated using following formula given by equation (8) and (9):

Lower Limit $= \mu - Zprob * \sigma$ $\tag{8}$

Upper Limit $= \mu + Zprob * \sigma$ \hfill (9)

where μ is mean, Zprob is the Z-score with certain probability, **σ** is standard deviation

Example: Assume that the price of the particular stock XYZ is normally distributed with a mean of 80 and standard deviation is 30. Determine the sampling distribution of the mean based on confidential interval with the sample size is 6.

Standard error of Mean $= 30/\sqrt{6} = 30/2.45 = 2.241$

Here we select the most common value of probability is 95%.
The critical value of Z = 1- 2 α

∴ 0.95 = 1 - 2α

∴ 2α = 1 – 0.95

∴ α = 0.05/2

∴ α = 0.025

Zprob = 1 – 0.025

∴Zprob=0.975= 1.96

Lower Limit = 80 – (1.96) * (12.24)= 56.01

Upper Limit= 80 + (1.96)*(12.24)=103.99

Figure 2. Confidence Interval of an example

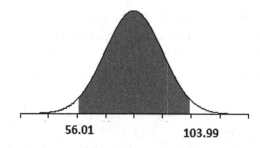

The figure-2 clearly describes the confidence interval of blue shaded area. The range is between 56.01 and 103.99.

Null Hypothesis Test of Significance Estimation

Test of significance is a technique that is used to support or reject the claims based on sample data(FigueiredoFilho D.B. et.al, 2014, Ronald K.Pearson,2018, Stolte, C. et al.,2002, Heer J. et al.,2012, Gratzl S. et al.,2013, Gratzl S. et al.,2014, Gartner. Inc.,Analytics and Business Intelligence Platform,2018). Every test of significance begins with null hypothesis H0. The researcher with the help of this null hypothesis will strive to negate or discredit. The alternate hypothesis (Hα) states that there is a statistically momentous association between two variables. The final conclusion of the test is always based on the null hypothesis. We either reject the null hypothesis or do not reject it. We neither accept nor reject the alternate hypothesis(Hα).

The hypothesis are always associated with population like mean μ. The alternate hypothesis having two forms: one or two sided. The formulas for one and two sided alternate hypothesis are equation (10) and (11) respectively as:

Null Hypothesis(H0): $\mu = n$

Alternate Hypothesis(Hα):$\mu > n$ or $\mu < n$ (One Sided) (10)

Alternate Hypothesis(Hα): $\mu \neq n$ (Two sided) (11)

Once the null and alternate hypothesis is determined, z-test is computed as given in equation (12):

$$z = (\bar{x} - \mu 0) / (\sigma / \sqrt{n})$$ (12)

where $\bar{x}$ is the sample mean, μ0 is the population means equals to μ, σ is standard deviation, nis the number of observations.

As a final point, the P-values for null hypothesis(H0) aligned with each of the probable alternative hypothesis are given by equation (13), equation (14), equation (15) respectively :

P(Z >= z) for Alternate Hypothesis(Hα): $\mu > \mu 0$ (One sided) (13)

P(Z <= z) for Alternate Hypothesis(Hα): $\mu < \mu 0$ (One sided) (14)

2P(Z>|z|) for Alternate Hypothesis(Hα): $\mu \neq \mu 0$ (Two sided) (15)

Example: Assume that in the election of one state, the mean vote cast is 70 and standard deviation equal to 20. Government official suspects that female voters cast votes with higher mean than men as from a random sample of 91 constituencies female voters mean is 75. Does it give strong indication that the by and large mean for female voters is higher than men? Assume that confidence interval is 95%.

The test statistic z= (75-70)/(20/9) = 2.25 // Computed Z value

There are three possibilities to reject or do not reject the null hypothesis.

Possibility: 1 (Two Sided Test i.e $\mu \neq n$)

The critical Z value = 1 – 2 α (As we have used Two sided test)

∴ α = 0.025

Z = 1 – 0.025

∴ Z=0.975= 1.96

The computed value of Z is far higher than the critical value of Z so we reject the null hypothesis. Refer the Figure 3 with $\mu \neq n$ criteria. According to the figure, the accepted region lies between -1.96 to +1.96. The computer Z value is 2.25 so it is more than +1.96 so we reject the null hypothesis.

Figure 3. Curve based on Two sided test i.e e $\mu \neq n$

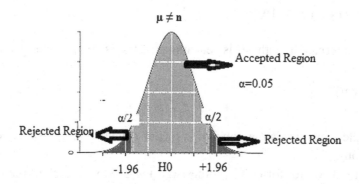

Possibility: 2 (One Sided Test i.e $\mu > n$)

The critical Z value = α

Here confidence level is 95% so level of significance = 0.05

∴ α = 0.05

Z = 1 – 0.05

∴ Z=0.95= 1.65

The computed value of Z is far higher than the critical value of Z so we reject the null hypothesis. Refer the Figure 4 with $\mu > n$ criteria. According to the figure, the accepted region lies less than or equal to +1.65. The computer Z value is 2.25 so it is more than +1.65 so we reject the null hypothesis.

Figure 4. Curve based on Two sided test i.e e μ > n

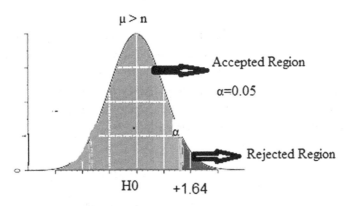

Possibility: 3 (One Sided Test i.e μ < n)

The critical Z value = α

    Here confidence level is 95% so level of significance = 0.05

∴ α = 0.05

Z = 1 – 0.05

∴ Z=0.95= 1.65

    The computed value of Z is 2.25 that is in the range of >-1.65 so we do not reject the null hypothesis. Refer the Figure 5 with μ < n criteria. According to the figure, the accepted region lies greater than or equal to -1.65. The computer Z value is 2.25 that are in the range so we do not reject the null hypothesis.

Figure 5. Curve based on Two sided test i.e e μ < n

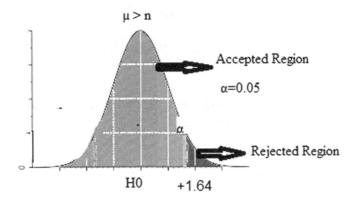

Visualization Techniques

Visualization techniques are essential in analysis in two stages of the life cycle of data science. The first stage is during the Exploratory Data Analysis and second stage during the final conclusion. During the Exploratory Data Analysis, visualization techniques are helpful to identify relationships among certain variables, correlation between features and efficiently display the data. Many visualization plots can be drawn during Exploratory Data Analysis, however the scope of this chapter is to deal with basic visualization plots such as Box Plot, Histogram and Scatter Plots.

Box Plot

The Box Plot undervalued plot in Exploratory Data Analysis (EDA) but very vital to display the distribution of data in a better way. The components of the Box Plot are depicted in the Figure-6. The rectangle area of the box has three lines: upper, middle and lower. Lower line considers as first quarter or 25%. Upper line considers as third quarter or 75% and the middle line considers as median or second quartile or 50%. Difference between third quartile and first quartile is known as Inter Quartile Range (IQR). The box is attached with one vertical line. The lower part of the vertical line is the minimum value while the upper part is the maximum value.

Figure 6. Components of Box Plot

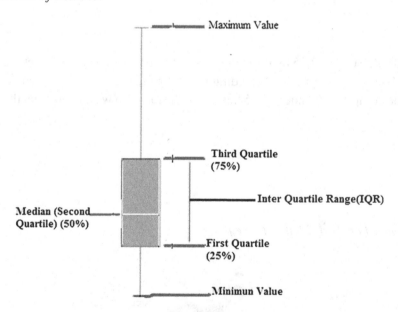

When you have multiple data sets from different independent sources that are related to each other Box Plot plays a vital representation. Examples are:

- Unit test marks among number of tests
- Vote cast between state and Centre election

- Change detection before and after a particular event
- Performance comparison of different manufacturing machines

Histograms

In an exploratory data analysis histogram acts as a more wide-ranging tool for displaying the distribution of quantitative data. Histogram answers the several questions such as:

- What is the type of distribution? Symmetric, left skewed or right skewed?
- How many frequent value distribution have?
- Are there any data not fit in the range means Outliers?

Visualization is the best way to present analysis of the data in efficient way (C. Ware, 2004, P.s. Lee et.al.,2015). Visualization is very important in exploratory data analysis for different kind of graphs.

Different kinds of histograms are possible for the same data set based on a number of parameters such as different numbers of intervals, starting and ending points and rules for assigning the data. There are common steps to construct the histogram efficiently:

Step 1: Calculate the Range given by equation (16).

Range (R) = Maximum Value – Minimum Value = R(n) – R(0)   (16)

Step 2: Divide the range with some random number (m) based on how many bins you want in histogram.
Step 3: Use the values R and m to set the interval boundaries.
Step 4: Choose the rules, how to assign the data into the intervals.

Figure 7 describes the marks of unit test of a specific student.

Figure 7. Histogram of Unit test marks

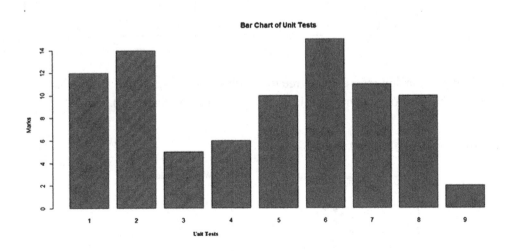

Scatter Plots

Exploratory Data Analysis uses the scatter plot extensively to show the relationship between two or more variables. These charts propose the correlations such as perfect positive, high positive, low positive, no correlation, low negative, high negative and perfect negative between variables based on the confidence interval. Scatter plot comprises of the horizontal axis say X, and the vertical axis say Y, and a progression of spots. Each spot on the scatter plot represents one observation from a data set. The location of the spot on the scatter plot characterizes its X and Y values. In exploratory data analysis, it is used to analyze the patterns between data. These patterns are described in terms of number of aspects. The main aspects are: linearity, slope and strength.

- Linearity describes the pattern as straight or curved shown by figure.8.

Figure 8. Linearity of scatter plot

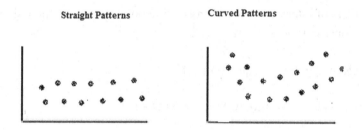

- Slope refers to the direction of change of one variable when another variable value changes. When variable Y becomes high the slope is positive else slop is negative(represented by figure.9).

Figure 9. Positive vs. negative slope

- Strength of relationship refers the degree to which spots are scattered. If spots are concentrated, the relationship is strong otherwise it is weak(represented by figure.10).

Figure 10. Strong vs. weak relationship

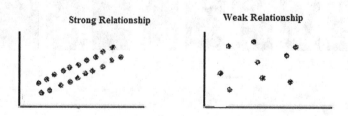

PLOTTING SYSTEM USING R

R programming language provides astonishing functions and packages for data visualization. It provides the obvious indulgent of patterns in data. It also has an ability to detect hidden structure in the data. The Table-2 describes the several components of plotting system in R.

Table 2. Components of Plotting System in R

Sr.No	Component	Description
1.	Packages	It includes functions for statistical plots. For example lattice,ggplot2
2.	Devices	The function is to produce the output. The screen is the default device. Several graphical devices like pdf, jpeg etc.
3.	Grammar of graphics	It includes the data, aesthetic mapping, geometric objects, coordinate system, scales, faceting.

Base Plotting System

The plot() function is the simplest technique to generate different kinds of plot in R. It is available as a part of the basic installation of R. It has the following syntax in R:

```
plot(x,y,parameter,main,sub,xlab,ylab)
```

"p" for point, "l" for line, "b" for both,"s" for stair step,"h" for histogram

main,subdepicts the main and sub title of the plot respectively, xlab,ylabare the labels of X and Y axes respectively.

Example: One data set of women contains 15 records of height and weight of different women. General different plots using a base plotting system of R.

```
>women                                    // Women data-
set
height weight
1      58     115
2      59     117
3      60     120
4      61     123
5      62     126
6      63     129
7      64     132
8      65     135
9      66     139
```

```
10     67     142
11     68     146
12     69     150
13     70     154
14     71     159
15     72     164
plot(women$height,women$weight,"p",main="Corealation of Height and Weight",
sub="Height Vs.Weight",xlab="Height",ylab="Weight")                    // R
code to generate the plot
```

Figure 11 represents the R coding and corresponding graph generated.

Figure 11. Left: Plot using R base plotting system; right: R code to generate the plot

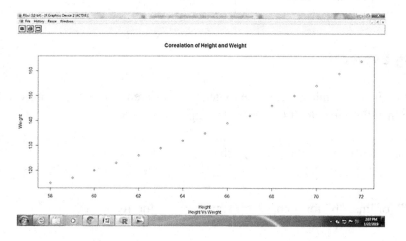

ggplot2 System

The ggplot2 package is the most powerful package for the visualization in R. The ggplot2 package is extensively used for the Exploratory Data Analysis. It is an explicit package so not available with the basic installation of R. The install.packages() command is used to install the ggplot2 package.

```
>install.packages("ggplot2")                    // R code to install an ex-
plicit package
```

Once it is installed, the library function is used to load it, so we can use its function.

```
>library(ggplot2)                               // Load the ggplot2 package
```

The main advantage of ggplot2 in comparison of base system is that we can assign the result of ggplot2 function into an object and can modify it whereas in base system there is no object that holds the results.

There are two main functions of ggplot2 for effective visualization:qplot() and ggplot(). The qplot() is an elementary function to generate different kind of visualizations very quickly while gplot() is a sophisticated function used for more complex plots.

The simplified syntax of the qplot() function is:

```
qplot(x,y,data,facet,geom,xlim,ylim,main,xlab,ylab)
```

Example:

```
qplot(women$height,women$weight,main="Scatter Plot of Height and Weight",xlab="
Height",ylab="Weight",colour=I("red"),size=I(5))          // Scatter Plot using
qplot() function
```

Figure 12 represents the R coding and corresponding graph generated.

Figure 12. Left: Scatter Plot using qplot()function of ggplot2; right: R code to generate the plot

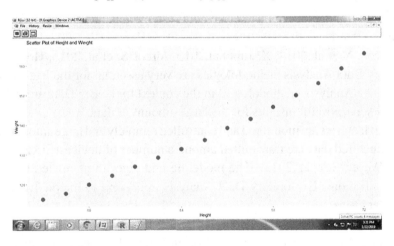

The basic syntax of ggplot() function is:

```
ggplot(data,aes (x,y,....)) + geometric objects
```

It also checks the several encoding aspects of the graph such assize,shape,color etc.
+ is the operator to append several features in the graph.
geometric objectsare objects that we want to draw in the graph such as geom.
Example:

```
ggplot(women,aes(x=height,y=weight))+geom_point()          // Scatter Plot using
ggplot()
```

Figure 13 represents the R coding and corresponding graph generated.

Figure 13. Left: Scatter Plot using ggplot()function of ggplot2; right: R code to generate the plot

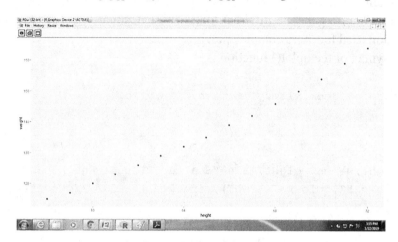

RESEARCH AND FUTURE TRENDS OF EDA

The modern technologies and information management systems generate abundant data of different types in nature and size and that have given the birth of Big Data(Keim D.A.,2002, Lynch C.,2008, Im J.-F. et al.,2013, Changwon. Y et al.,2014, X. Jin et.al,2015, Idreos S. et al.,2015, Hourieh R. et al.,2016). Effective Exploratory Data Analysis methodologies are very essential for the Big Data related applications. Exploratory Data Analysis methodologies in the context to the Big Data are in the demand of the modern era to provide purposeful insights for the large amount of data.

The modern smart factories are monitored and controlled remotely and large amount of structured,semi structured and unstructured data are transmitted among a number of devices(Erickson, B. H., &Nosanchuk, T. A.,1992, Weyer, S. et.al.,2015). The modeling and algorithms related to Indusry4.0 require the interactive visualizations. Exploratory Data Analysis in context to interactive visualization is the demand of the modern time.

The Fog Computing is another rapidly emerging research area that is concerned with decentralized computing between the data source and the cloud(Borkin MA et al.,2016, Dastjerdi, A.V.,2016).Exploratory Data Analysis is very essential for the efficient modeling of such kind of problems. Both statistical and visualization revolutions in terms of Exploratory Data Analysis are very vital for the Fog Computing.

The Scientific Computing (Michael L. Overton, 2004, Naureen Moon et al.,2006, Gao H. et al.,2019) deal with the numbers that is extremely large to fit into standard integer types. The Scientific Computing needs the model that solves the scientific problems efficiently. Exploratory Data Analysis plays a crucial role in order to design a perfect kind of model for the scientific computing. The traditional methodologies for Exploratory Data Analysis need to be changed in order to accommodate Scientific Computing concepts in order to design the efficient model.

CONCLUSION

The Exploratory Data Analysis is vital in the modern era to design efficient modeling of any applications as it does the analysis first and then design the model accordingly. Statistics and visualization techniques are the core components of the Exploratory Data Analysis. This chapter has discussed about the Interval Estimation, which is very essential for the determining the estimation of the interval of unfamiliar population. The chapter concluded that the confidence interval is the right approach to decide the interval of unfamiliar population. The chapter also dealt with the null hypothesis test of significance estimation in details. The researcher with the help of the null hypothesis will try to reject or not reject the hypothesis based on the alternate hypothesis. The chapter discussed the one sided and two sided tests and concluded that based on the different type of situations, the proper test must be designed to take appropriate decision about null hypothesis. The chapter discussed the basic visualization techniques in the context to the Exploratory Data Analysis. The chapter has used R programming for the visualization techniques related to Exploratory Data Analysis. The chapter has demonstrated both base plotting and ggplot2 package for visualizations inR. The chapter concluded that the R programming is the right tool to achieve appropriate visualizations for the Exploratory Data Analysis. The chapter concluded that the base plotting is better for the basic visualizations whereas ggplot2 is for sophisticated visualizations.

REFERENCES

Anderson, A., & Semmelroth, D. (2015). *Statistics for Big Data For Dummies* (1st ed.). For Dummies.

Bonomi, F., Milito, R., Zhu, J., & Addepalli, S. (2012).Fog computing and its role in the internet of things. In *Proc. of MCC*. ACM. 10.1145/2342509.2342513

Borkin, M. A., Bylinskii, Z., Kim, N. W., Bainbridge, C. M., Yeh, C. S., Borkin, D., Pfister, H., &Oliva, A. (2016). Beyond Memorability: Visualization Recognition and Recall. *IEEE Transactions on Visualization and Computer Graphics, 22*(1), 519–528. doi:10.1109/TVCG.2015.2467732

Bradley, E. S., Toomey, M. P., Still, C. J., & Roberts, D. A. (2010, December). Multi-Scale Sensor Fusion With an Online Application: Integrating GOES, MODIS, and Webcam Imagery for Environmental Monitoring. *IEEE Journal of Selected Topics in Applied Earth Observations and Remote Sensing, 3*(4), 497–506. doi:10.1109/JSTARS.2010.2048419

Dastjerdi, A. V., & Buyya, R. (2016). Fog Computing: Helping the Internet of Things Realize Its Potential. *Computer, 49*(8), 112–116. doi:10.1109/MC.2016.245

Di Blas, N., Mazuran, M., Paolini, P., Quintarelli, E., & Tanca, L. (2017). Exploratory computing: A comprehensive approach to data sensemaking. *International Journal of Data Science and Analytics, 3*(1), 61–77. doi:10.100741060-016-0039-5

Duffy, F. H., Jones, K., Bartels, P., Albert, M., McAnulty, G. B., & Als, H. (1990). Quantified Neurophysiology with mapping: Statistical inference, Exploratory and Confirmatory data analysis. *Brain Topography, 3*(1), 3–12. doi:10.1007/BF01128856 PMID:2094310

Erbacher, R. F. (2007). Panel Position Statement: The Future of CMV. *Fifth International Conference on Coordinated and Multiple Views in Exploratory Visualization (CMV 2007)*, 75-75, 10.1109/CMV.2007.17

Erickson, B. H., & Nosanchuk, T. A. (1992). *Understanding data* (2nd ed.). University of Toronto Press.

Figueiredo Filho, D. B., Rocha, E. C., Batista, M., Paranhos, R., & Silva, J. A. Jr. (2014). Reply on the Comments on When is Statistical Significance not Significant? *Brazilian Political Science Review*, 8(3), 141–150. doi:10.1590/1981-38212014000100024

Furmanova, K., Gratzl, S., Stitz, H., Zichner, T., Jaresova, M., Ennemoser, M., . . . Streit, M. (2017). *Taggle: Scalable visualization of tabular data through aggregation.* arXiv preprint arXiv:1712.05944.

Gao, H., Nie, H., & Li, K. (2019). Visualisation of Pareto Front Approximation: A Short Survey and Empirical Comparisons. *2019 IEEE Congress on Evolutionary Computation (CEC)*, 1750-1757. 10.1109/CEC.2019.8790298

Gartner. Inc. (n.d.). *Gartner Customer Choice Awards - Analytics and Business Intelligence Platform.* https://www.gartner.com/reviews/customer-choice-awards/analytics-business-intelligence-platforms//

Gibson, D., & de Freitas, S. (2016). Exploratory Analysis in Learning Analytics. *Tech Know Learn*, 21(1), 5–19. doi:10.100710758-015-9249-5

Godfrey, P., Gryz, J., & Lasek, P. (2016). Interactive Visualization of Large Data Sets. *IEEE Transactions on Knowledge and Data Engineering*, 28(8), 2142–2157. doi:10.1109/TKDE.2016.2557324

Gratzl, S., Gehlenborg, N., Lex, A., Pfister, H., & Streit, M. (2014). Domino: Extracting, comparing, and manipulating subsets across multiple tabular datasets. *IEEE Transactions on Visualization and Computer Graphics*, 20(1), 2023–2032. Advance online publication. doi:10.1109/TVCG.2014.2346260 PMID:26356916

Gratzl, S., Lex, A., Gehlenborg, N., Pfister, H., & Streit, M. (2013). LineUp: Visual Analysis of Multi-Attribute Rankings. *IEEE Transactions on Visualization and Computer Graphics*, 19(12), 2277–2286. doi:10.1109/TVCG.2013.173 PMID:24051794

Haddara, M., & Elragal, A. (2015). The Readiness of ERP Systems for the Factory of the Future. *Procedia Computer Science*, 64, 721–728. doi:10.1016/j.procs.2015.08.598

Heer, J., & Shneiderman, B. (2012). Interactive dynamics for visual analysis. *Queue, 10*(2), 30. doi:10.1145/2133416.2146416

High, R. (2012). *The era of Cognitive Systems: An Inside look at IBM Watson and how it Works.* IBM Corporation.

Huang, Ward, & Rundensteiner. (2005). Exploration of dimensionality reduction for text visualization. *Coordinated and Multiple Views in Exploratory Visualization (CMV'05)*, 63-74. . doi:10.1109/CMV.2005.8

Idreos, S., Papaemmanouil, O., & Chaudhuri, S. (2015). Overview of data exploration techniques. In *Proceedings of the 2015 ACM SIGMOD International Conference on Management of Data.* ACM. 10.1145/2723372.2731084

Im, J.-F., Villegas, F. G., & McGuffin, M. J. (2013). VisReduce: Fast and responsive incremental information visualization of large datasets. *IEEE International Conference on Big Data*, 25-32.

Information Visualization, Second Edition: Perception for Design (Interactive Technologies) by Colin Ware (2004–04-21). (1783). Morgan Kaufmann.

Isaacs, E., Damico, K., Ahern, S., Bart, E., & Singhal, M. (2014). Footprints: A visual search tool that supports discovery and coverage tracking. *IEEE Transactions on Visualization and Computer Graphics, 20*(12), 1793–1802. doi:10.1109/TVCG.2014.2346743 PMID:26356893

Iyer, G., DuttaDuwarah, S., & Sharma, A. (2012). DataScope: Interactive visual exploratory dashboards for large multidimensional data. *IEEE Workshop on Visual Analytics in Healthcare (VAHC),* 17-23.

Javed, W., & Elmqvistm N. (2013). ExPlates: spatializing interactive analysis to scaffold visual exploration. *Computer Graphics Forum, 32,* 441-450. doi:10.1111/cgf.12131

Jhon, W. T. (1961). The future of Data Analysis. Research Sponsored by the Army Research Office. Princeton University.

Jin, X., Wah, B. W., Cheng, X., & Wang, Y. (2015). Significance and Challenges of Big Data Research. *Big Data Research, 2*(2), 59–64. doi:10.1016/j.bdr.2015.01.006

Johnstone, I. M., & Titterington, D. M. (2009). Statistical challenges of high-dimensional data. *Philos. Trans. R. Soc., 367*(1906), 4237–4253. doi:10.1098/rsta.2009.0159

Jöreskog, K. G., & Sörbom, D. (1993). *LISREL 8: Structural equation modeling with the SIMPLIS command language. Scientific Software International.* Lawrence Erlbaum Associates, Inc.

Kaisler, S., Armour, F., Espinosa, J. A., & Money, W. (2013). Big data: Issues and challenges moving forward. *46th Hawaii International Conference on System Sciences (HICSS),* 995-1004. 10.1109/HICSS.2013.645

Kamat, N., & Nandi, A. (2014). *InfiniViz: Interactive Visual Exploration using Progressive Bin Refinement.* arXiv preprint arXiv:1710.01854.

Keim, D. A. (2002). Information visualization and visual data mining. *IEEE Transactions on Visualization and Computer Graphics, 8*(1), 1–8. doi:10.1109/2945.981847

Kelly, J. E. (2015). Computing, cognition and the future of knowing. Whitepaper IBM Res.

Kolawole, E. B. (2001). *Tests and Measurement, AdoEkiti.* Yemi Printing Services.

Komorowski, M., Marshall, D. C., Salciccioli, J. D., & Crutain, Y. (2016). Exploratory Data Analysis. In *Secondary Analysis of Electronic Health Records.* Springer. doi:10.1007/978-3-319-43742-2_15

Konopka, B. M., Lwow, F., Owczarz, M., & Łaczmański, Ł. (2018). Exploratory data analysis of a clinical study group: Development of a procedure for exploring multidimensional data. *PLoS One, 13*(8), e0201950. doi:10.1371/journal.pone.0201950 PMID:30138442

Krekhov, A., Cmentowski, S., Waschk, A., & Krüger, J. (2020, January). Deadeye Visualization Revisited: Investigation of Preattentiveness and Applicability in Virtual Environments. *IEEE Transactions on Visualization and Computer Graphics, 26*(1), 547–557. doi:10.1109/TVCG.2019.2934370 PMID:31425106

Lee, P. S., & Howe, B. (2015). Dismantling composite visualizations in the scientific literature. In *International Conference on Pattern Recognition Applications and Methods.* ICPRAM. 10.5220/0005213100790091

Lee, P. S., West, J. D., & Howe, B. (2018). Viziometrics: Analyzing Visual Information in the Scientific Literature. *IEEE Transactions on Big Data, 4*(1), 117–129. doi:10.1109/TBDATA.2017.2689038

Lewis, J. M., Hull, P. M., Weinberger, K. Q., & Saul, L. K. (2008). Mapping Uncharted Waters: Exploratory Analysis, Visualization, and Clustering of Oceanographic Data. *2008 Seventh International Conference on Machine Learning and Applications*, 388-395, 10.1109/ICMLA.2008.125

Lynch, C. (2008). Big data: How do your data grow? *Nature, 455*(7209), 28–29. doi:10.1038/455028a PMID:18769419

Mao, J., & Jain, A. K. (1995, March). Artificial neural networks for feature extraction and multivariate data projection. *IEEE Transactions on Neural Networks, 6*(2), 296–317. doi:10.1109/72.363467 PMID:18263314

Meyer-Baese, A., Wismueller, A., & Lange, O. (2004, September). Comparison of two exploratory data analysis methods for fMRI: Unsupervised clustering versus independent component analysis. *IEEE Transactions on Information Technology in Biomedicine, 8*(3), 387–398. doi:10.1109/TITB.2004.834406 PMID:15484444

Michael, L. O. (2004). Numerical Computing with IEEE Floating Point Arithmetic. SIAM Publications.

Moon, N., Hsu, Y.-W., & Singh, R. (2006). A Multiple-Perspective, Interactive Approach for Web Information Extraction and Exploration. *22nd International Conference on Data Engineering Workshops (ICDEW'06)*, 41-41. 10.1109/ICDEW.2006.11

Mulaik, S. A. (1984). Empiricism and exploratory statistics. *Philosophy of Science, 52*, 410–430. doi:10.1086/289258

Munzner, T., Johnson, C., Moorhead, R., Pfister, H., Rheingans, P., & Yoo, T. (2006). NIH-NSF visualization research challenges report summary. *IEEE Computer Graphics and Applications, 26*(2), 20–24. doi:10.1109/MCG.2006.44 PMID:16548457

Natrella, M. (2010). *NIST/SEMATECHe-Handbook of Statistical Methods*. NIST/SEMATECH.

Oliveira, S., & Stewart, D. (2006). *Writing Scientific Software: A Guide to Good Style*. Cambridge University Press., doi:10.1017/CBO9780511617973

Pearson, R. K. (2018). *Exploratory Data Analysis using R*. CRC Press. doi:10.1201/9781315382111

Peng, R. (2016). *Exploratory Data Analysis with R*. lulu.com.

Perin, C., Vuillemot, R., & Fekete, J. (2013, December). SoccerStories: A Kick-off for Visual Soccer Analysis. *IEEE Transactions on Visualization and Computer Graphics, 19*(12), 2506–2515. doi:10.1109/TVCG.2013.192 PMID:24051817

Piringer, H., Berger, W., & Hauser, H. (2008). Quantifying and Comparing Features in High-Dimensional Datasets. *2008 12th International Conference Information Visualisation*, 240-245. 10.1109/IV.2008.17

Puolamäki, K., Oikarinen, E., Kang, B., Lijffijt, J., & De Bie, T. (2020). Interactive visual data exploration with subjective feedback: An information-theoretic approach. *Data Mining and Knowledge Discovery, 34*(1), 21–49. doi:10.100710618-019-00655-x

Rodríguez-Ibáñez, M., Muñoz-Romero, S., Soguero-Ruiz, C., Gimeno-Blanes, F., & Rojo-Álvarez, J. L. (2019). Towards Organization Management Using Exploratory Screening and Big Data Tests: A Case Study of the Spanish Red Cross. *IEEE Access: Practical Innovations, Open Solutions*, *7*, 80661–80674. doi:10.1109/ACCESS.2019.2923533

Roger, D. (2016). *Exploratory Data Analysis with R.* Leanpub.

Shlomo, S. A. W. I. L. O. W. S. K. Y. (2003). Deconstructing Arguments From The Case Against Hypothesis Testing. *Journal of Modern Applied Statistical Methods; JMASM*, *2*(2), 467–474. doi:10.22237/jmasm/1067645940

Stolte, C., Tang, D., & Hanrahan, P. (2002). Polaris: A system for query, analysis, and visualization of multidimensional relational databases. *IEEE Transactions on Visualization and Computer Graphics*, *8*(1), 52–65. doi:10.1109/2945.981851

Thompson, J. M. (1992). Visual Representation of Data Including Graphical Exploratory Data Analysis. In C. N. Hewitt (Ed.), *Methods of Environmental Data Analysis. Environmental Management Series.* Springer. doi:10.1007/978-94-011-2920-6_6

Weyer, S., Schmitt, M., Ohmer, M., & Gorecky, D. (2015). Towards Industry 4.0 - Standardization as the crucial challenge for highly modular, multi-vendor production systems. *IFAC-PapersOnLine*, *48*(3), 579–584. doi:10.1016/j.ifacol.2015.06.143

Yan, L. (2011). Global exploratory analysis of massive neuroimaging collections using Microsoft Silverlight PivotViewer. *Proceedings of the 2011 Biomedical Sciences and Engineering Conference: Image Informatics and Analytics in Biomedicine*, 1-4. 10.1109/BSEC.2011.5872323

Yang, W., Wang, X., Lu, J., Dou, W., & Liu, S. (2020). Interactive Steering of Hierarchical Clustering. *IEEE Transactions on Visualization and Computer Graphics*, 1. Advance online publication. doi:10.1109/TVCG.2020.2995100 PMID:32746252

Yoo, C., Ramirez, L., & Liuzzi, J. (2014). Big data analysis using modern statistical and machine learning methods in medicine. *International Neurourology Journal*, *18*(2), 50–57. doi:10.5213/inj.2014.18.2.50

Chapter 7

Continuous Autoregressive Moving Average Models:
From Discrete AR to Lévy–Driven CARMA Models

Yakup Ari

🆔 https://orcid.org/0000-0002-5666-5365

Alanya Alaaddin Keykubat Üniversitesi, Turkey

ABSTRACT

The financial time series have a high frequency and the difference between their observations is not regular. Therefore, continuous models can be used instead of discrete-time series models. The purpose of this chapter is to define Lévy-driven continuous autoregressive moving average (CARMA) models and their applications. The CARMA model is an explicit solution to stochastic differential equations, and also, it is analogue to the discrete ARMA models. In order to form a basis for CARMA processes, the structures of discrete-time processes models are examined. Then stochastic differential equations, Lévy processes, compound Poisson processes, and variance gamma processes are defined. Finally, the parameter estimation of CARMA(2,1) is discussed as an example. The most common method for the parameter estimation of the CARMA process is the pseudo maximum likelihood estimation (PMLE) method by mapping the ARMA coefficients to the corresponding estimates of the CARMA coefficients. Furthermore, a simulation study and a real data application are given as examples.

INTRODUCTION

Time series analysis, which includes many methods, is one of the most important tools used to understand the existing and predict the unknown. The most important material of this tool is the observed data over time. When the data of an observation in a certain time unit are examined, it is observed that they are under the influence of some fluctuations. These effects, which can also be defined as the components of time series; are a trend, seasonal fluctuations, cyclical fluctuations, and irregular movements. The

DOI: 10.4018/978-1-7998-7701-1.ch007

effects of each of these factors on the event can be in different directions and intensity, as well as in the same direction and intensity. Therefore, when analyzing with time series, it is necessary to investigate the effects of these factors.

The purposes of analyzing time series can be listed as follows: (i) To reveal the properties of the time series by separating it into its components. It is an important step to separate the time series into components in order to eliminate some effects and to find out what effects they are under. (ii) Another purpose of time series analysis is to control the system and try to understand whether the system is developing in the desired direction by investigating the mechanism of the event that constitutes the series. (iii) The most important purpose of time series analysis is to determine the relationship between time series and to make predictions for the future. In order for the relationships between variables to be meaningful, the time series used must show stationary properties. Otherwise, it may seem that there is a relationship between the two variables, although there are no significant relationships.

The effects of time series factors can be temporary or permanent. These permanent effects cause the series to deviate from stationarity and prevent the series from approaching a certain value. In addition, although some effects are effective in the short term, they lose their effects after a certain period. In the literature, series that are not affected by time and of which mean, variance, and covariance are fixed are called weak stationary (or covariance stationary) series.

When the time series are not stationary, the relationship between variables only covers the period in which the data is valid, and this prevents general inferences about the population. This situation is called spurious correlation and it is known that there may be a high correlation between two unrelated variables that do not actually affect each other. It is stated that this high correlation, if it is not a coincidental situation, may result from a third variable that affects both. In order for the standard errors calculated for the correlation coefficient to be meaningful and reliable, especially two conditions must be met. First one; The sample data used should be taken from the whole population with equal probability. Second, each sample data must be independent of the previous and next observation. However, in the financial time series, these two assumptions are not fulfilled because they are not independent of each other. More importantly, correlation coefficients and their standard errors are closely related to the autocorrelation structure of time series variables.

In most time series consecutive observation values are highly auto correlated. However, it is not very useful in terms of application since it contains many measurements. As the number of scales increases, their reliability in the calculation from the available sample decreases. Considering this situation, researchers named Box and Jenkins (1970) have developed models with very little but appropriate number of parameters. These models used in analyzing time series are called Box-Jenkins models. In Box-Jenkins models, the value of time series in any period comes from the fact that the same series is a linear combination of observation values and / or error terms in the previous periods. Therefore, the Box-Jenkins method is also called as the Autoregressive Moving Average (ARMA) method. Box-Jenkins method, which is used in the prospective estimation of univariate time series, shows a systematic approach in establishing the forecasting models of the discrete and stationary time series consisting of the observation values obtained in these equal time intervals and making the estimations. Discrete and stationary series consisting of observation values obtained with equal time intervals are important assumptions of the method. Box-Jenkins estimation method differs from other estimation methods.

It does not require any prior knowledge about the structure or general development trend of the series. In addition, the method can also be applied to complex time series. An important advantage of the method is that it uses past observation values as an explanatory variable. Unlike econometric models, the Box-Jenkins method does not provide a behavioral explanation for the variable studied, so it does not fit into the theoretical framework. It considers the internal dynamics of the time series.

In real life, financial time series have a high frequency and the difference between observations is not regular. Therefore, continuous models can be used instead of discrete-time series models. The foundations of the Continuous Autoregressive Moving Average (CARMA) models were laid by the studies (Doob,1944) and (Brockwell & Hyndman, 1992). But, Brockwell (1994) introduced the linear continuous-time model for the first time. CARMA models have particularly advantageous for dealing with irregularly spaced data as continuous-time threshold ARMA(p, q) process with $0 \leq q < p$. These models are a continuous version of the well-known ARMA models. It is assumed that the error terms are normally distributed in both models. Many of the studies in the theoretical framework of modern finance have been carried out using continuous time models since Merton's (1969) study of optimal portfolio selection under uncertainty. Continuous time methods are used in many areas such as derivative pricing, structure of interest rates and asset pricing. Moreover, these models are used in many fields of engineering. CARMA models, whose second moments allow asymmetry and are derived fromLévy processes, have been extended by Brockwell (2001, 2004) and Brockwell et al. (2009, 2010) and as a class of the CARMA model. The CARMA models derived from the Lévy processes allow asymmetry as well as heavy-tailed increments that often occur in time series. The studies of Stramer(1996),Barndorff-Nielsen and Shephard (2001), Todorov and Tauchen (2006) and Brockwell & Marquardt (2005) are among the most important examples of this model.

The purpose of this chapter is to define CARMA models and their applications. CARMA model is an explicit solution to stochastic differential equations and also, it is analogue of the discrete ARMA models. Because of these mentioned properties, the outline of the chapter is organized as follows; in the following sections, the stationarity conditions and features of discrete AR, MA and ARMA model will be discussed. Tsay (2012) is one of the best sources for detailed examination of discrete AR, MA and ARMA processes and practical examples of these processes. Tsay (2012) has been used in the sections where these processes are examined. In the fifth section, the stochastic differential equations are defined briefly. The definition and the properties of Lévy processes are given. Then, the Compound Poisson and Variance Gamma processes are discussed since the increments of the Lévy processes are based on these processes. The main subject of the chapter, CARMA models, is defined with its features in the subsection. The last section, aCARMA model is applied for illustrative purposes.

AUTOREGRESSIVE MODELS

In time series modelling, the information available in the lagged values of a variable such as X_t is very useful in predicting the future values of the X_t variable. In particular, a statistical model that reflects this idea has been tried to be estimated by autoregressive processes. The model example describing the dependence on such delay values can be specified as the AR (p) process. In order to understand the general features of autoregressive processes, it is necessary to understand the features of the AR (1) process. The AR (1) process is as given in equation (1).

$$X_t = \varnothing_0 + \varnothing_1 X_{t-1} + a_t \ for \ t = 1, 2, \dots, T \tag{1}$$

In the model where a_t is a white noise process with a mean of zero and a variance of σ_a^2, $\varnothing_0$ is a constant parameter, and $\varnothing_1$ is another parameter that shows the relationship between X_t and its lag-value. a_t is also considered as an error term without correlation. Equation 5.1 is a first-order autoregressive time series model. Because X_t only depends on its previous lag-value and a random residual such as a_t. X_t can be depended not only to X_{t-1}, but also X_{t-2}, X_{t-3} ... Therefore, it is possible to define a p-order autoregressive process. The $AR(p)$ process given in equation (2) is then expressed as the generalized version of the $AR(1)$ process.

$$X_t = \varnothing_0 + \varnothing_1 X_{t-1} + \varnothing_2 X_{t-2} + ... + \varnothing_p X_{t-p} + a_t = \varnothing_0 + \sum_{i=1}^{p} \varnothing_i X_{t-i} + a_t \tag{2}$$

where $p \in \mathbb{Z}^+$ and $a_t \sim d\left(0, \sigma_a^2\right)$. It can be concluded from the mentioned model that, given the past data, the first p lagged variables $X_{t-i} \left(i = 1, ..., p\right)$ jointly determine the conditional expectation of X_t. The $AR(p)$ model has the same form as a multiple linear regression model with lagged values serving as the explanatory variables.

In the time series analysis, it is important to calculate the expected value namely the mean, variance and covariance of X_t, in order to determine the characteristics of the AR process. Therefore, the properties of the model will be explained through the AR (1) process. From the assumption that the dependent variable has the same probability density function for all periods in the AR process, it is concluded that the mean and variance are the same for all periods. So, it means that

$$E\left[X_t\right] = ... = E\left[X_{t-p}\right] = \mu .$$

Then, the expectation of Equation 1 is

$$E\left[X_t\right] = \varnothing_0 + \varnothing_1 E\left[X_{t-1}\right].$$

Hence;

$$\mu = \frac{\varnothing_0}{1 - \varnothing_1} \tag{3}$$

Equation 3 shows that $\varnothing_1 \neq 1$ for the process to be stationary. If $\varnothing_1 = 1$, the variance of the process is not constant and increases continuously. Let suppose that $\varnothing_0 = 0$ which does not affect the variance and covariance of the process. Then, rewrite the equation 1.1 as $X_t = \varnothing_1 X_{t-1} + a_t$. Because of the assumption that equality of variances for all periods,

$$\text{var}\left[X_t\right] = \ldots = \text{var}\left[X_{t-p}\right] = \sigma_X^2.$$

The variance of the rewritten form of the AR(1) process is

$$\text{var}\left[X_t\right] = \text{var}\left[\varnothing_1 X_{t-1} + a_t\right].$$

Hence;

$$\sigma_X^2 = \frac{\sigma_a^2}{1 - \varnothing_1^2} = \gamma_0 \tag{4}$$

From equation 4 the following conclusion is reached; for the AR(1) process to be stationary, it must be $-1 < \varnothing_1 < 1$ in addition to the $\varnothing_1 \neq 1$ condition. In the AR processes, the mean and variance are identical for all periods, as well as the covariances, more precisely, the auto covariance is assumed to be constant. This situation can be explained by obtaining first-degree auto covariance.

$$\gamma_1 = Cov\left(X_t, X_{t-1}\right) = E\left[X_t - E\left[X_t\right]\right]E\left[X_{t-1} - E\left[X_{t-1}\right]\right] \tag{5}$$

Let suppose that $E\left[X_t\right] = 0$. Then,

$$\gamma_1 = E\left[X_t\right]E\left[X_{t-1}\right] = E\left[\varnothing_1 X_{t-1} + a_t\right]E\left[X_{t-1}\right] = \varnothing_1 E\left[X_{t-1}^2\right]E\left[a_t X_{t-1}\right]$$

Since a_t and X_{t-1} are independent, $E\left[a_t X_{t-1}\right] = 0$. So,

$$\gamma_1 = \varnothing_1 \sigma_X^2 \tag{6}$$

Based on the result in Equation 6, a generalization as follows can be made for k-lagged autocovariance between X_t and X_{t-k}.

$$\gamma_k = \varnothing_1^k \sigma_X^2 \; for \; k = 0, 1, 2, \ldots$$

Using the result in Equation 4, k-lagged auto covariance is expressed as in equation (7).

$$\gamma_k = \varnothing_1^k \gamma_0 \tag{7}$$

In this case, the correlation between X_t and X_{t-k} is obtained by equation (8) as

$$Cor\left(X_t, X_{t-1}\right) = \rho_k = \frac{Cov\left(X_t, X_{t-k}\right)}{\sqrt{\text{var}\left[X_t\right]\text{var}\left[X_{t-k}\right]}} = \frac{\gamma_k}{\gamma_0} = \varnothing_1^k \tag{8}$$

The results given from Equation 3 through Equation 8 can be extended for the AR(p) process and used for parameter estimation of the AR (p) model in addition to determining the order of the model. In application, the order p of an AR time series, which must be specified empirically, is unknown. There are two general ways for order specification, which are using partial autocorrelation function (PACF) and applying some information criteria such as Akaike, Schwarz and Bayesian.

MOVING AVERAGE MODELS

The relationships in the structure shown by autoregressive models are frequently encountered in the time series. As an example of these structures, data on many fields ranging from economics to hydrology can be shown. Another time series structure form, called the moving average process, is also used to explain many data. For example, if the price of a stock on any day t is P_t, the change in price can be expressed as follows:

$$X_t = P_t - P_{t-1} = a_t \, for \, t=1,2,...,T \tag{9}$$

where e_t is a random variable with known properties and refers to the change in price. The reasons why meat is included in the model can be listed as follows; learning new information, sudden increases and decreases in prices, change in risk perception, developments and response to them, negative news.

There are two ways to introduce MA models. The first way is to treat the model as a simple extension of the white noise series. The second one is to extend the model as an AR (∞) model with constraints on parameters. Let the following equation (10) be an AR (∞)

$$X_t = \varnothing_0 + \varnothing_1 X_{t-1} + \varnothing_2 X_{t-2} + ... + a_t = \varnothing_0 + \sum_{i=1}^{\infty} \varnothing_i X_{t-i} + a_t \tag{10}$$

Equation 10 is not realistic since it has an infinite number of parameters which makes the estimation of the model impossible. Therefore one can put some constraints on the parameters such as $\varnothing_i = -\theta_1^i$ for $i \geq 1$ that satisfies the stationarity of the model by $|\theta_1| < 1$. So;

$$X_t + \sum_{i=1}^{\infty} \theta_1^i X_{t-i} = \varnothing_0 + a_t \tag{11}$$

If the Equation 11 is rewritten for X_{t-1}, then it becomes

$$X_{t-1} + \sum_{i=1}^{\infty} \theta_1^i X_{t-1-i} = \varnothing_0 + a_{t-1} \tag{12}$$

Multiplying Equation 12 by θ_1 and subtracting the result from Equation 13, the following is obtained

$$X_t = \varnothing_0 \left(1 - \theta_1\right) + a_t - \theta_1 a_{t-1} \tag{13}$$

So, Equation 13 is a process which is called MA(1) model shows that X_t is a weighted average of error terms a_t and a_{t-1} except for the constant term. Let $c_0 = \varnothing_0 \left(1 - \theta_1\right)$, then the general form of the MA(1) model is given by equation 14 as

$$X_t = c_0 + a_t - \theta_1 a_{t-1} \tag{14}$$

where c_0 is a constant and $\{a_t\}$ is a white noise series. Equation 14 can be generalized for MA(q) model as the following equation 15

$$X_t = c_0 + a_t - \theta_1 a_{t-1} - \theta_2 a_{t-2} - \ldots - \theta_q a_{t-q} = c_0 + a_t - \sum_{i=1}^{q} \theta_i a_{t-i} \tag{15}$$

Since MA models are a finite linear combination of stationary series (white noise series) they are weakly stationary. It is known that the first two moments of white noise series are time-invariant. Moreover, a_t and its lagged values are not correlated. To determine the properties of MA process MA (1) model is considered. So, the first moment of MA(1) model given in Equation 15 is obtained by taking the expectation of the model given by equation 16 as

$$E\left[X_t\right] = c_0 \tag{16}$$

The variance of MA(1) model is given by equation 17

$$\mathrm{var}\left[X_t\right] = \sigma_a^2 \left(1 + \theta_1^2\right) = \gamma_0 \tag{17}$$

The auto covariance between X_t and X_{t-1} is

$$\gamma_1 = Cov\left(X_t, X_{t-1}\right) = E\left[X_t - E\left[X_t\right]\right] E\left[X_{t-1} - E\left[X_{t-1}\right]\right] \tag{18}$$

Let suppose that $E\left[X_t\right] = c_0 = 0$. Then,

$$\gamma_1 = E\big[X_t\big]E\big[X_{t-1}\big]$$

$$= E\big[a_t - \theta_1 a_{t-1}\big]E\big[a_{t-1} - \theta_1 a_{t-2}\big]$$

$$= E\big[a_t a_{t-1}\big] - \theta_1 E\big[a_t a_{t-2}\big] - \theta_1 E\big[a_{t-1}^2\big] - \theta_1^2 E\big[a_{t-1} a_{t-2}\big] \tag{19}$$

Since a_t, a_{t-1} and a_{t-2} are independent. So, equation 20 gives the relation as

$$\gamma_1 = -\theta_1 \sigma_a^2 \tag{20}$$

The auto covariance between X_t and X_{t-2} is

$$\gamma_2 = Cov\big(X_t, X_{t-2}\big) = E\big[X_t\big]E\big[X_{t-2}\big] = 0 \tag{21}$$

The extended form of the auto covariance between X_t and X_{t-k} is

$$\gamma_k = Cov\big(X_t, X_{t-k}\big) = E\big[X_t\big]E\big[X_{t-k}\big] = 0 \tag{22}$$

It is concluded that $\gamma_k = 0$ for $k > 1$ in MA(1) models. In this case, the MA (1) process has one term memory. The autocorrelation function of the MA (1) process is given in equation 23 as

$$\rho_k = \frac{\gamma_k}{\gamma_0} = \begin{cases} \dfrac{-\theta_1}{1+\theta_1^2} & k = 1 \\ 0 & k > 1 \end{cases} \tag{23}$$

That is, the autocorrelation function for the MA (1) process is interrupted after the first lag. If one can extend the results obtained from Equation 20 through Equation 24, the following conclusions can be reached for MA(q) model

- The mean of an MA(q) model is $E\big[X_t\big] = c_0$ which is time invariant and
- The variance of an MA(q) model is

$$\gamma_0 = \sigma_a^2\big(1 + \theta_1^2 + \theta_2^2 + \dots + \theta_q^2\big) \tag{24}$$

which is time invariant and the auto covariance given in equation 25 as

$$\gamma_k = \begin{cases} \sigma_a^2 \left(\theta_k \pm \theta_{k+1}\theta_1 \pm \theta_{k+2}\theta_2 \pm \ldots \pm \theta_q\theta_{q-k} \right) & k = 1, 2, \ldots, q \\ 0 & k > qz \end{cases} \tag{25}$$

- The autocorrelation function of the MA (1) process is given in equation 26 as

$$\rho_k = \begin{cases} \dfrac{\theta_k \pm \theta_{k+1}\theta_1 \pm \theta_{k+2}\theta_2 \pm \ldots \pm \theta_q\theta_{q-k}}{1 + \theta_1^2 + \theta_2^2 + \ldots + \theta_q^2} & k = 1, 2, \ldots, q \\ 0 & k > q \end{cases} \tag{26}$$

As can be seen from the results, the autocorrelation function is zero after the q-lag. This situation also helps in identifying the order of the MA(q) process.

ARMA MODELS

Separate autoregressive processes and moving average processes fail to explain the dynamic structure of time series and face various difficulties. Because these models need many parameters when explaining the dynamic structure of the time series. In this case, it makes the models cumbersome and makes parameter estimation difficult. Box et al. (1994) introduced ARMA models to overcome this challenge. Thus, autoregressive processes and moving average processes were collected in a single compact structure and the number of parameters was kept to a minimum. The model is useful for modelling business, economic and engineering time series. It also provides a basis for the generalized autoregressive conditional heteroscedastic (GARCH) model used in volatility modelling.

A general ARMA(p, q) model is in the form given by equation 27 as

$$X_t = \varnothing_0 + \sum_{i=1}^{p} \varnothing_i X_{t-i} + a_t - \sum_{i=1}^{q} \theta_i a_{t-i} \tag{27}$$

where a_t is a white noise series and $p, q \in \mathbb{Z}^+$. It can be concluded from this point of view that AR and MA processes are special cases of the ARMA(p, q) model (Tsay, 2012). The model can be written as following using backshift operator given by equation 28 as

$$\left(1 - \varnothing_1 B - \varnothing_2 B^2 - \ldots - \varnothing_p B^p \right) X_t = \varnothing_0 + \left(1 - \theta_1 B - \theta_2 B^2 - \ldots - \theta_q B^q \right) a_t \tag{28}$$

The polynomials

$$\left(1 - \varnothing_1 B - \varnothing_2 B^2 - \ldots - \varnothing_p B^p \right)$$

and

$$\left(1 - \theta_1 B - \theta_2 B^2 - \ldots - \theta_q B^q\right)$$

in equation (29) are AR polynomial and MA polynomial of the model respectively.

As with the AR and MA models, the properties of the ARMA (1,1) model will be used to reveal the properties of the ARMA (p, q) process. A time series X_t follows an ARMA(1,1) model if it satisfies equation 3.3

$$X_t - \varnothing_1 X_{t-1} = \varnothing_0 + a_t - \theta_1 a_{t-1} \tag{30}$$

where $\{a_t\}$ is a white noise series. The expected value, that is the mean, of the ARMA (1,1) model as given equation 31.

$$E[X_t] = \mu = \frac{\varnothing_0}{1 - \varnothing_1} \tag{31}$$

Under the conditions of $|\theta_1| < 1$ which provides stationarity and where the mean is zero, in other words $\varnothing_0 = 0$, the variance of the ARMA (1,1) model is given in the following equation 32.

$$\gamma_0 = \left(\frac{1 + \theta_1^2 - 2\varnothing_1 \theta_1}{1 - \varnothing_1^2}\right) \sigma_a^2 \tag{32}$$

Then, the general form of the autocovariances are given in equation 33 as

$$\gamma_k = \varnothing_1 \gamma_{k-1} \, for \, k \geq 2 \tag{33}$$

CONTINUOUS ARMA MODELS

In this section, first, the general definitions of Stochastic Differential Equations and Levy Processes are given and then CARMA models are introduced.

Definition 1: Stochastic Differential Equations (SDE)

The general form of an SDE is given in equation 34 as

$$dX\left(t,\omega\right) = f\left(t, X\left(t,\omega\right)\right)dt + g\left(t, X\left(t,\omega\right)\right)dW\left(t,\omega\right) \tag{34}$$

where ω denotes that $X = X\left(t,\omega\right)$ is a random variable and having an initial condition $X\left(0,\omega\right) = X_0$ with probability one. The eqn (1) can be written in the integral in equation 35 as

$$X\left(t,\omega\right) = X_0 + \int_0^t f\left(s, X\left(s,\omega\right)\right)ds + \int_0^t g\left(s, X\left(s,\omega\right)\right)dW\left(s,\omega\right) \tag{35}$$

where $f\left(t, X\left(t,\omega\right)\right) \in \mathbb{R}$, $g\left(t, X\left(t,\omega\right)\right) \in \mathbb{R}$ and $W\left(t,\omega\right) \in \mathbb{R}$.

Definition 2: Lévy Processes

A Lévy process is a stochastic process $\left(L_t\right)_{t \geq 0}$ that satisfies the following properties:

- The starting value of Lévy process is $L_0 = 0$
- It has independent and stationary increments.
- It is stochastically continuous.

Definition 3: Compound Poisson Process

A Poisson process N for $t \geq 0$ and parameter $\lambda > 0$, If a Poisson process is independent of an i.i.d. (independent and identically distributed) sequence of random variables $\left(Y_i\right)_{i \in \mathbb{N}}$, then a compound Poisson process L is defined in equation 36 as

$$L_t = \sum_{i=1}^{N_t} Y_i, t \geq 0 \tag{36}$$

The compound Poisson (CP) process has jumps with random size instead of the constant jumps of size 1 of a Poisson process.

Definition 4: Variance Gamma Process

A Variance Gamma (VG) process is obtained by evaluating Brownian motion with drift at a random time given by a Gamma process. Let $B\left(t\right)$ be a standart Brownian motion then Brownian motion with drift can be defined in equation 37 as

$$b\left(t, \theta, \sigma\right) = \theta t + \sigma B\left(t\right); t \geq 0 \tag{37}$$

where $\theta \in \mathbb{R}$ is the drift term and σ is variance. The time change of the Brownian motion is done with respect to a Gamma process $\left(H_t\right)_{t \geq 0}$ with parameters $a, b > 0$, such that each of the i.i.d increments is Gamma distributed with density defined in equation 38 as

$$f_{H_t}\left(x\right) = \frac{b^{at}}{\Gamma\left(at\right)} e^{at-1} e^{-bx} \; for \; x \geq 0 \tag{38}$$

where $\Gamma\left(.\right)$ is Gamma function. Thus, the VG process $\left(V_t\right)_{t \geq 0}$ can be obtained by equation 39 as

$$V_t = \theta H_t + \sigma B_{H_t}; t \geq 0 \tag{39}$$

CAR Models

The definition of the Continuous AR (CAR) process is given depending on the studies of Brockwell et al. (1992) and Hyndman (1992). The p-order continuous-time autoregressive process can be shown as in equation 40 as

$$X_{t_{(p)}} + a_{p-1} X_{t_{(p-1)}} + \ldots + a_0 X_t + b_0 = \sigma Z_t \tag{40}$$

where Z_t is continuous time Gaussian white noise. From equation 40, CAR(1) processcan be defined as a stationary solution of the stochastic differential equation 41 given below

$$dX_t + a_0 X_t dt + b_0 dt = \sigma dW_t, for \; t \geq 0 \tag{41}$$

where W_t represents standart Brownian motion, X_0 converges in distribution to $N\left(-b_0 / a_0, \sigma^2 / 2a_0\right)$, $a_0 > 0$ and $\sigma > 0$. Under these conditions, equation 4.8 has the unique stationary solution given by equation 42 as

$$X_t = e^{-a_0 t} X_0 - \frac{b_0}{a_0}\left[1 - e^{-a_0 t}\right] + \sigma \int_0^t e^{-a_0(t-u)} dW_u, for \; t > 0 \tag{42}$$

If $b_0 = 0$, the process reduces to a stationary Ornstein–Uhlenbeck process. The mean and the variance of the process are as in equation 43 and 44 as

$$m\left(x, t\right) = e^{-a_0 t} x - \frac{b_0}{a_0}\left[1 - e^{-a_0 t}\right] \tag{43}$$

$$v\left(x,t\right) = \sigma^2 \int_0^t e^{-2a_0(t-u)} du = \frac{\sigma^2}{2a_0}\left[1 - e^{-2a_0 t}\right] \tag{44}$$

Lévy-Driven CARMA Models

In this subsection, definition of Brockwell (2001) is applied to define Lévy-driven CARMA(p,q) models and the properties of the model is discussed. The CARMA(p,q) process is defined as in equation 45 with $p, q \in \mathbb{Z}^+$

$$a\left(D\right)Y_t = b\left(D\right)DL_t \tag{45}$$

where a and b are polynomials

$$a\left(z\right) = z^p + a_1 z^{p-1} + a_2 z^{p-2} + \ldots + a_p \tag{46}$$

$$b\left(z\right) = b_0 + b_1 z^1 + b_2 z^{p-2} + \ldots + b_{p-1} z^{p-1} \tag{47}$$

with $a_1, \ldots, a_p$ and $b_1, \ldots, b_{p-1}$ are coefficients of the auto-regressive part and moving average part, respectively such that $b_q \neq 0$ and $b_j = 0$ for $\forall j > q$ and D is the differentiation operator with respect to t. Finally, $p, q \in \mathbb{N}$ are the auto-regressive and moving average order.

So, the convenient state-space representation of the CARMA(p,q) model is $-a_{p-1}$ given in equation 48 and 49

$$the\ observation\ equation : Y_t = b^T X_t \tag{48}$$

where X_t is p-dimensional stationary process solution to stochastic differential equation

$$the\ state\ equation : dX_t = AX_t dt + edL_t \tag{49}$$

where $X_t = \begin{bmatrix} X_t & X_t^{(1)} & \ldots & X_t^{(p-2)} & X_t^{(p-1)} \end{bmatrix}^T$.

A is a $p \times p$ matrix given in Equation 4.14 and

$$A = \begin{bmatrix} 0 & 1 & 0 & \cdots & 0 \\ 0 & 1 & 0 & \cdots & 0 \\ \vdots & \vdots & \vdots & \ddots & \vdots \\ 0 & 0 & 0 & \cdots & 1 \\ -a_p & -a_{p-1} & -a_{p-2} & \cdots & -a_1 \end{bmatrix}$$

e is a $pX1$ unit vector such as $e = \begin{bmatrix} 0 & \cdots & 0 & 1 \end{bmatrix}^T$ and the $pX1$ coefficient vector $b = \begin{bmatrix} b_0 & \cdots & b_{p-2} & b_{p-1} \end{bmatrix}^T$. So, X and Y can be called the state process and the output process, respectively. The state equation is an Itô differential equation. Since the linearity, the solution given in equation 50 as

$$X_t = e^{A(t-s)} X_s + \int_s^t e^{A(t-u)} e dL_u, \, for \, all \, t > s \geq 0 \tag{50}$$

where the Taylor expansion of $e^A = \sum_{h=0}^{\infty} \frac{1}{h!} A^k$. Let $\lambda_1, \ldots, \lambda_p$ be the real part of the eigenvalues of the matrix A. If the $\lambda_1, \ldots, \lambda_p$ have negative real parts, the covariance stationary solution (Iacus et al., 2018) of X_t is given in equation 51 as

$$X_t = \int_{-\infty}^t e^{A(t-u)} e dL_u = \int_0^{\infty} e^{Au} e dL_u \tag{51}$$

where the solution is convergence in distribution with $E[X_t] = \frac{\mu}{a_p} e$ and

Covariance is given by equation 52 as

$$Cov[X_{t+h}, X_t] = \sigma^2 e^{Ah} \int_0^{\infty} e^{Au} e e^T e^{A^T u} du \, for \, h \geq 0 \tag{52}$$

where $\mu = E[L_1]$ and $\sigma^2 = Var[L_1]$.

If all eigen values of the matrix A are not equal to each other and the real parts of the eigen values are negative then the CARMA(p,q) model can be written as a summation of a finite number of continuous autoregressive models of order one as given in equation 53 as

$$Y_t = b^T e^{A(t-s)} X_s + \int_0^{\infty} \sum_{i=1}^p \left(\frac{b(\lambda_i)}{a'(\lambda_i)} e^{A(t-u)\lambda_i} \right) dL_u \tag{53}$$

and $a'(.)$ denotes the derivative of $a(.)$. One can see the study of Hitaj et al. (2019) for the proof of the above representation.

During the definition of the process, one can note that the differential operators on the auto-regressive side of act like integration operators on the moving average side. Thus, the differential operators on the auto-regressive side offset the differential operators of the moving average side acting on the Lévy process. Since Lévy processes are not differentiable, a necessary condition ensuring the proper existence and stationarity of CARMA processes is $p > q$.

The most common method for the parameter estimation of the CARMA process is the Pseudo Maximum Likelihood Estimation (PMLE) method. PMLE method finds the parameter values of the discrete ARMA model which maximizes the Likelihood Function of the sampled data. The ARMA coefficients can be transferred to the estimates of the CARMA coefficients. But, not all ARMA processes can be embedded in a CARMA process. It can be observed that ARMA processes, which cannot occur as equidistant sampled CARMA processes, can exist. The appropriate way is to estimate discrete ARMA coefficients in the CARMA parameter space. Thornton et al.(2017) and Chambers et al.(2011) examined new state space, and exact discrete-time representations taking the general case of mixed stock and flow variables. They compare the likelihood evaluations based on an exact discrete-time representation, state-space representation, and the Kalman–Bucy filter by applying real data taking account of the mixed frequencies. A Whittle likelihood estimator is applied by obtaining discrete or continuous-time spectral densities using equidistant samples of the output (Gillberg&Ljung, 2005). They applied the estimation procedure in the cases of low sampling rates and rapid sampling. Larsson et al. (2006) showed the problems in parameter estimation of CARMA models from the discrete ARMA model by comparing the direct and indirect methods. Moreover, they showed the impact of the sampling interval on the estimation results. In the study of Chen et al. (2017), they proposed an algorithm to obtain MLEs utilizing irregularly spaced sampled data which is based on the expectation-maximization algorithm. They compared the proposed method with the prediction error method and the results indicate that the expectation-maximization algorithm has high robustness and convergence rate.

Another approach to estimating CARMA parameters is Bayesian inference. Ji et al. (2018) proposed an innovative Bayesian inference method without the requirement of likelihood evaluation by applying an algorithm to approximate Bayesian computation distance correlation for computing the dependence between generated samples. They also conducted a simulation study of which results showed that this approach is straightforward and effective in inferring CAR model parameters. Müller et al. (2019) developed a Bayesian estimation procedure for the electricity spot price model in Benth et al. (2014). Their MCMC algorithm produces samples from the full posterior distribution over all parameters. So, the estimations are obtained much more accurately. In this chapter, the PMLE method is applied in the illustration part.

Example 1: CARMA(2,1) Process

For the example of CARMA (2,1) process, it is recommended to refer to the studies of (Brockwell et al., 2013, 2014), (Benth et al.,2014), (Müller et al.,2019), and (Jónsdóttir et al., 2018). The difference of the Müller et al. (2019) study from others is that the parameter estimation is made using the MCMC algorithm with a Bayesian approach.

Let CARMA(2,1) process be given by equation 54,55 and 56 as

$$\left(D^2 + a_1 D + a_2\right) Y_t = \left(b_0 + D\right) DL_t, t \in \mathbb{R} \tag{54}$$

$$b(z) = b_0 + z \tag{55}$$

and

$$a(z) = z^2 + a_1 z + a_2 = \left(z - \lambda_1\right)\left(z - \lambda_2\right) \tag{56}$$

where $\lambda_1 \neq \lambda_2$ and both have negative real parts.

ARMA(2,1)

$$Y_t = \varnothing_0 + \varnothing_1 Y_{t-1} + + \varnothing_2 Y_{t-1} + a_t + \theta_1 a_{t-1} \tag{57}$$

The transformation of dynamics parameters from CARMA to ARMA $\left(a_1, a_2, b_0\right) \rightarrow \left(\varnothing_1, \varnothing_2, \theta_1\right)$ is the main structure of this part.

So, the stationarity condition for CARMA(2,1) is that the λ_1, λ_2 roots of the characteristic polynomial $a(z)$ should have negative real parts. If the CARMA(2,1) model is stationary an equivalent discrete-time ARMA(2,1) model can be found. The autoregressive parameters of both discrete and continuous processes are directly connected. The eigenvalues of matrix A in the state equation are

$$\lambda_{1,2} = -\frac{1}{2} a_1 \pm \frac{1}{2}\sqrt{a_1^2 - 4a_2} \ .$$

Thus, AR parameters can be computed as $\varnothing_1 = e^{\lambda_1 h} + e^{\lambda_2 h}$ and $\varnothing_2 = -e^{(\lambda_1 + \lambda_2) h}$ where h denotes the grid size. The kernel of the process is given by equation 58 as

$$g(h) = \left(\frac{b_0 + \lambda_1}{\lambda_1 - \lambda_2}\right) e^{\lambda_1 h} + \left(\frac{b_0 + \lambda_2}{\lambda_2 - \lambda_1}\right) e^{\lambda_2 h} \tag{58}$$

Consuquently, Y_t has the following representation in equation 59

$$Y_t = \left(\frac{b_0 + \lambda_1}{\lambda_1 - \lambda_2}\right) \int_{-\infty}^t e^{\lambda_1(t-u)} dL_u + \left(\frac{b_0 + \lambda_2}{\lambda_2 - \lambda_1}\right) \int_{-\infty}^t e^{\lambda_2(t-u)} dL_u \tag{59}$$

The autocovariance function (acf) of the CARMA process is given in equation 60 as

$$\gamma_Y\left(k\right) = 2\sigma^2 b' e^{A|k|}\Sigma b \tag{60}$$

with

$$\Sigma = \int_0^\infty e^{Au}ee^T e^{A^T u} du = \begin{bmatrix} 1/2a_1 a_2 & 0 \\ 0 & 1/2a_1 \end{bmatrix}.$$

The random variables

$$u_t = Y_t - \varnothing_1 Y_{t-1} - \varnothing_2 Y_{t-1} \tag{61}$$

The acf of the corresponding MA process at lag 0 and 1 has following values

$$\gamma_u\left(0\right) = \left(1 + \varnothing_1^2 + \varnothing_2^2\right)\gamma_Y\left(0\right) + \left(2\varnothing_1\varnothing_2 - 2\varnothing_1\right)\gamma_Y\left(1\right) - 2\varnothing_2\gamma_Y\left(2\right) \tag{62}$$

$$\gamma_u\left(1\right) = -2\varnothing_2\gamma_Y\left(3\right) + \varnothing_1\left(\varnothing_2 - 1\right)\gamma_Y\left(2\right) + \left(1 + \varnothing_1^2 + \varnothing_2^2 - \varnothing_2\right)\gamma_Y\left(1\right) + \varnothing_1\left(\varnothing_2 - 1\right)\gamma_Y\left(0\right) \tag{63}$$

the autocorrelation function at lag 1 of an arbitrary MA(1) process with coeficient θ is given by equation 64 as

$$\frac{\gamma_u\left(1\right)}{\gamma_u\left(0\right)} = \frac{\theta}{1 + \theta^2} \tag{64}$$

Above quadratic equation has two explicit solutions. The solution that leads to an invertible MA process should be chosen, which is characterized by $|\theta| < 1$. After the parameter estimation of the autoregressive process is made, the error terms obtained from this model, in other words, the noise process are obtained. Garcia et al. (2010) applied the parameter estimation method for the error terms of the independent stable ARMA process and showed reliable results in the simulation study. These results are justified in the work of Varin et al. (2011)that has been concluded based on the statistics in estimation approach of (Wald, 1949).One can see how the parameters of the α-stable distribution are determined in the study of Benth et al. (2014) at Table 1.

Example 2: Simulation of CARMA(2,1) Process

In this section, the R package "yuima" developed by Iacus et al. (2017, 2018) is used for illustrative purposes. One can use "yuimaGUI" software, which is an interface application used for the "yuima" package developed by the Yuima Project Team (2018, 2020) to perform simulation studies and stochas-

Table 1. CARMA(2,1) Model Output for Log-Returns of IBM Stock Prices

Coef	True	Compound Poisson CARMA(2,1)		Variance Gamma CARMA(2,1)	
		Estimate	**Std. Error**	**Estimate**	**Std. Error**
b1	2.00	2.02	0.03	2.74	0.05
b0	1.00	0.55	0.27	1.05	0.34
a2	0.10	0.06	0.05	0.03	0.03
a1	1.50	1.35	0.23	1.26	0.22

tic continuous model applications on real data sets. The web application of "yuimaGUI" software can be accessed via https://yuimaproject.com/yuimagui/. The program can be downloaded or used online.

This simulation study includes CARMA(2,1) process driven by the Compound Poisson (CP) process and Variance Gamma (VG) process. The jumps of the compound Poisson process increments follow a normal distribution and the constant intensity is one. The sample size is chosen 2000 with delta 200.

Table 1 shows the CARMA(2,1) Model Output for Log-Returns of IBM Stock Prices. The corresponding eigenvalues of matrix A for the estimated parameters in the auto regression in the state equation are $\lambda_1 = -1.43$ and $\lambda_2 = -0.07$ which real and strictly negative. The simulations are given in Figure 1 and Figure 2, respectively.

Figure 1. CP – CARMA (2,1)

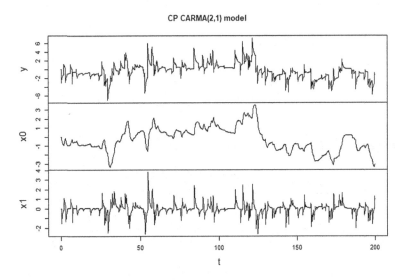

Figure 2. VG – CARMA(2,1)

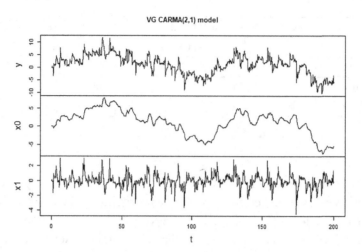

Table 2. The Descriptive Statistics of Increments of CP and VG

Statistics of Increments	Compound Poisson	Variance Gamma
Min.	-2.29	-3.12
1st Qu	-0.01	-0.02
Median	0.00	0.00
Mean	-0.01	0.00
3rd Qu.	0.01	0.01
Max.	3.79	2.52
Average	-0.01	0.00
Sd. of Inc	0.30	0.30
loglikelihood	-3667.13	-4941.15

Figure 3. The increments of CP and VG Processes

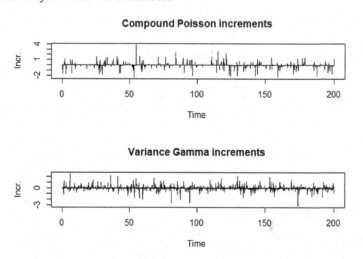

The descriptive statistics and plot of increments of CP and VG processes are given in Table 2 and Figure 3, respectively.

According to the mean absolute error of both processes, the CP-CARMA(2,1) estimates are slightly better than VG-CARMA(2,1) estimates with errors of 0.147, where the error of VG-CARMA(2,1) is 0.158.

Example 3: The Real Data Application of CARMA(2,1) Process

For the illustration, the daily IBM stock price data for the period 2015-2020 is used. The dataset is downloaded using "quantmod" package of R software (Ryan and Ulrich, 2020). Summary descriptive statistics of stock price data and return data are given in Table 3. At the same time, the time series graphs of both series are given in Figure 4.

Figure 4. The closing prices and log-returns of IBM stocks

Table 3. Descriptive statistics of data

Statistics	IBM Stock Prices	IBM Returns
min	94.77	-0.14
max	181.95	0.11
range	87.18	0.24
sum	211885.30	0.00
median	146.04	0.00
mean	145.73	0.00
SE.mean	0.41	0.00
CI.mean.0.95	0.80	0.00
var	241.90	0.00
std.dev	15.55	0.02

In this example, Gaussian and Variance Gamma processes are used as the Lévy noise process. The difference between two consecutive observations is taken as $\Delta = 1/252$, since the data set used is financial time series, the time horizon for daily observations is one year, and the average working days in a year is 252. The parameters of the mentioned models are estimated by using the Quasi Maximum Likelihood Estimation (QMLE) method. One can also make parameter estimation with the Least Square Estimation (LSE) method for CARMA processes. The parameter estimation results for both processes are given in Table 4.

Table 4. CARMA(2,1) Model Output for Log-Returns of IBM Stock Prices

Coefficients	Gaussian CARMA(2,1)		Variance Gamma CARMA(2,1)	
	Estimate	Std. Error	Estimate	Std. Error
b1	-0.71	0.05	0.00	0.11
b0	-0.01	0.13	-0.78	0.07
a2	50.10	NA	27.05	390.52
a1	999.00	140.00	1208.92	200.50
mu	0.00	0.00	0.00	0.00
lambda	-	-	246.78	20.55
alpha	-	-	19.28	1.13
beta	-	-	1.13	0.58
mu0	-	-	-1.51	0.59

In the parameter estimation of the VG-CARMA (2,1) model, lambda, alpha, beta and mu0 parameters are the scale and shape parameters. The descriptive statistics of Lévy increments derived from Gaussian and Variance Gamma distributions used in the models can be seen in Table 5. Moreover, the loglikelihood values of the models are given in the same table. One can use the R package "Variance-Gamma" (Scott and Dong, 2018) to examine whether the increments of the VG-CARMA (2,1) model

Table 5. The Descriptive Statistics of Increments of CARMA(2,1) Processes

Statistics of Increments	Gaussian	Variance Gamma
Min.	-0.45	-0.53
1st Qu	-0.03	-0.04
Median	0.00	0.00
Mean	0.00	0.00
3rd Qu.	0.03	0.03
Max.	0.50	0.61
Average	0.00	0.00
Standart dev.	0.07	0.08
Loglikelihood	3956.24	1850.48

fit the Variance Gamma distribution. After all, it should not be forgotten that this data set is used only for exemplary purposes.

Probability densities were checked to see which Lévy increments fit better. The red-striped distribution shows the VG distribution and the blue one shows the normal distribution. Figure 5 shows that the VG increments fit better.

Figure 5. The rescaled log-densities of Gaussian and VG processes

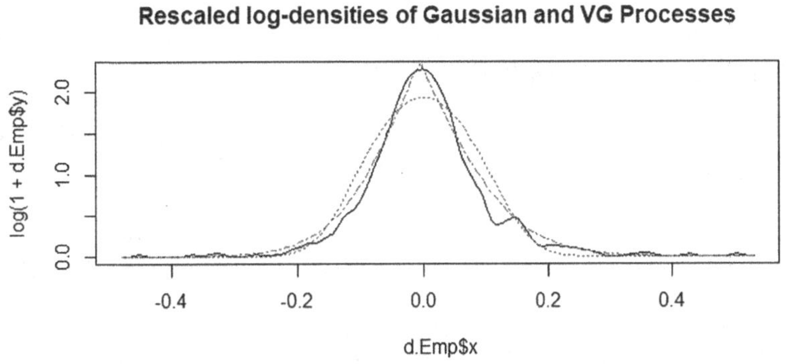

As in the examples given in this section, it is understood that the distributions of Lévy increments are important both in parameter estimation and in fitting the distribution of the variable. Lévy CARMA models are used in many engineering and finance fields from wind studies to electric spot price modeling. Since it was first put forward by Brockwell (2001), frequentist and Bayesian methods have been developed on the parameter estimates of the model. Especially the "yuima" R package developed by Iacuset al. (2017, 2018) allows the model to be implemented and simulated on real data sets. Not every discrete ARMA model is suitable can be accepted as a constraint while estimating the parameters of the model. This model is especially used in pricing, volatility and signaling studies, as well as producing a numerical solution for stochastic equations, indicating that it can be applied in many different areas.

REFERENCES

Barndorff-Nielsen, O. E., & Shephard, N. (2001). Non-Gaussian Ornstein-Uhlenbeck based models and some of their uses in financial economics (with discussion). *Journal of the Royal Statistical Society. Series B. Methodological, 63*(2), 167–241. doi:10.1111/1467-9868.00282

Benth, F. E., Klüppelberg, C., Müller, G., & Vos, L. (2014). Futures pricing in electricity markets based on stable CARMA spot models. *Energy Economics, 44*, 392–406. doi:10.1016/j.eneco.2014.03.020

Box, G. E. P., & Jenkins, G. M. (1970). *Time series analysis: Forecasting and Control.* Holden-Day.

Box, G. E. P., Jenkins, G. M., & Reinsel, G. C. (1994). *Time Series Analysis: Forecasting and Control* (3rd ed.). Prentice Hall.

Brockwell, P. J., & Hyndman, R. J. (1992). On continuous-time threshold autoregression. *International Journal of Forecasting, 18*(3), 439–454.

Brockwell, P. J. (1994). On continuous-time threshold ARMA processes. *Journal of Statistical Planning and Inference, 39*(2), 291–303. doi:10.1016/0378-3758(94)90210-0

Brockwell, P. J., & Davis, R. A. (2001). Discussion of Levy-driven Ornstein–Uhlenbeck processes and some of their applications in financial economics. Shephard. *Journal of the Royal Statistical Society. Series B. Methodological, 63,* 218–219.

Brockwell, P. J. (2004). Representations of continuous-time ARMA processes. *Journal of Applied Probability, 41*(A), 375–382. doi:10.1239/jap/1082552212

Brockwell, P. J., & Marquardt, T. (2005). Lévy-driven and fractionally integrated ARMA processes with continuous time parameter. *Statistica Sinica, 15,* 477–494.

Brockwell, P. J., & Lindner, A. (2009). Existence and uniqueness of stationary Lévy-driven CARMA processes. *Stochastic Processes and Their Applications, 119*(8), 2660–2681. doi:10.1016/j.spa.2009.01.006

Brockwell, P. J., & Hannig, J. (2010). CARMA(p, q) generalized random processes. *Journal of Statistical Planning and Inference, 140*(12), 3613–3618. doi:10.1016/j.jspi.2010.04.028

Brockwell, P. J., & Lindner, A. (2013). Integration of CARMA processes and spot volatility modelling. *Journal of Time Series Analysis, 34*(2), 156–167. doi:10.1111/jtsa.12011

Brockwell, P.J., &Lindner, A. (2014). Prediction of stationary Lévy-driven CARMA processes. *Journal of Econometrics, 189*(2), 263-271

Chambers, M. J., & Thornton, M. A. (2011). Discrete Time Representation of Continuous Time Arma Processes. *Econometric Theory, 28*(01), 219–238. doi:10.1017/S0266466611000181

Chen, F., Agüero, J. C., Gilson, M., Garnier, H., & Liu, T. (2017). EM-based identification of continuous-time ARMA Models from irregularly sampled data. *Automatica, 77,* 293–301. doi:10.1016/j.automatica.2016.11.020

Doob, J. L. (1944). The elementary Gaussian processes. *Annals of Mathematical Statistics, 25*(3), 229–282. doi:10.1214/aoms/1177731234

Garcia, I., Klüppelberg, C., & Müller, G. (2010). Estimation of stable CARMA models withan application to electricity spot prices. *Statistical Modelling, 11*(5), 447–470. doi:10.1177/1471082X1001100504

Gillberg, J., & Ljung, L. (2005). Frequency-Domain Identification of Continuous-Time Arma Models From Sampled Data. *IFAC Proceedings Volumes, 38*(1), 225–230. 10.3182/20050703-6-CZ-1902.00038

Hyndman, R. J. (1992). *Continuous-time threshold autoregressive modelling.* Unpublished doctoral dissertation]. University of Melbourne. https://robjhyndman.com/papers/PhDThesis.pdf

Hitaj, A., Mercuri, L., & Rroji, E. (2019). Lévy CARMA models for shocks in mortality. *Decisions Econ Finan, 42*(1), 205–227. doi:10.100710203-019-00248-9

Iacus, S. M., Mercuri, L., & Rroji, E. (2017). COGARCH(p,q): Simulation and Inference with the yuima Package. *Journal of Statistical Software, 80*(4), 1–49. doi:10.18637/jss.v080.i04 PMID:30220889

Iacus, S. M., & Yoshida, N. (2018). *Simulation and Inference for Stochastic Processes with YUIMA: A Comprehensive R Framework for SDEs and other Stochastic Processes.* Springer. doi:10.1007/978-3-319-55569-0

Ji, C., Yang, L., Zhu, W., Liu, Y., & Deng, K. (2018). On. Bayesian Inference for Continuous-time Autoregressive Models without Likelihood. *21st International Conference on Information Fusion,* 2137-214. 10.23919/ICIF.2018.8455660

Jónsdóttir, G. M., Hayes, B., & Milano, F. (2018).Continuous-Time ARMA Models for Data-Based Wind Speed Models. *Power Systems Computation Conference (PSCC),* 1-7. 10.23919/PSCC.2018.8442659

Larsson, E. K., Mossberg, M., & Soderstrom, T. (2006). An Overview of Important Practical Aspects of Continuous-Time ARMA System Identification. *Circuits, Systems, and Signal Processing, 25*(1), 17–46. doi:10.100700034-004-0423-6

Müller, G., & Seibert, A. (2019). Bayesian estimation of stable CARMA spot models for electricity prices. *Energy Economics, 78,* 267–277. doi:10.1016/j.eneco.2018.10.016

Ryan, A. J., & Ulrich, M. J. (2020). *quantmod: Quantitative Financial Modelling Framework. R package version 0.4.17.* https://CRAN.R-project.org/package=quantmod

Scott, D., & Dong, C.Y. (2018). *VarianceGamma: The Variance Gamma Distribution. R package version 0.4-0.* CRAN.R-project.org/package=VarianceGamma

Stramer, O. (1996). On the Approximation of Moments For Continuous Time Threshold ARMA Processes. *Journal of Time Series Analysis, 17*(2), 189–202. doi:10.1111/j.1467-9892.1996.tb00272.x

Todorov, V., & Tauchen, G. (2006). Simulation methods for Lévy-driven CARMA stochastic volatility models. *Journal of Business & Economic Statistics, 24,* 455–469. doi:10.1198/073500106000000260

Thornton, M. A., & Chambers, M. J. (2017). Continuous time ARMA processes: Discrete time representation and likelihood evaluation. *Journal of Economic Dynamics & Control, 79,* 48–65. doi:10.1016/j.jedc.2017.03.012

Tsay, R. S. (2012). *Analysis of Financial Time Series.* Wiley and Sons.

Varin, C., Reid, N. M., & Firth, D. (2011). An overview of composite likelihoodmethods. *Statistica Sinica, 21*(1), 5–42.

Wald, A. (1949). Note on the consistency of the maximum likelihood estimate. *Annals of Mathematical Statistics, 20*(4), 595601. doi:10.1214/aoms/1177729952

YUIMA Project Team. (2018). *yuimaGUI: A Graphical User Interface for the 'yuima' Package. R package version 1.3.0.* CRAN.R-project.org/package=yuimaGUI

YUIMA Project Team. (2020, October 2). *YUIMAGUI:Web Application.* yuimaproject.com/yuimagui/

Chapter 8

A Computational Statistics Review for Low Complexity Clutter Cancellation for Passive Bi-Static Radar

Venu D.

Osmania University, India

N. V. Koteswara Rao

Chaitanya Bharathi Institute of Technology, India

ABSTRACT

Direct signal, clutter, and multipath echoes are received along with surveillance signal in passive bi-static radars. These signals degrade the target detection capability of the radar processing algorithm and thus require additional processing to achieve a decent performance. Different clutter and multipath cancellation algorithms are devised for removal of unwanted signals. These algorithms require different computational complexity to provide different level of clutter cancellation. This chapter reviews different clutter cancellation techniques and compares their performance based on the computational complexity. This performance comparison allows understanding the computation load put up by different clutter cancellation techniques and ultimately the response rate of the radar system while maintaining decent target detection.

INTRODUCTION

Passive Bi-static radars (PBR) have received renewed interest in recent years for its application as surveillance radar. PBR are low cost radar as they do not require any spectrum allocation. These radars do not emit any signal on their own due to absence to dedicated transmitters and use existing transmitters as illuminators of opportunity. This makes PBR completely inconspicuous and covert in nature (Colone, F. et al,2009). Widely available signals like FM (frequency modulation), Digital TV (DTV), Digital Audio/

DOI: 10.4018/978-1-7998-7701-1.ch008

Figure 1. Passive Bi-static radar geometry

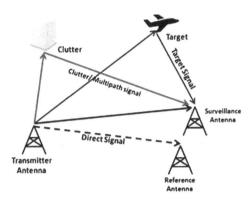

Video broadcast (DAB/DVB), Global system for mobile communication (GSM) are used are illuminator of opportunity in PBR. The selection of a suitable transmitter depends on transmitter's coverage area, transmitted power, carrier frequency and bandwidth. Broadcast transmitters like a commercial FM station make a suitable transmitter for PBR due to their high coverage and high transmitted power(Howland P.E.et al.,2005). This makes the passive radars suitable candidate for continuous surveillance of stealth targets, hence, PBR is widely used in homeland security applications (Celik, N. et al.,2011).

As oppose to active radars where time reference information is available, in case of passive radars time reference is not available. Therefore, two antennas, reference antenna and surveillance antenna are used in PBR receiver circuitry as shown in fig. [1]. The reference antenna receives the signal emitted by the already deployed transmitters and is used to provide timing reference. This signal is not controlled of the radar designer. The surveillance antenna receives the surveillance signal after reflection from the target. Both the received signals are then cross correlated and a cross ambiguity function is solved to detect the targets range and doppler.

However, surveillance antenna along with target echoes also receives multipath signals as well as reflected signal from nearby clutter like buildings, ground, and tree leaves etc. The reference signal also leaks into the surveillance channel. Thus, the surveillance antenna, along with target echo also receives clutter and multipath signal. The direct signal leaked into the surveillance antenna has a comparatively higher power of approximately 60 dB - 90 dB to those of the targets echo power, ultimately degrading radar target detection capacity. These disturbance signals are mostly concentrated near zero doppler therefore a possible solution is to reject the zero doppler result of the cross ambiguity range doppler map. However, this zero doppler signal leads to high side lobes in frequency domain due to the application of fast fourier transform (FFT). These side lobes often mask the target echoes due to their much higher power. Targets with high doppler frequency are also masked by these side lobes. Thus, these unwanted signals and non anticipatory behaviour of the transmitted signal leads to time varying side lobes in the ambiguity function. This masking of target limits the radar's detection capability and degrades its performance.

Many physical and processing techniques are used in the past to reduce the radar performance degradation due to direct path interference (DPI). With an aim to prevent the DPI from entering the receiver antenna, the transmitter and the receiver were shielded from each other. Another important physical method used to reduce DPI is beam forming where a null is formed in the receiving antenna pattern towards the transmitted signal.

Many temporal and spatial filters are designed which aims to remove the interference and increase the dynamic range of radar. Spatial filters uses beam forming for clutter cancellation(Di Lallo, A. et al.,2008). A spatial correlation matrix is formed and the eigen structure was utilize for clutter cancellation in Tao, R. et al., 2010 while Villano, M. et al.,2013 investigates the null steering methods and non adaptive beam steering methods.

1. The temporal filters uses the reference signal to suppress the clutter. Conventional techniques like moving target indicator (MTI) cannot remove side lobes due to spreading in range-doppler map. Thus, adaptive cancellation filters are used for target detection in PBR. Many disturbance cancellation algorithms are designed for clutter cancellation and to retrieve target signatures. These temporal algorithms solve weiner filtering problems. These algorithms either employ iterative filters to cancel the clutter and multipath (Palmer, J. E. et al.,2012; Xiaode, L. et al.,2014) or projects the received signal on clutter and pre detected target subspaces (Colone, F. et al.,2006;Ansari, F. et al.,2013).

Thus, target detection using PBR has 4 main stages: Pre processing, clutter cancellation using adaptive filters, cross ambiguity function (CAF) calculation yielding range-doppler map and lastly target tracing and detection (Howland, P. E. et al.,2005). Second step, i.e. clutter cancellation using adaptive filter marks the most computationally expensive part of the radar system. Thus, a significant time and effort is put up by radar designer in designing these algorithms to minimize the effects of clutter and multipath signal without reducing the desired target echoes while decreasing the computational complexity. A trade-off is however seen in between the processing gain and the range resolution. Increasing the processing gain for good target detection is possible by increasing the coherent processing interval (CPI) which may result in target migration. Also, CPI cannot be increased much for time sensitive applications, like security and defence, as they require a high response rate.

Thus, this paper discusses different disturbance cancellation algorithms and a comparative study is being done based on computational complexity while maintaining decent target detection. Popular clutter cancellation algorithm like Least mean square (LMS), Normalized least mean square (NLMS), recursive least square (RLS), fast block least mean square (FBLMS) are studied and compared with sequential cancellation algorithm (SCA), extensive cancellation algorithm (ECA), and their variants. These algorithms minimises the least mean square (LMS) error between the weighted reference signal and the target echo using either iterative method or block processing.

LMS, NLMS and RLS are iterative algorithms discussed in (Howland, P. E. et al.,2005; Jichuan, L. et al.,2013) where NLMS provides improvement over LMS in terms of greater stability and faster rate of convergence whereas RLS provide better cancellation capacity but with an increased computational complexity. These algorithms function has high pass filter and provides better cancellation capability for static clutter and multipath signal. Block algorithms like ECA and SCA algorithm, discussed in (Colone, F. et al.,2006), provides significant performance improvement by minimizing the effect of clutter on received target echoes when compare to transversal filters, LMS and RLS. ECA is found to not able to detect weak targets, thus, different variants of ECA, ECA-B and ECA-S discussed in [(Colone, F. et al,2009;Colone, F. et al,2016; Cardinali, R. et al.,2007) which provides a better target detection capacity. Jarrah, A. et al.,2016 implements the ECA on a GPU while utilizing parallel processing and due to its high computational load, renders unfit for use in time sensitive applications. SCA algorithm provides better possibility to detect weak targets by sequentially cancelling strong targets and has a reduced computational complexity when compared to ECA (Colone, F. et al.,2006; Garry, J. L. et al.,2015). NLMS is modified in work by Zhao, Y. D., et al.,2013 for low complexity and faster convergence in real time implementation while sparsity of weight matrix is discussed in (Xiaode, L. et al.,2014; Ma, Y. et al.,2016). Shan, T. et al.,2014 discusses the MCNLMS (multi channel NLMS) for sea clutter cancellation. It uses multiple modulated reference signal and is accompanied with an increase in processing time. Attalah, M. A. et al.,2019 and Attalah, M. A. et al.,2016, discusses the range-doppler fast block least mean square (RD-FBLMS), an improvement over FBLMS for a better cross ambiguity function and faster convergence rate target detection using DVB-T signal(Cardinali, R. et al.,2007).

This paper discusses few of the above mentioned algorithm, namely, LMS, NLMS, RLS, FEBLMS, ECA and SCA and their variants and compares the performance of NLMS, ECA, FBLMS and RLS using a simulated FM signal. The comparison is performed based on computational complexity while maintaining a decent clutter cancellation capability thus decreasing the computation load on the radar system and thereby increasing the response time of the radar. The paper is organised as follows. Section 2 discusses the signal model while different clutter cancellation algorithms are discussed in section 3. Section 4 compares the performance of the different algorithms based on computational complexity and section 5 marks the conclusion of the paper.

SIGNAL MODEL

A typical bi-static radar geometry consisting of targets clutter, a surveillance antenna and a reference antenna is shown in fig. [1]. In bi-static radar, the direct signal (or reference signal) is collected by the reference antenna while surveillance antenna collects the target echo. It is assumed for reference antenna that the signal received by it is not corrupted with multipath signals and clutter echoes as the reference antenna is pointed straight towards the transmitter of opportunity. Practically, however, this isn't the case as the reference antenna also receives multipath and clutter echoes and its is discussed in Colone, F. et al.,2009. Thus, with above assumptions, the reference signal is modelled as shown in equation 1

$$S_{ref}(t) = A_{ref}d(t) + n_{ref}(t) \tag{1}$$

Where, A_{ref} and $d(t)$ are the complex amplitude and complex envelope of the received direct signal and the thermal noise contribution is given by $n_{ref}(t)$.

The surveillance antenna is pointed directly towards the area under surveillance and receives reference signal, multipath and direct path interference (MPI, DPI) signal along with target echoes. Clutters are modelled as multiple discrete scattering centres where a collection of such multiple scattering centres represents the continuous clutter backscattered and received at the surveillance antenna. Thus, the surveillance signal received at the receiver given by Colone, F. et al.,2006 is shown in Eq.2 as

$$S_{surv}(t) = A_{surv}d(t) + \sum_{m=1}^{N_T} a_m d(t - \tau_m)e^{j2\pi f_{dm}t} + \sum_{i=1}^{N_c} c_i d(t - \tau_{ci})e^{j2\pi f_{dci}t} + n_{surv}(t) \tag{2}$$

Where, A_{surv} is the complex amplitude and $d(t)$ represents the envelope of the direct signal. a_m, f_{dm} and t_m are the complex amplitude, doppler shift and delay of the m-th target. Complex amplitude, doppler frequency and delay of the i-th scatterer is represented by c_i, f_{dci} and τ_{ci}. n_{surv} represents the thermal noise contribution. For static scatterers $f_{dci} = 0$ and these static scatteres causes zero doppler interference (ZDI). And for non stationary clutters, f_{dci} is nearly zero as they represent slowly varying clutter. The clutters are assumed to be slowly moving with time, therefore, the clutter doppler frequency, f_{dci} is considered to be nearly zero. N samples of surveillance signal is collected to form a $N \times 1$ vector represented in Eq. 3 as

$$S_{surv} = \left[S_{surv}(t_0), S_{surv}(t_1), S_{surv}(t_2), \ldots\ldots, S_{surv}(t_{N-1}) \right]^T \tag{3}$$

And, for reference signal, N+R-1 samples are collected resulting $(N + R - 1) \times 1$ vector in S_{ref} as shown in Eq. 4

$$S_{ref} = \left[S_{ref}(t_{-R+1}), \ldots\ldots, S_{surv}(t_0), \ldots\ldots, S_{surv}(t_{N-1}) \right]^T \tag{4}$$

The collected reference signal in Eq. 1 and surveillance signal in Eq. 2 are fed as input to the matched filter for further processing and target detection. The reference signal and the surveillance signal are further cross correlated and a cross ambiguity function as shown in Eq. 5 is obtained, to detect the target signatures by measuring the bi-static range (distance between the source of illumination-target-receiver antenna) and doppler.

$$\psi(\tau, f_d) = \int_0^{T_0} S_{surv}(t) S_{ref}^*(t - \tau) e^{-j2\pi f_d t} dt \tag{5}$$

As the collected signals are digitized, the discrete implementation of cross ambiguity function defined in Eq.5 for target detection is given by Eq. 6 as

$$\xi[l, p] = \sum_{i=0}^{N-1} s_{surv}[i] s_{ref}^*[i - l] e^{-j2\pi pi/N} \tag{6}$$

Where, l, p represents the time bin and doppler bin respectively for calculating the delay (or distance) and velocity of the target. Time bin is given by $l = 0, \ldots, R - 1$ and doppler bin by with $f_D[p] = p / NT_s$, where, T_s is the sampling time. Also, the doppler resolution achieve able is directly proportional to the integration time, therefore, a trade-off exists between the desired doppler resolution for processing gain and Doppler- range migration of the targets.

Surveillance antenna receives the direct signal with a magnitude of approx 90 dB greater than the target echoes. Also, a strong peak is seen at $\Delta R = 0\,and\,\Delta v = 0$ due to the presence of stationary or low moving clutters. As PBR uses illuminator of opportunity, the transmitted waveform is not controlled/designed by the radar designer, this result in range-doppler map obtained using Eq.5 to have an inherent time varying structure. Thus clutters echoes due to spreading in the filter domain leads to masking of the target echoes. This makes the application of clutter cancellation algorithm a crucial step in target detection for passive radars where surveillance signal is filtered to remove direct signal and clutter before it is cross correlated with reference signal to yield target signatures.

ADAPTIVE DISTURBANCE CANCELLATION TECHNIQUE

Interference caused due to direct path and multipath signal as well as clutter echoes in passive radars needs to be cancelled before detecting target using matched filtering operations. In temporal filtering, reference signal is used to construct a zero doppler interference signaland then this constructed signal is subtracted from the surveillance signal thus yielding a surveillance signal free of clutters and multipath. It tends to minimize the mean square error (MMSE) by solving for weiner hopf equations. The basic operation of adaptive filters is shown in Fig. 2 where the reference signal is considered as input and filter tap weights are adapted according to a pre-defined algorithm to result in a surveillance signal free of multipath signal and clutter echoes. This newly obtained clean signal is then used to detect target signatures y cross correlating it with reference signal. This section discussed the different block algorithms like ECA and ECA-B and iterative algorithms like LMS, NLMS, FBLMS, RLS clutter cancellation algorithms in brief where the surveillance signal is progressively cleaned by cancelling direct path and multipath signal.

Figure 2. Adaptive cancellation techniques

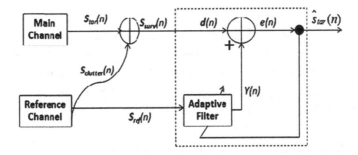

<u>Least mean square algorithm:</u> In LMS a gradient descent algorithm is solved. In this algorithm, the aim is to reduce the least mean square error between the surveillance signal and reference signal. Filter tap weights are calculated iteratively which minimises this LMS error. Fig. [3] shows the basic block diagram for LMS algorithm where it uses a transversal filter to produce the filtered signal and to update weight.

It is assumed that the clutter echoes and the multipath signal are present in the first K range bins in the range doppler ambiguity function, therefore, the filter length in LMS filter is selected such that $L > K$. The error function using LMS is given in eq. 7. $e\big[n\big]$ is the error vector output which represents

Figure 3. Least mean square transversal filter

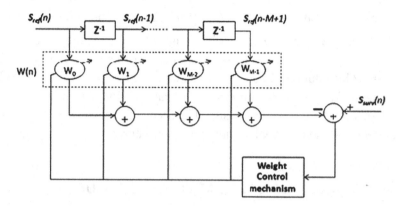

the difference between the surveillance signal and the product of the filter tap weights and the reference signal resulting in suppression of direct path and multipath signals (Garry, J. L. et al.,2015; Haykin, S.,2002).

$$e\left[n\right] = s_{tar}\left[n\right] = s_{surv}\left[n\right] - w\left[n\right]^{H} s_{ref}\left[n\right] \tag{7}$$

Where, $w\left[n\right]^{H}$ is the hermitian transpose of the weight vector. The filter tap weights are updated according Eq.8 to

$$w\left[n+1\right] = w\left[n\right] + \mu s_{ref}\left[n\right] s_{tar}\left[n\right]^{*} \tag{8}$$

Where (*) represents the complex conjugate and μ is the step size. Step size, μ, controls the accuracy and the convergence rate of the algorithm. A higher μ leads to faster response with decrease in clutter attenuation (CA) capability while a lower value leads to a more sluggish filter response. The computational complexity in LMS filters scales linearly with the dimensionality of the weight vector w[n]. LMS algorithm works well for stationary clutter but has a reduced clutter cancellation capacity in presence of time varying clutter. LMS algorithm is also subjected to gradient noise amplification for large weight vector.

Normalized least mean square algorithm: As mentioned in the previous section, LMS algorithm encounters the problem of gradient noise amplification when the input weight vector used has large values. This issue is resolved using NLMS algorithm where the step size, μ, is normalized by calculating its Euclidean norm. This normalizes the weight vector by its energy and making the algorithm independent of the magnitude of the reference signal(Jichuan, L. et al.,2013). For filter designing using LMS and NLMS, the maximum delay due to the disturbance signal is assumed to be less than the filter length, MT_{s}, where, T_{s} is the signal sampling time. The error function for NLMS algorithm is given by Eq.9 as:

$$e\big[n\big] = s_{tar}\big[n\big] = s_{surv}\big[n\big] - w\big[n\big]^H_{NLMS}\, s_{ref}\big[n\big] \tag{9}$$

The weight updation for NLMS filter is given in Eq. 10:

$$w\big[n+1\big] = w\big[n\big] + s_{ref}\big[n\big] s_{tar}\big[n\big]^* \frac{\mu}{a + s_{ref}\big(n\big)^2} \tag{10}$$

Where, a is used to prevent instability. Step size, μ ranges between 0 and 2. Normalizing the step size leads to an increase in stability and faster rate of convergence of NLMS algorithm as compared to LMS algorithm however its long impulse response leads to increased computational complexity as the instantaneous energy of the reference signal needs to be calculated after each iteration. NLMS also works well for static clutter however its complexity scales with the linearly with the dimension of the weight vector.

Recursive Least Square algorithm: The RLS algorithm, in contrast with the LMS algorithm, is a deterministic algorithm having transversal structure. RLS algorithm considers instantaneous as well as the weighted sum of the previous error. The error cost function of the RLS algorithm minimises the least square difference between the received target echoes and a weighted reference signal and is given by Eq. 11 as

$$e_{RLS}\big[n\big] = \sum_{i=0}^{n} \lambda^{n-1}\Big(s_{surv}\big[i\big] - w_{RLS}^H\big[n\big] s_{ref}\big[n\big]\Big)^2 \tag{11}$$

Where, λ represents the filter forgetting factor, is a measure of algorithm memory. It ranges between $0 < \lambda < 1$. λ decreases the weight of the previously estimated error over time and determines the rate of convergence of the algorithm. For $\lambda = 1$, RLS algorithm converges to LMS algorithm. The filter output is given by Eq.12 as

$$y_{RLS}\big[n\big] = s_{surv}\big[n\big] - w\big[n-1\big]^H s_{ref}\big[n\big] \tag{12}$$

Where, $w\big[n-1\big]$ is the filter weight vector. The following equations eq. 13 and Eq.14 are used for updating the weight.

$$w\big[n+1\big] = w\big[n\big] + k\big[n+1\big] y_{RLS}^*\big[n+1\big] \tag{13}$$

Where, $k\big[n+1\big]$ represents the weight factor given by Eq.14 as

$$k\big[n\big] = \frac{\lambda^{-1}\phi^{-1}\big(n-1\big) s_{ref}\big[n\big]}{1 + \lambda^{-1} s_{ref}^H\big[n\big] \phi^{-1}\big(n-1\big) s_{ref}\big[n\big]} \tag{14}$$

And, ϕ, which represents the time average correlation matrix of the input vector $s_{ref}(i)$ is given by eq.15 as

$$\phi(n) = \sum_{i=0}^{n} \lambda^{n-1} s_{ref}(i) s_{ref}^{H}[i] \tag{15}$$

The RLS algorithm is computationally more complex while having a faster convergence rate as compared to both LMS and NLMS. Thus the real time implementation of RLS algorithm is difficult. RLS algorithm also does not have a good tracing capacity and hence deems unsuitable for use in case of moving clutter like high speed wind, moving water waves etc. Also, at times target echoes can also be removed if the target delay is smaller than the filter length.

Fast Block Least Mean Square Algorithm: LMS and NLMS are time domain algorithms leading to a high computation load thus requiring higher memory consumption. In FBLMS algorithm, it takes the advantage of the fact that the convolution in time domain is changed to point wise multiplication in frequency domain. Thus the received signal is converted into frequency domain(Attalah M.A. et al.,2019). To gain maximum speed-up, the block size is chosen in multiples of 2. Also, in FBLMS, filter weights are updated block by block unlike LMS and NLMS where weights are updated after each sample. The weights are updated according to Eq. 16

$$w^{B}(n+M) = w^{B}(n) + \mu^{B} \sum_{m=0}^{M-1} \varepsilon(n+m) s_{ref}(n+m) \tag{16}$$

FBLMS algorithm solves the steepest descent algorithm for minimizing the squared error function. In FBLMS, error is accumulated for a period equivalent to M samples before weights are again updated. The error function is defined in Eq. 17 as

$$\varepsilon(n+m) = s_{surv}(n+m) - \left[w^{B}(n)\right]^{H} s_{ref}(n+m) \tag{17}$$

Where Eq. 18 gives

$$\mu^{B} = \frac{\mu}{L\left[\gamma + x_{ref}^{H} x_{ref}\right]} \tag{18}$$

The application of fourier transform and block updating of filter weight improves both the computational complexity and the convergence rate. FBLMS algorithm has same steady state but lower mid-adjustment errors as compared to that of NLMS. It exhibits a much broader notch in doppler frequency resulting in the loss of target detection with lower radial velocities. FBLMS algorithm improves the radar dynamic range along with weak target detection capability. FBLMS algorithm also provides improvement over NLMS algorithm both in terms of clutter attenuation and computational complexity. It also provides a quicker processing and a faster convergence rate. FBLMS algorithm is also suitable for use while using wide band illuminator of opportunity as transmitters. Also, FBLMS algorithm can be easily implemented on GPU using the advantage offered by parallel computation of FFT. Thus, this algorithm deems suitable for use in target detection in real time for stationary platform.

Excessive Cancellation Algorithm and Its Variants

1. 1. *Excessive Cancellation Algorithm:* Introduced by Colone et.al.,2006, it is assumed in this algorithm that the multipath clutter and the ZDI are present in the first K doppler frequency bins i.e. It assumes the backscattered multipath and clutter echoes to be present in the first K range bins. This is caused by the almost static slowly varying environment. ECA uses the least square estimation for clutter cancellation by minimizing the power available at the output of the filter(Haykins S.,2002). ECA, unlike previously discussed algorithm is a batch algorithm operating on N samples. It also does not yield a transient response. The objective function of the ECA is given by Eq. 19 where the similarity between the surveillance signal and the weighted direct signal is reduced thus cancelling the disturbance signal.

$$y_{ECA} == \min \left\{ s_{surv} - X\alpha \right\}^2 \tag{19}$$

Where, X forms a N-by-M matrix which contains the delayed replica of reference signal and α is a weight vector having a dimension of M and is defined in Eq.20 as:

$$\alpha = \left(X^H X \right)^{-1} X^H s_{surv} \tag{20}$$

making the surveillance signal after disturbance cancellation to be as in Eq.21

$$y_{ECA} = s_{surv} - X\alpha = \left[I_N - X \left(\left(X^H X \right)^{-1} X^H \right) \right] s_{surv} = P s_{surv} \tag{21}$$

Where, P is a projection matrix which projects the received signal s_{surv} in an orthogonal subspace, orthogonal to both, the pre detected targets and the disturbance subspace. It results in a notch in the filter domain. A wider notch results in high clutter cancellation leading to better target detection. It also leads to performance degradation while detecting slow moving targets. After the cancellation of disturbance from the direct signal and ground echoes, the signal model is exploited and cross ambiguity function is evaluated as given in Eq.22 by:

$$\xi_{ECA}[l, p] = \sum_{i=0}^{N-1} y_{ECA}[i] s_{ref}^*[i-l] e^{-j2\pi pi/N} \tag{22}$$

Strong targets were still not detected using ECA algorithm, hence, Cell averaging constant false alarm rate (CA-CFAR) is used on the cross ambiguity range doppler map to detect the strong targets easily. Weak targets are still not detected in ECA algorithm with a possible reason being residual from the disturbance cancellation and the side lobes of strong targets masking the weaker targets. Many false targets are also detected mainly due to the side lobes of stronger targets (Colone F. et al.,2009).

The selection of the value of M affects the computational complexity of the algorithm as it needs to calculate the inverse of the M-by-M matrix $X^H X$ which has a complexity of $O\left[NM^2\right]$. ECA due to its batch processing requires a higher data storage capacity. Also, ECA algorithms are computationally expensive, requiring a total of $O\left[NM^2 + M^2 \log M\right]$. Thus, the main limitation of ECA algorithm is its inability to detect weaker targets and high computational complexity.

Figure 4. Batch Excessive Cancellation Algorithm

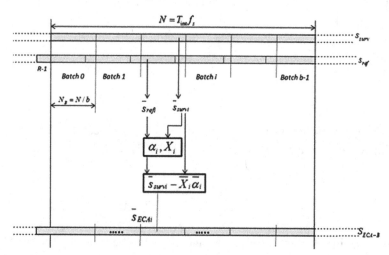

2. <u>Batch Excessive Cancellation Algorithm:</u> ECA filtering deploys all the signal samples at once. This implies that in ECA, filter coefficients are computed once for the whole data. While in a non-stationary environment, optimal filters coefficients changes rapidly. ECA being a slow algorithm is not able to adapt fast for non stationary environment. Thus, for better performance in non stationary environment, batch version of ECA is proposed by Colone F. et al.,2009 for removal of static and dynamic clutter. In ECA-B, the received input signal is divided into a number of batches B and ECA is performed on all the batches separately. The different batch responses achieved after the application of ECA, are then combined to form the resultant signal free of clutter and multipath disturbance as shown in figure [4] and represented by Eq.23 as

$$S_{ECA-B} = \left[S^T_{ECA_0} \quad S^T_{ECA_1} \quad S^T_{ECA_{b-1}}\right]^T \tag{23}$$

Where, Eq.24 represents

$$S_{ECA_i} = S_{surv_i} - X_i \alpha_i \tag{24}$$

The zero doppler notch is inversely proportional to the batch size. Thus, ECA-B has a comparatively wider notch around zero doppler than that of ECA as the number of samples in each batch has reduced. This results in better removal of the direct signal disturbance while also makes the algorithm robust to slow moving targets. As the sample size is reduced, ECA-B may also lead to inaccurate estimation and increase in unwanted side lobes. This leads to the presence to false targets in the range doppler map.

Weaker targets are not detected in ECA-B as well, however, unlike ECA, they are limited to only masking effect of stronger targets and no longer the residual from disturbance cancellation. Also, ECA-B algorithms are marginally more computationally expensive compared to ECA, requiring a total of $O\left[NM^2 + bM^2 \log M\right]$ complex multiplications. It is also seen in Colene F.,et al.,2009 that the dynamic behaviour of the transmitted waveform impacts the overall detection capability of this algorithm.

However, as the data to be processed for each batch is reduced, the memory requirement is significantly reduced in ECA-B. Moreover, the algorithm can be paralleled with data acquisition system, hence the data processing can be started as soon as the data for the first batch is received. This reduces the pre-processing rate which, in real time applications, turns out to be a limiting factor for ECA algorithm(Colene F.,et al.,2009) .On the other hand, it also makes the algorithm more robust to the slowly varying characteristics of the slow moving typical broadcast radios.

3. Sliding window ECA: To remove the effect of the doppler side lobes in ECA-B algorithm, ECA-S is discussed in the work by Jarrah A. et al.,2016. It is seen that the doppler ambiguities are separated by a duration of T_B, where, T_B is the signal duration. Thus, by reducing the batch duration, side lobes can be moved out of the region of interest. This, however, leads to further widening of notch in doppler domain resulting in performance degradation while detecting slow moving targets. This issue is resolved by using different sizes of window for both, estimating filter coefficients and actual filtering.

In ECA-S, filter coefficients are calculated using overlapped signal samples resulting in a smoothed estimate and improving the filter performance. Thus, the width of the zero doppler notch remain the same as in ECA-B, however, this algorithm is more computationally complex.

4. ECA batches and stages: A multistage version of ECA-B is discussed in Colene F.,et al.,2009 where ECA-B is used consecutively over multiple stages. In this algorithm, ECA-B is applied on the initially collected reference and surveillance signal to remove the major contributing disturbances. The cancellation process is then refined to remove the strongest targets thus increasing the detection possibility of weaker targets.

The flow diagram for ECA-Batches and stages is shown in fig. [5]. First, ECA-B is applied on the collected data. This removes the direct and multipath signal and ground clutter and allows the detection of strong targets. Reference signal is then auto correlated resulting in 2-D auto ambiguity function of the reference signal. This gives information about doppler and range resolution, peak to side lobe ratio (PSLR) and peak to noise floor ratio (PNFR) in the collected reference signal. PNFR and PSLR are further compared. A smaller value to PNFR as compared to PSLR implies that target detection capability is limited by the noise floor. In this case, no further processing is required and the algorithm coincides with ECA-B

However, if PSLF < PNFR, a threshold (η) is applied on range-doppler map of ECA-B between range of [0, PSLF] which allows detection of strong peaks due to strong targets and residual disturbance if any. The cancellation mask is then computed which removes the strong target and ECA-B is performed on the newly formulated data . The PNFG value is then updated after evaluating 2-D CCF. This process is repeated until PSLF > PNFR.

This algorithm, first removes the strongest clutter and direct signal and then targets are detected in decreasing order of their strength resulting in a more robust algorithm. This joint exploration of multiple batches and stages improves the overall target detection due to stronger clutter cancellation capability and the ability to detect weak targets as well.

This algorithm is able to detect all the targets irrespective of their strength, however, the overall computational complexity increases as it requires the usage of ECA-B multiple times. Also, the value of η defines the overall convergence rate of the algorithm. A low value of η results in more stages while a higher value of η decreases the weaker target detection capability.

Figure 5. Successive cancellation algorithm

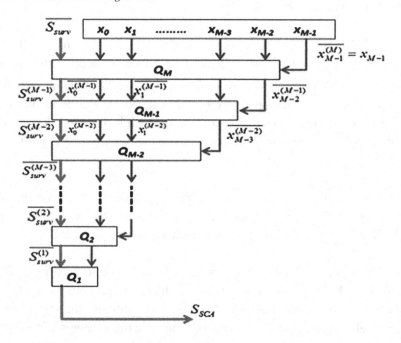

Successive Cancellation Algorithm and Its Variants

1. *Successive Cancellation Algorithm:* ECA and ECA-B are computationally very complex and isn't practical for use where high response rate is required. Therefore, SCA is devised which is the sequential version of the same as shown in figure [5]. In SCA, for interference cancellation, the strongest clutter is progressively detected and removed to decrease its effect on the received signal.

For this, the coefficient matrix of the input matrix is projected on a subspace orthogonal to clutter subspace and previously detected target subspace. The projection matrix is computed using Eq.25 given as

$$P_0 = P_1 - \frac{P_1 x_0 x_0^H P_1}{x_0^H P_1 x_0} \tag{25}$$

Where, P_1 is defined as given in Eq. 26 as

$$P_1 = I_N - X_1 \left(x_1^H x_1 \right)^{-1} x_1^H \tag{26}$$

Thus, the strongest clutter is detected and removed. The received signal is improved after each recursive step i.e. after each recursive step; the signal received has reduced clutter and multipath interference as compared to the signal received in the last recursion. The complexity of SCA algorithm is given by $O\left[NM^2\right]$. However, it is possible to restrict the computational cost after S stages where, $S \leq M$. This is done when a predefined level of clutter cancellation has been achieved. Then the computational complexity of $O\left[NMS\right]$ is achieved for SCA algorithm restricted after S stages, which is much smaller than the computational cost required for the complete algorithm.

It is however possible that the weak target echoes still remain undetected even after the application of SCA as they are masked by the side lobes of the strong targets. Therefore, to detect the weak targets, range-doppler map is refined by detecting the strongest targets and a properly scaled reference signal is produced corresponding to that target. This signal is subtracted from the matched filter output resulting in the detection of weaker targets. A continuous application of this procedure results in all targets being detected but it also increases the computational complexity. But as the number of targets is limited, even though the complexity is increased, it is still lower than that of ECA.

Figure 6. Batch successive cancellation algorithm

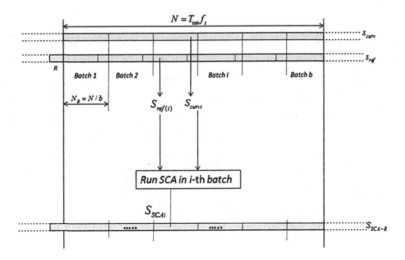

2. Batch Successive Cancellation Algorithm: Similar to ECA-B, SCA-B is the batch version of SCA used primarily for non stationary environment where the clutter cancellation performance is increased(Ansari, F. et al.,2016). In this algorithm, the signal is divided into batches. SCA algorithm is then applied individually on each of these batches. Each batch obtained at the output of SCA algorithm has almost negligible clutter and direct signal. The complete surveillance signal to be use further for match filtering is obtained by the union of these packets as shown in figure [6]. This output is then used for cross correlation ambiguity function analysis for target detection as shown in Eq.27.

$$S_{SCA-B} = \begin{bmatrix} S^T_{SCA_0} & S^T_1 & S^T_{SCA_{b-1}} \end{bmatrix}^T \tag{27}$$

A projection matrix is formed similar to SCA, projecting the received signal on a subspace orthogonal to clutters and disturbance subspace and previously detected targets. Thus SCA-B employs multiple batches which allows for strong target detection along with strong clutter cancellation. Weak targets are not detected using SCA-B, hence this algorithm can be clubbed with clean algorithm discussed in Kulpa, K. (2008) to detect weak targets as well. Thus, SCA-B allows cancellation of both stationary and non stationary clutters as well as direct signal cancellation thus allowing better target detection capability than that of ECA, ECA-B and SCA.

The sequential approach of the algorithm renders it suitable for use in both stationary and non stationary clutter cancellation. The computational complexity for SCA-B algorithm for each individual batch is $O[N_B MS]$. Since the algorithm is used over b batches, the overall complexity for SCA-B becomes $O[bN_B MS]$ which is equal to the computation complexity of SCA algorithm. However, the use of clean algorithm increases the complexity marginally. As, this algorithm can also be paralleled like ECA-B and the processing starts when the samples corresponding to the first batch is received, thus using less memory as compared to the SCA algorithm. Hence, SCA-B is less computationally complex as compared to ECA and ECA-B along with having lesser memory requirement as compared to SCA.

RESULTS

This section compares the performance of few of the above mentioned algorithm. An innovated weight matrix disturbance attenuation cancellation discussed in work by Lauri, A. et al., 2007 is used on a simulated FM signal operating in 88 MHz – 108 MHz band, by constructing a weight matrix for LMS, RLS, ECA-B and FBLMS consisting of useful clutter information. Signal simulator discussed in work by Lauri, A. et al., 2007 is used to generate a FM radio stereo signal. This increases the clutter cancellation capability especially for non stationary environment. A error vector is computed first consisting of only the target signatures. This error vector is then used in the ambiguity function for target detection. Also, it was observed that the element wise product of the i-th column of generator matrix and weight matrix gives clutter signatures. As the number of clutters in an environment is limited, a sparse weight matrix is obtained. The average of each column of the weight matrix is then computed and checked for non trivial entries. The columns with non trivial entries are then identified as clutter delay bins. The

Table 1. Target scenario for Weight matrix based Target Detection

Targets	T1	T2	T3
Delay (ms)	0.3	0.6	0.5
CNR	4	-10	2
Doppler (Hz)	-50	50	100

Table 2. Clutter scenario for Weight matrix based target Detection

Clutters	C1	C2	C3	C4	C5	C6
Delay (ms)	0.1	0.2	0.07	0.15	0.25	0.17
CNR	30	10	27	20	5	18
Doppler (Hz)	3	1	-2	2	-1	-3

clutters are then reproduced with the available information of clutter amplitude, range and doppler. After that, the computed clutter signatures are subtracted from the surveillance signal, thus, resulting in a clean signal. This computed clean signal is called the weight matrix based clean signal and is further used for target detection.

The efficacy of the algorithm is demonstrated using a simulated data generated using work by Venu-Dunde et al., 2019. Table 1 and 2 shows the target and clutter doppler, delay and clutter to noise ratio (CNR) information. It assumes 6 clutters and 3 targets. To simulate a more realistic scenario a relatively larger doppler band is considered for clutters. This allows for emulation of clutters in non stationary environment. The above mentioned algorithms are then compared in terms of computational complexity and clutter attenuation where clutter attenuation is defined as the ratio of the power of the original surveillance signal and the filtered surveillance signal as shown in Eq.27 as

$$CA = \frac{\sum_{n=0}^{N-1} \left| x_f[n] \right|^2}{\sum_{n=0}^{N-1} \left| x_s[n] \right|^2} \tag{27}$$

In RLS algorithm, it is observed that the evaluated amplitude and Doppler are closely tracking the factual amplitude and Doppler for weight matrix based clutter cancellation technique. The clean signal is then computed by subtracting the clutter signal modelled using the factual amplitude, doppler and delay from the surveillance signal. The ambiguity function is then plotted in Fig. 7. It is noted that the weak target signatures are also detected in weight matrix based RLS cancellation algorithm. False targets are also seen which due to the small mismatch error in factual and evaluated amplitudes and doppler.

In clutter cancellation using LMS algorithm, it is seen that the LMS requires a higher number of initial filter taps as compared to that of RLS due to the slow convergence of LMS algorithm. The ambiguity function is then plotted in Fig. [8]. It is seen that the weak target signatures are also detected in weight matrix based cancellation algorithm for LMS algorithm. False targets are seen due to the small mismatch error in factual and evaluated amplitudes and doppler.

Figure 7. Ambiguity function for RLS algorithm

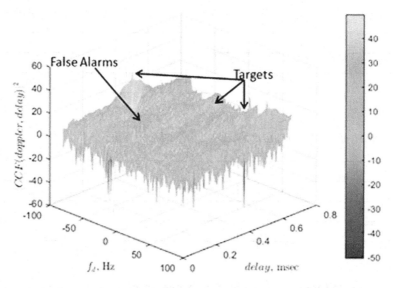

Figure 8. Ambiguity function for LMS algorithm

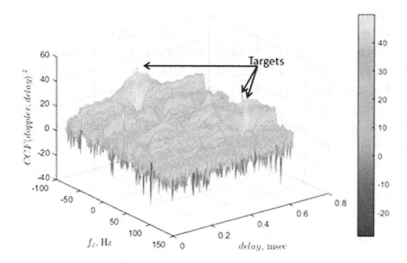

LMS and RLS uses transversal filters thus having transient response. LMS is a stochastic algorithm while RLS is a deterministic algorithm. The performance of these filters depends on filter parameters μ, λ which controls the overall clutter cancellation capability and the convergence rate. It is however seen that the convergence rate of RLS algorithm is faster than the LMS algorithm but has a higher computational complexity than RLS.

Range – doppler map obtained for FBLMS algorithm is shown in fig. [9]. FBLMS algorithm requires a larger values of filter repetition taps than that of RLS and LMS due to more slower convergence rate. Clutter ranges are than computed in a similar fashion. False targets are detected in FBLMS algorithm also due to the coarse approximation of Doppler and amplitude in weight matrix based methods. FBLMS

Figure 9. Ambiguity function for FBLMS algorithm

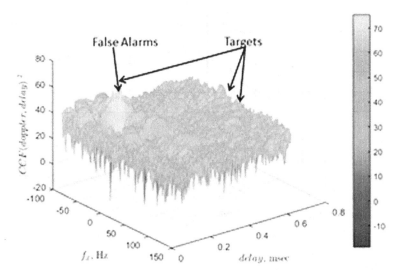

Figure 10. Ambiguity function for ECA algorithm

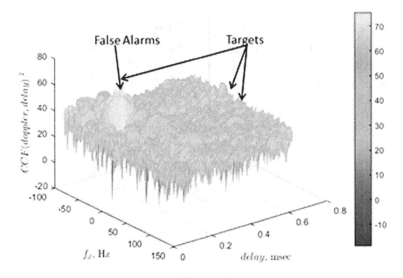

algorithm has a faster response rate and provides a much lower computational complexity as compared to RLS and LMS methods.

Next, weight matrix based method is examined to evaluate the computational complexity and clutter properties for ECA-B. The proposed method no longer requires the augmentation of H matrix. Therefore, it has a column dimension of M. Clutter amplitude, doppler and delay are then evaluated in similar way as described in above sections. The ambiguity function is then plotted in Fig. 10. As seen, a wide null is present at zero doppler allowing better cancellation of disturbances. Strong targets are very well detected. However, weaker targets are not detected as they are masked by the residual disturbance and the side lobes of the stronger targets.

Table 3. Clutter attenuation and complexity

Algorithm	Clutter Attenuation	Complex Products
RLS	54.4657	1.3235e9
LMS	54.4563	10.71e6
FBLMS	49.1314	3.8612e6
ECA-B	49.1165	111.78e6

Table 3 compares the clutter attenuation and computational complexity based on the number of required complex products for different algorithms. As seen from table 3, RLS algorithm provides the highest clutter attenuation while ECA-B has the lowest clutter attenuation. However, RLS algorithm has the highest computational complexity while LMS has the lowest computational complexity. However, the computation complexity of LMS increases with increase in sample size.

It is observed from table 3 that ECA-B and FBLMS has a lower clutter evaluation precision as compared to other algorithms resulting in false targets. This is because, in both FBLMS and ECA-B, the rows of weight matrices are updated after a batch in comparison to RLS and LMS where the weight matrix is updated after every step. Therefore, it can be inferred that ECA-B and FBLMS clutter cancellation capability decreases for moving clutters.

Table 4 tabulates the complexity of LMS, RLS, ECA-B and FBLMS for error based clean signal in terms of number of complex multiplications required. As seen, due to logarithmic dependence, the computational complexity increases the least for FBLMS algorithm with increase in sample size while it increases the most for RLS algorithm.

Table 4. Computational Complexity for error based algorithms

Algorithm	Complex Products
RLS	$\left(5M^2 + 4M\right)N$
LMS	$\left(2M + 1\right)N$
FBLMS	$\left(16M + 12\log\left(2M\right)\right)N_m$
ECA-B	$\left(2M_a^2 b + M_a b + M_a^3\right)N_b$

CONCLUSION

Usage of illuminator of opportunity for target detection and localization in passive radars includes many challenges like reference signal leakage, clutter and multipath signals received by the surveillance antenna along with target echo. Suppression of these unwanted signals becomes crucial for maintaining a reasonable detection performance and to maximize the radar dynamic range. This amounts to huge computational complexity and increases the radar response time. However, in time sensitive application, like homeland security and defense, which requires swift action, importance is given to faster response time while maintaining a decent target detection.

Different clutter cancellation algorithms are discussed along with their benefits and limitations. LMS and NLMS are iterative algorithm based on stochastic gradient and have an easy implementation. The step size is NLMS is normalized to provide more stable and has a faster rate of convergence as compared to that of LMS algorithm. However the computational complexity in both LMS and NLMS scales linearly with the dimensionality of the weight vector w[n]. An improvement over NLMS algorithm, RLS is discussed. It has a faster rate of convergence and produces zero mis-adjustment error but is more computationally complex compared to LMS and NLMS.

FBNLMS algorithm uses the fourier transform property to convert convolution to multiplication and thus provides easy implementation. It also provides better clutter cancellation capability and computation efficiency and has lower mis-adjustment error. For application in stationary environment, ECA and SCA algorithms were discussed. In these algorithms, a projection matrix is formulated which projects the input signal in a subspace orthogonal to both the clutter subspace and the signal subspace. ECA and SCA algorithm is computationally intensive and requires high memory for its operation. For decreasing the required memory and to decrease the processing time, the batches version of both ECA and SCA, ECA-B and SCA-B are discussed. As this algorithm performs ECA and SCA on smaller batches of data, it decreases the memory requirements. However both algorithms, ECA and its batch version and SCA and its batch version have same computational complexity.

Therefore, it can be said that the effect of different cancellation algorithms are used for different scenarios. When low computational complexity is needed, its better to use FBLMS while RLS provides the best performance in terms of clutter cancellation. In case of time varying environment, RLS algorithm deems unfit as it has poor target tracing capacity while LMS can be used in such scenario. ECA and SCA works well for moving target detection but has a reduced performance for weak targets. Thus, there is always a trade off between the computational complexity and clutter attenuation and the usage of a specific algorithm is based on its application.

REFERENCES

Ansari, F., & Taban, M. R. (2013, May). Implementation of sequential algorithm in batch processing for clutter and direct signal cancellation in passive bistatic radars. In *2013 21st Iranian Conference on Electrical Engineering (ICEE)* (pp. 1-6). IEEE.

Ansari, F., Taban, M. R., & Gazor, S. (2016). A novel sequential algorithm for clutter and direct signal cancellation in passive bistatic radars. *EURASIP Journal on Advances in Signal Processing, 2016*(1), 1–11. doi:10.118613634-016-0431-2

Attalah, M. A., Laroussi, T., Aouane, A., & Mehanaoui, A. (2016, December). Adaptive filters for direct path and multipath interference cancellation: Application to FM-RTL-SDR based Passive Bistatic Radar. In *2016 7th International Conference on Sciences of Electronics, Technologies of Information and Telecommunications (SETIT)* (pp. 461-465). IEEE.

Attalah, M. A., Laroussi, T., Gini, F., & Greco, M. S. (2019). Range-Doppler fast block LMS algorithm for a DVB-T-based passive bistatic radar. *Signal, Image and Video Processing*, *13*(1), 27–34. doi:10.100711760-018-1324-7

Cardinali, R., Colone, F., Ferretti, C., & Lombardo, P. (2007, April). Comparison of clutter and multipath cancellation techniques for passive radar. In 2007 IEEE Radar Conference (pp. 469-474). IEEE.

Cardinali, R., Colone, F., Lombardo, P., Crognale, O., Cosmi, A., & Lauri, A. (2007). *Multipath cancellation on reference antenna for passive radar which exploits FM transmission*. Academic Press.

Celik, N., Youn, H. S., Omaki, N., Lee, Y. L., Gagarin, R., & Iskander, M. F. (2011, July). Experimental evaluation of passive radar approach for homeland security applications. In *2011 IEEE International Symposium on Antennas and Propagation (APSURSI)* (pp. 224-227). IEEE.

Colone, F., Cardinali, R., & Lombardo, P. (2006, April). Cancellation of clutter and multipath in passive radar using a sequential approach. In *2006 IEEE Conference on Radar* (pp. 1-7). IEEE.

Colone, F., O'hagan, D. W., Lombardo, P., & Baker, C. J. (2009). A multistage processing algorithm for disturbance removal and target detection in passive bistatic radar. *IEEE Transactions on Aerospace and Electronic Systems*, *45*(2), 698–722. doi:10.1109/TAES.2009.5089551

Colone, F., Palmarini, C., Martelli, T., & Tilli, E. (2016). Sliding extensive cancellation algorithm for disturbance removal in passive radar. *IEEE Transactions on Aerospace and Electronic Systems*, *52*(3), 1309–1326. doi:10.1109/TAES.2016.150477

Di Lallo, A., Fulcoli, R., & Timmoneri, L. (2008, September). Adaptive spatial processing applied to a prototype passive covert radar: Test with real data. In *2008 International Conference on Radar* (pp. 139-143). IEEE. 10.1109/RADAR.2008.4653906

Dunde & Koteswara Rao. (2019). Weight matrix- based representation of sub-optimum disturbance cancellation filters. *International Journal of Intelligent Systems and Applications, 11*(10), 15-24.

Garry, J. L., Smith, G. E., & Baker, C. J. (2015, June). Direct signal suppression schemes for passive radar. In *2015 Signal Processing Symposium (SPSympo)* (pp. 1-5). IEEE. 10.1109/SPS.2015.7168278

Haykin, S. (2002). *Adaptive filter theory*. Prentice Hall.

Howland, P. E., Maksimiuk, D., & Reitsma, G. (2005). FM radio based bistatic radar. *IEE Proceedings. Radar, Sonar and Navigation*, *152*(3), 107–115. doi:10.1049/ip-rsn:20045077

Jarrah, A. A., & Jamali, M. M. (2016). A parallel implementation of extensive cancellation algorithm (ECA) for passive bistatic radar (PBR) on a GPU. *Journal of Signal Processing Systems for Signal, Image, and Video Technology*, *85*(2), 201–209. doi:10.100711265-015-1066-5

Jichuan, L., Yaodong, Z., Yongke, Z., & Xiaode, L. (2013, August). Direct path wave purification for passive radar with normalized least mean square algorithm. In *2013 IEEE International Conference on Signal Processing, Communication and Computing (ICSPCC 2013)* (pp. 1-4). IEEE. 10.1109/IC-SPCC.2013.6663987

Kulpa, K. (2008, September). The CLEAN type algorithms for radar signal processing. In *2008 Microwaves, Radar and Remote Sensing Symposium* (pp. 152-157). IEEE. 10.1109/MRRS.2008.4669567

Lauri, A., Colone, F., Cardinali, R., Bongioanni, C., & Lombardo, P. (2007, March). *Analysis and emulation of FM radio signals for passive radar. In 2007 IEEE Aerospace Conference.* IEEE.

Ma, Y., Shan, T., Zhang, Y. D., Amin, M. G., Tao, R., & Feng, Y. (2016). A novel two-dimensional sparse-weight NLMS filtering scheme for passive bistatic radar. *IEEE Geoscience and Remote Sensing Letters*, *13*(5), 676–680.

Palmer, J. E., & Searle, S. J. (2012, May). Evaluation of adaptive filter algorithms for clutter cancellation in passive bistatic radar. In *2012 IEEE Radar Conference* (pp. 493-498). IEEE.

Shan, T., Ma, Y., Tao, R., & Liu, S. (2014). Multi-channel NLMS-based sea clutter cancellation in passive bistatic radar. *IEICE Electronics Express*, *11*(20), 11–20140872. doi:10.1587/elex.11.20140872

Tao, R., Wu, H. Z., & Shan, T. (2010). Direct-path suppression by spatial filtering in digital television terrestrial broadcasting-based passive radar. *IET Radar, Sonar & Navigation*, *4*(6), 791–805. doi:10.1049/iet-rsn.2009.0138

Villano, M., Colone, F., & Lombardo, P. (2013). Antenna array for passive radar: Configuration design and adaptive approaches to disturbance cancellation. *International Journal of Antennas and Propagation*.

Xiaode, L., Jichuan, L., Kuan, L., Daojing, L., & Yi, Z. (2014, October). Range-Doppler NLMS (RDNLMS) algorithm for cancellation of strong moving targets in passive coherent location (PCL) radar. In *2014 International Radar Conference* (pp. 1-5). IEEE. 10.1109/RADAR.2014.7060444

Zhao, Y. D., Zhao, Y. K., Lu, X. D., & Xiang, M. S. (2013). *Block NLMS cancellation algorithm and its real-time implementation for passive radar*. Academic Press.

Chapter 9

Machine Intelligence–Based Trend Analysis of COVID–19 for Total Daily Confirmed Cases in Asia and Africa

Yibeltal Meslie
Mekdela Amba University, Ethiopia

Wegayehu Enbeyle
https://orcid.org/0000-0002-0788-6137
Department of Statistics, Mizan-Tepi University, Ethiopia

Binay Kumar Pandey
Govind Ballabh Pant University of Agriculture and Technology, Pantnagar, India

Sabyasachi Pramanik
https://orcid.org/0000-0002-9431-8751
Haldia Institute of Technology, India

Digvijay Pandey
https://orcid.org/0000-0003-0353-174X
Department of Technical Education, Institute of Engineering and Technology, Lucknow, India

Pankaj Dadeech
https://orcid.org/0000-0001-5783-1989
Swami Keshvanand Institute of Technology, Management, and Gramothan (SKIT), Jaipur, India

Assaye Belay
Department of Statistics, Mizan-Tepi University, Ethiopia

Ashwini Saini
RPS Group of Institution, Haryana, India

ABSTRACT

COVID-19 is likely to pose a significant threat to healthcare, especially for disadvantaged populations due to the inadequate condition of public health services with people's lack of financial ways to obtain healthcare. The primary intention of such research was to investigate trend analysis for total daily confirmed cases with new corona virus (i.e., COVID-19) in the countries of Africa and Asia. The study utilized the daily recorded time series observed for two weeks (52 observations) in which the data is obtained from the world health organization (WHO) and world meter website. Univariate ARIMA models were employed. STATA 14.2 and Minitab 14 statistical software were used for the analysis at 5% significance level for testing hypothesis. Throughout time frame studied, because all four series are

DOI: 10.4018/978-1-7998-7701-1.ch009

non-stationary at level, they became static after the first variation. The result revealed the appropriate time series model (ARIMA) for Ethiopia, Pakistan, India, and Nigeria were Moving Average order 2, ARIMA(1, 1, 1), ARIMA(2, 1, 1), and ARIMA (1, 1, 2), respectively.

INTRODUCTION

COVID-19 is the class of corona (Crown) viruses. They are so-called because of the crown-like appearance under a microscope (Otom,R., 2020). Coronavirus disease 2019, cases endure to rise rapidly transversely the African continent (W, H, O., 2020). It has the potential to cause tremendous social disruptions, economic loss, and political and security (Pramanik, S. and Singh, R. P., 2017) crises and to reverse the health and socio-economic development gains (Pandey, D., et.al., 2020). A lack of state-run healthcare services and everyone's insufficient resources to access health-care systems, COVID-19 seems to be likely to pose a serious threat, particularly to vulnerable groups. Due to transportation but also immigration controls, the outbreak has an impact on the delivery and cost of humanitarian assistance. At the same time, if resources are sidetracked to sustenance national COVID-19 attempts, expenditures are likely to be cut(Parmeshwar, U. et. al., 2020). This might have serious ramifications for population groups that heavily trust on philanthropic help to live and/or defend their maintenances. Similarly, support approaches would consider major obstacles in coordinating face to face-based evaluations and post-delivery tracking (Emmanuel, S., et. al., 2020).

COVID-19's global collateral effects may include a rise in food costs due to agricultural manpower scarcity, as well as the detrimental influence of protectionist measures (Parmeshwar, U. et. al., 2020). Ethiopia is among the most vulnerable developing countries in the world, with COVID-19 spreading at an exponential rate. The scarcity of emergency care services, the extensive use of public transportation, the scarcity of hygiene resources, including water, the concealment of suspicious cases, the absence of necessary safety devices for healthcare professionals, and the presence of immune-compromised individuals all are significant motivators (Birhanu, A.et. al., 2020). Even if COVID-19 is contained in Africa, the disease's economic consequences will be unavoidable and irreversible (Yakubu, L., 2021). COVID-19 is perhaps the most recent humanitarian difficult task that Ethiopia's government and humanistic partner organizations have confronted (Mekonnen, H, Z. 2020).

The COVID-19 pandemic is wreaking havoc on global supply chains and disrupting manufacturing operations (Daniel B. et. al., 2021). Closing schools would then reduce food intake and nutrition, possibly increase school dropouts, and have a negative effect on social economic growth. Long-term consequences of interrupted education but also impoverished childhood development nutrition would then disproportionately affect poor families, limiting their human capital development and able to earn growth in the field (Lee, H. et. al., 2020).

Flight cancellations, hotel reservation cancellations, and cancelled local and international events totaling more than $200 billion were all on the rise (Ozili,P, K., 2020). In Ethiopia, public transportation is highly vulnerable in boosting outbreaks such as COVID-19 pandemic. Heavy focus on hygiene, sanitation, temperature screening at entry sights, and a limited number of bus seats and onboard cameras to enforce these rules is the need of the hour (Birhanu, A.et. al., 2020). Beginning with biggest employers, flower and horticulture farms utilize 150,000 people in such a number of focused areas. Garment suppliers, other significantly impacted export group, hires between 50,000 and 70,000 people (Ababulgu, A, N. and Wana, F, H., 2021).

The main objective of this article would be to examine pattern analysis in Ethiopia, India, Pakistan and Nigeria of the cumulative confirmatory cases of recent corona virus (Pandey, D. et. al., 2021). Which article aimed specifically? To choose appropriate model for the total daily confirmed (COVID-19) case (s) for each country. To predict (Pramanik, S. and Bandyopadhyay, S. K. 2013) total daily confirmed cases with (COVID-19) based on appropriate time series model.

Since December 2019, there has been a severe outbreaks of corona virus disease 2019 triggered via contamination with serious critical lungs disorder corona virus 2 (SARS-CoV-2) virus in wuhan city, a well-known city center in the China whereby movement remains exceptionally easy for linking all the other locations across China and overseas (Emmanuel, S. et. al., 2020).There have been 80 813 recorded cases throughout all Chinese cities since March 7, 2020, and 21 110 case registered in 93 nations around 6 continents. Algeria, Egypt, and Nigeria are among the African nations where infections have also been recorded. Since 1949, when the Peoples of China was established that was China's biggest infectious disease crisis. This is the most difficult challenge because the disease is spreading so quickly and has such a high prevalence, and preventing propagation has enlisted the help of everybody in the country.

In view of the both healthcare and socio - economics, the new corona virus disease is experiencing the unforeseen devastating impact. Although COVID-19 doesn't really differentiate, the 55 states and regions which are homes to 135 million people who are extremely nutritional poor and also in urgent need for humanitarian health and nutrition aid are most sensitive to a pandemic's repercussions because they have hardly any capacity to deal with both the health or socioeconomic dimensions of an outbreak. Furthermore, the disease outbreak could exacerbate severe food shortages in nations which rely on imported food, oil exports, tourist industry, and settlements for revenue, such as Small Island Developing States (SIDS).

COVID-19 is expected to have a significant impact on diet and wellness, especially among poorer populations, due to limited conditions in public health systems and people's lack of funds for service provision. Increased costs of chronic health problems, such as non-communicable deficiency, are expected to raise the likelihood of significant effects. COVID-19 could also have a negative impact on organizational and economic stability, promoting unrest, particularly in the most vulnerable countries affected by the food crisis. This could stymie global resolving conflicts mediation attempts as well as peacekeeping operations, with adverse repercussions for susceptible but also food-insecure strife people. Owing to transportation and restrictions on travel, the disease outbreak could have an impact on the delivery and cost of disaster relief. At the very same time, if possessions are sidetracked to provision nationwide COVID-19 effort in order, budgets are probable to have been reduced (Ashebir, Y. et. al., 2015).

This could have serious consequences for populations that rely heavily on humanitarian aid to survive and/or secure subsistence. Humanitarian groups also face greater difficulties in conducting face-to-face assessments and post-distribution tracking (Dong, E.et. al., 2020) . The planet has been struggling to contain a global public health pandemic COVID-19 that is expanding extremely rapidly to diverse numbers of deaths throughout various regions of the globe. COVID-19, that is initially emerged as a deadly virus in late 2019 in China and it suddenly blowout to worldwide as a disease outbreak. Currently, Europe, especially Italy, Spain, the United Kingdom, France, and the United States, have been stiffest hinted in relations to contagions and deaths, considering the increasingly advanced heath care systems. It has triggered significant apprehension aimed at the African continent, owing to the continent's reasonably meager healthcare arrangements in contrast to Europe and the Global North, and also as the continent's enormous expanse happens in the health care systems.

COVID-19 remained stated a disease outbreak as declared by World Health Organization on March 11, 2020. Viral infections have varying consequences for women, children, adults, teenagers, and people of all races, never to mention the many at so that vulnerable populations. Children and women, such as at and disadvantaged populations, could be disproportionately affected by the challenges in growth and humanitarian perspectives. COVID-19's broad unintended consequences may result in an increase in food costs due to agricultural labour shortages and the detrimental consequences to protectionist measures policies. A combination of travel information constraints as well as the anticipated downturn throughout the world economy threatens to hinder agricultural production, manufacturing, and marketing activities, as well as livestock moves. 5th. Corona virus disease 2019 (COVID-19) cases are increasingly increasing throughout the African continent. After our last situation update on April 8, 2020, no new countries have been impacted. In the WHO African Region, In the last week, there seems to be a 51% growth in the number of case scenarios and then a 60% spike in the proportion of fatalities. Since about 14 April 2020 (epidemiological week 16), a total of 10 759 confirmed COVID-19 cases with 520 deaths (case fatality ratio CFR: 4.8%) had already been registered around the continent's 45 impacted countries (Yakubu, L., 2021).

A new landmark has hit the COVID-19 pandemic in the WHO African Region, with over 10 000 reported cases in all Member States, other than in Comoros and Lesotho, and over 500 deaths. Two-thirds of cases in the country are recorded in six nations (Algeria, Cameroon, Côte d'Ivoire, Ghana, Niger, and South Africa). COVID-19 is spreading rapidly around the world. Many African nations have now confirmed cases, and indeed the number of fatalities is increasing. While unchecked, its effect to African people and economies will be significant (Mekonnen, H, Z., 2020).

Ethiopia, the east African underprivileged nation, does have a extremely less well-being care staff concentration of around 0.96 per 1000 population, which is exacerbated by either a shortage of facilities, widespread need for public transit, an inadequate sanitation materials, particularly water, the concealment of cases reported, a deficiency of individual protecting equipment for health care workers, as well as the existence among immune-compromised people. COVID-19(Manne, R., Kantheti,S., 2020) is the most recent acute threat that Ethiopia's government and diplomatic allies have encountered. The very first case was confirmed on March 13, 2020, and as of March 31, there were 26 confirmed cases, with 20 in Addis Abeba, two in Amhara region (Bahir Dar and Addis Kidam), three in Oromia region (Adama), and one in Dire -Dawa(Lee, H. et. al., 2020).

In wellbeing and in socio-economic contexts a new corona virus infection (COVID-19) does have an unparalleled health effects worldwide. Though perhaps not discriminating against, COVID-19 how has greatest vulnerability to a effects of such a disease outbreak to 55 regions and countries of 135 million extremely nutritional poor people in most need of immediate diet and health aid, since they are very small or unable to cope with either the economic and social dimensions of such a contagion. A comparison, in countries which rely on food imports and energy production, tourists, and incomes remittances, as SIDS (Small Islands Developing States), the disease outbreak will raise the level of acute food insecurity (Emmanuel, Sekyere. et. al., 2020).

Circumstances of instability in its most vulnerable countries of the food shortage may have detrimental impacts on political stability. Global mediation and peacekeeping attempts could obstruct resolving conflicts attempts, associated with adverse impacts on fragile and food-insect infected individuals. This could lead to resolving conflict. Due to transport and immigration bans, a pandemic could influence a delivery and costs of providing humanitarian assistance. Around the same time, budgets will be cut when finances were altered to sustain their COVID-19 activities around national level. This could have big implications to people who depend heavily on humanitarian aid for survive and/or safeguard the lives.

LITERATURE REVIEW

`The ongoing COVID-19 problem has been finding it tough to provide basic resources to some of the most vulnerable segments of the population. The COVID-19 (Bokam, Y. et. al., 2021) infection is initially thought to be restricted to China. As a product of human action, it then spread across the world. Once individuals are asked to sit at home, the financial damage increases significantly, and the effect is felt throughout the board, like travel bans in the transportation sector, sporting activity outages in the sports industry, and the prohibition of major gatherings in the events and entertainment companies.

As per(Mekonnen, H, Z., 2020)UNECA, with a projected 3.2% down to 1.8%, this pandemic will hit economic growth. It might undo the promising development of a recent decade and affect regions where Africa has gradually improved, whether it's in the fight against malaria or hunger, when it is not handled collectively and organized. It can also spread beyond the economies towards challenging certain countries' structural fragile nature by fueling additional disputes, including insecurity (Pramanik, S. and Bandyopadhyay, S. K., 2014).

According to (Ozili, P, K., 2020), Africa's average growth rate in 2020 would decrease by 1.4 percentage points, from 3.2 percent to 1.8 percent. In the worst situation, analysts estimate that Africa's gdp will decline by close to 2.6% during 2020. COVID-19 is a significant impediment to Africa's development. That virus's instability, and also policy responses like physical separation and lockdowns, has caused a drop in demand for African products, owing to a dramatic drop in global industrial production, that has been exacerbated by a drop-in economic activity on the continent as the work force remains at home to fight a virus.

As seen by (Ababulgu, A, N. and Wana, F, H. 2021), culture is already pushing for just a relaxation of a lockout strategy. To conform to societal disassociating conditions, buses and minibuses, locally known as "kombi," are now only allowed to carry a small percentage of the population on board as well as operate for a short time span to transport essential personnel. As a result of all of this, a carrier could become impossible to afford. Also as consequence of the problems faced both by drivers and riders, amended guidelines have been enforced, allowing bus services to board approximately to 70% with to their capacity while still requiring everybody to wear masks.

In (Ashebir, Y. et. al., 2015) mathematical analysis performed to analyses that patterns of a COVID-19 outbreak in asian country. The idea of researching an effect of public distance on demographic characteristics between many patient populations was presented. This evaluated the statistics of many developing and developed country and identified a most disadvantaged age ranges as well as different genders across all three countries. That study even projected an increase of infected cases in India (Singh, R., Adhikari, R., 2020) with various lockdown times. Likewise, one of the experiments used a network structure approach to see whether some unique node nodes are forming. However, the researchers assumed just migrate data nodes in order to ascertain that what prominent regions have an effect in developed and emerging nations.

A research (Daniel B. et. al., 2021). has addressed the effort of health professionals and frontline health care workers. When compared to developed regions like Italy, Spain, and the United States. It was found that health-care employee's positions are far less stressful as this propagation process of a corona outbreak was stage wise, or even the stage of regional transmission instead of global spread. Even so, it was also reported that many countries' medical facilities is not so good. According to world health organization norms, and that if crowd spreading occurs, the government of the country will struggle to handle that spread. Research has provided a thorough analysis mostly on the existence of a Corona Virus. Moreover, the study presented the SIR model to evaluate the frequency of corona virus transmission between infected peoples in a country. Earlier authors have provided the analysis of the testing laboratories and facilities.

As mentioned in (Gujarati, D, N., and Sangeetha, S., 2007), when COVID-19 incidents increase in Africa (currently 91 reported cases) and spread across various states, domestic demand is expected to fall. This really is due to a domestic policy goal to restricting movement and closing enterprises. This could escalate when businesses adjust to new economic realities by lying off staff, raising the rate of unemployment, which was recorded as 23.3 percent in its most recent data available.

In(Box, George E. P., and Gwilym M. Jenkins., 1976)the financial effects of a 2019-20 corona virus disease outbreak in India seems to have been tremendously devastating. A world bank and credit rating agencies have marked decrease India progress for financial year 2021 to a minimum level shown in three decades, dating back to the 1990s when India's economy was liberalized. Though, the International Monetary Fund's estimate of 1.9 percent Economic growth of India during fiscal year 2021-22 was its maximum amongst G-20 countries.

The corona virus, as reported in (Bhatnagar, V. et al., 2020), wreaked havoc on the $600 billion higher education industry. Instructors and learners all around the world felt the effects of a corona virus, which forced several universities into close after virus was declared a public health emergency in many nations.

In (Khongsai, L. et.al., 2020)states that mostly during the COVID-19 disaster, there were many significant difficulties in providing equal access to learning for all children. To delay the spread of the disease, many authorities therefore temporarily closed schools and universities. According to UNESCO, 1.6 billion learners (nearly 9 out of 10 children) are out of school worldwide related to school disruptions because of the first day of April 2020. Since about mid-March 2020, a disease outbreak has impacted nearly every African school student and university student. For instance, in Ethiopia, schools have closed since March 16, 2020, and nearly 25 million pre-primary, primary, secondary, and tertiary students have also been made to remain at home. While it would be too initial to determine how the COVID-19 institute closings are impacting especially vulnerable people in low-income countries like Ethiopia, there are many indications that they could have a long-term effect on income disparity. Over the last months in Ethiopia, I've noticed that the less wealthy and technologically illiterate families are the further off their children. Prior to COVID-19, there were already disparities in access to quality education between children from urban and rural areas, as well as students from low and high socioeconomic backgrounds.

In (Singh, V. et. al., 2020) provides more information. Due to the closure among all classes, both school-aged children enrolled in schools have been impacted throughout terms of daily training. Through school openings, home-schooling would not help a most vulnerable and poor students, increasing disparities of learning between the lowest and highest quintiles. Closing schools would result in a decrease in food intake and nutrition, and also a rise of school dropouts as well as a negative effect on human capital growth. Furthermore, school closures can worsen long-term drop - outs. Long-term consequences of interrupted education and inadequate early childhood nutrition would disproportionately affect disadvantaged families, limiting their human capital growth and earnings opportunities in the future.

As per (Jembere, G, B. et. al., 2018), the tourism sector has been impacted because Foreign visitors, that spend billions yearly, will have their transport options seriously constrained. Enhanced cancelled flights, resort cancellations, as well as domestic and international events worth more than $200 billion all were postponed. South Africa will have proclaimed a "nation of devastation" after holds the greatest COVID-19 figures (62) in the country (2nd largest on the region). The proportion of travel bans have been implemented, including the closure of half of the country's territory terminals. To prevent and control the spread of a disease outbreak as well as protect their people and financial systems, more nations in the area are anticipated to enforce similar travel restrictions on foreign citizens.

A state line closing will also have a detrimental effect on not just the tourism business as well as freight flows (due to increased delays), and also merchants, particularly unofficial brokers, across the region, to South Africa becoming a primary component of products bought by merchants: In South Africa, unofficial inter trade is indeed a notable factor of provincial trade and foreign maneuverability. Even so, tiny commerce creates greater revenue throughout the continent, enabling vulnerable groups to obtain products and services essential to their social and economic rehabilitation, thus playing an important role in alleviating poverty, food safety, but also domestic livelihood opportunities (Tanne, J. H. et. al., 2020).

In (Ayenew, B. et al., 2020) the growing number of jobs to countries currently facing problems, like conflicts and/or different economic and social instability, would demonstrate a decline of spending power and also the degree of poverty. Incredibly dangerous from the far effects of such an illness were also displaced people choosing to live in encampments and displaced people in large cities but rather older persons, young children, pregnant and lactating women and the disabled.

A statistic concerned can surpass a million, depending on mostly use, by concentrating on the potential size of eligible workers. 150 000 employees are noted to be employed at many concentrating places from of the biggest employer groups, flower and horticultural farmlands. Its other production team, which is greatly affected by fabric export markets, has between 50 000 and 70 000 staff members. A Planning Department approximated that perhaps the disease outbreak would make between 12.3 million and 18.5 million of people unemployed. Unemployment rate rose from 6.7% on 15 March to 26% on 19 April in one couple of weeks. A number of employees lost job at the lockdown was approximated to be 14 crore (140 million). Compared to the previous year, over 45% of households all throughout global community confirmed the decreases in earnings.

DATA AND METHODOLOGY

The whole article provides a thorough analysis of a field of research, an information sources, and also the interpretation underneath the many methods and procedures for both the analysis of time series.

Study Area

The study area was four developing countries namely: Ethiopia, Pakistan, India and Nigeria.

Data Source

The data source for the study was the country's ministry of health, world meter website and world health organization (WHO). The data was obtained through time daily for each country.

Analysis of Data

A sequence of observation that is arranged in chronological order of time based on variables of interest is called time series data.

Stationary

A vector (stochastic process) $\{X_t\}$ can be broadly classified as weak (covariance) stationary and strong (strict) stationary process. A mean vector that is independent of time shown if equation (1) and covariance given by equation (2) as:

$$E\left(X_t\right) = E\left(X_s\right) = \mu, \text{for all } t \text{ \& } s \tag{1}$$

Covariance:

$$Cov\left(X_t, X_s\right) = Cov\left(X_{t+j}, X_{s+j}\right) = \Gamma_{t,s}; \forall_{t,s,j} \geq 1 \tag{2}$$

A stationary series shows the overall behavior remains the same over time. It fluctuates around a constant mean.

Stationarity of Time Series Data

When some of those stay stable over time, a series has been said to be steady (i.e. it is not a random walk/has no unit root), otherwise it will be described as a non - stationarity phase (e.g that is a random walk/has unit root).

The time series is said to have been constant when there is no systematic change in mean (no trend), hardly any systematic change of variance, but purely periodic fluctuations have also been excluded. The static process does have a fixed mean which defines the time rate at which something varies but a constant variance which estimates the variability around a certain time level. Because the entirety in time series probability and statistics was associated with stationary time series, time series analysis usually necessitates converting the non-stationary time - series data to a static one in order to apply that concept.

In static sequences, the variable among two-time intervals is determined solely by duration or delay among them, not by the period at which covariance becomes estimated. The probabilistic method of this type is defined as just a weak constant or covariance stationary of time series (Sritha Z, D, B. et. al., 2020).

A stochastic process (vector) $\{X_t\}$ can be broadly classified as weak (covariance) stationary and strong (strict) stationary process. $\{X_t\}$ is thought to be feebly (covariance) still, if the foremost and anotherinstant are time-invariant, i.e., stochastic process (vector) $\{X_t\}$ has:

A mean vector shown in equation(3) that is independent of time and covariance in equation (4):

$$E\left(X_t\right) = E\left(X_s\right) = \mu, \text{for all } t \text{ \& } s \tag{3}$$

Covariance:

$$Cov\left(X_t, X_s\right) = Cov\left(X_{t+j}, X_{s+j}\right) = \Gamma_{t,s}; \forall_{t,s,j} \geq 1 \qquad (4)$$

Stationary series: A series whose overall behavior remains the same over time. It fluctuates around a constant mean. A time-series appearance still, if the time plot of the series seems alike at different points along the time axis.

Testing for Stationarity

The condition of stationarity is somewhat unrealistic situation in most macroeconomic variables. Trivially, a no stationary process arises when one of the properties for stationarity does not hold. The nonstationary of a time series influence the behavior of the series and inferences made unless it is detected (with standard unit root tests) and properly handled.

In these processes, the effect of a shock never dies away and it leads to spurious regressions (i.e.,one can regress completely unrelated series as a result; it would get inflated t-ratio which suggests whether a coefficient of one variable is significant or not to explain the other and high R^2 which indicate show good one term is at predicting another) and forged results of standard tests. In order to ensure the condition of stationarity, a series must be combined with order of zero, *I (0)*. There are dissimilar techniques so as to test for stationarity in time series analysis. Among them the most popular procedures are: Time plot (visual inspections) and Unit root tests: such as Augmented Dickey-fuller (ADF) and Phillips-Perron (PP) test (Pandey, D. et.al., 2020).

Time Plot

A time plot would typically suggest whether any differencing was done to know series stationery or non-stationarity before performing formal tests (Chatfield, C., 2004).Time Plot is a plot of observations against time. This is the most crucial step in any time series analysis. This graph would highlight key aspects of a series like pattern, seasonality, outliers, structural changes, and so forth. A plot is essential both for representing the data and aiding throughout the formulation of a reasonable model.

Once provided with a time series, the first step in evaluation would be to plot the data and obtain basic descriptive measures of a series' main properties. The time plot would usually indicate whether or not variance is required to render a series stationary. Until running structured experiments, it is always a great idea and map the time - series data in consideration.

The plot like that really gives an important indication of a possible existence of the time series. For example, when a line graph of a time series demonstrates the upward trend, this could indicate that the data's mean has not been stable or is increasing. As a result, the series doesn't really satisfy the criteria of stationarity(Zhu, Na. et al., 2020).

Unit Root Tests

Testes initially proposed by Dickey and fuller in 1979. This theory is the cornerstone for the methodology used interesting non-stationarity or stationarity for time series data in the real-world process. Nowadays, many of the procedures are standard offerings in econometric software packages like STATA. Augmented

Dickey-Fuller (ADF) and Phillip-Perron (PP) tests were to done (Pandey, D. et.al., 2020). These test procedures are developed for models with and without intercept as well as trend terms. Pre-tests for unit roots are often required to check the stationarity of a series (Pandey, D. et. al., 2021).

Differencing of Time Series

It is a useful technique for removing a trend, until a series becomes stationary. Differencing is an integral part Box and Jenkins methodology (Box, George E. P., and Gwilym M. Jenkins., 1976). First order differencing (Period-to-period change) is frequently used for non-seasonal data sets. 1st order difference is defined by equation(5) as:

$$\nabla X_t = X_t - X_{t-1} = \left(1 - B\right) X_t \tag{5}$$

Where, *B* is the backshift operator.

Univariate Models

These models consider only one variable measured over time. A univariate prototypical aimed at a given series depends only on past values of series. The Box-Jenkins have endowed modeling developed serves not only to explain the underlying process generating the series but as a basis for forecasting. The appropriate ARIMA processes were done for this study.

ARIMA Models

An auto-regressive of an order *P*, *AR (P)* is given in equation (6) as

$$x_t = \mu + \emptyset_1 x_{t-1} + \emptyset_2 x_{t-2} + \ldots + \emptyset_p x_{t-p} + \varepsilon_t \tag{6}$$

Where, x_t is the series; μ is the mean; $\emptyset_1, \emptyset_2 \ldots \emptyset_p$ are the parameters and ε_t is the residual.

A moving average of order *q*, *MA (q)* is written in equation (7) as:

$$x_t = \mu + \theta_1 \varepsilon_{t-1} + \theta_2 \varepsilon_{t-2} + \ldots + \theta_q \varepsilon_{t-q} + \varepsilon_t \tag{7}$$

Whereas, x_t is series; μ is mean and $\theta_1, \theta_2 \ldots \theta_q$ are the parameters and ε_t is error term.

Autoregressive Moving Average models were formed as combination of Autoregressive and Moving Average processes shown in equation (8) as

$$x_t = \mu + \emptyset_1 x_{t-1} + \ldots + \emptyset_p x_{t-p} + \theta_1 \varepsilon_{t-1} + \ldots + \theta_q \varepsilon_{t-q} + \varepsilon_t \tag{8}$$

Parameter estimation is a way estimating the values of the model coefficient (Like: $\varnothing_1, \varnothing_2 \ldots \varnothing_p; \theta_1, \theta_2 \ldots \theta_q$). The maximum likelihood estimation procedure was used in this study.

Model Diagnostic Checking

Model-checking is an obligatory activity to investigate the validity and reliability of all inference procedures made by ARIMA before one is going to use these models to forecast future patterns of series. Autocorrelation and Partial Autocorrelation residuals plot, NPP of Residuals, Normality of residuals through histogram was applied to check the underlined assumptions of residuals.

Forecasting from Time Series Models

Predictions for just a given frame are based on models that's only been adapted for current and previous observations of a time series(Box, George E. P., and Gwilym M. Jenkins., 1976).

Statistical Software

In this paper, STATA 14.2 and Minitab 14 Statistical Software were used for the analysis of total confirmed Novel coronavirus (COVID-19) case data.

RESULTS

The data used for this study was observed from February 26, 2020, to April 17. 2020. Hence, the total number of observations is 52. In this chapter, the results of univariate AR, MA, and ARIMA model specifications that can be used for forecasting the study variables would be presented. The discussion begins by describing the nature of the series and results from the model selection procedure. Then, the results would be interpreted (see table 1).

Table 1. Summary Statistics total confirmed COVID-19 case (s) for each Country

	Ethiopia	Pakistan	India	Nigeria
Mean	1.846154	135.0962	257.3846	9.384615
Median	1.000000	101.0000	51.00000	5.000000
Maximum	8.000000	931.0000	1211.000	51.00000
Standard Deviation	2.395949	181.5153	362.3784	11.98881
Skewness	1.262898	2.271362	1.260622	1.420308
Kurtosis	3.603789	9.304786	3.182509	4.544940
Jarque-Bera	14.61246	130.8378	13.84497	22.65453
p-value	**0.000671**	**0.000000**	**0.000985**	**0.000012**
Sum	96.00000	7025.000	13384.00	488.0000
Observations	52	52	52	52

Description of Data

The descriptive statistics in Table 1 show that 52 observations are included and found the positive mean of all confirmed case (s). That is, on average the total number of confirmed cases for Ethiopia, Pakistan, India, and Nigeria were: 2, 136, 258 & 10 respectively per day. Since the statistic values were the number of individuals face with COVID-19 they should round up always.

As normality is an essential property for time series data to obtain accurate and feasible results. The Jarque-Bera test statistic revealed the normal distribution of the total confirmed COVID-19 case (s), as having a p-value greater than 0.05(5%) for all countries.

Time Plot of Total Confirmed Daily Case (s)

From the time plot below (see figure. 1) one can observe that the total confirmed COVID-19 case (s)of India & Pakistan shows an upward (increasing) trend whereas for Ethiopia and Nigeria the increment is too slow even though it rises from day-to-day.

Figure 1. The Total Daily Confirmed Case (S) With COVID-19 at level

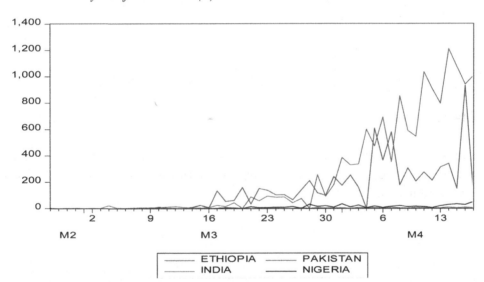

The time plot of all the study variables after taking their first difference are shown below in (see figure 2), the trend component (Pramanik, S. and Raja, S. S. 2020) is removed and all series looks weak (covariance) stationary as the plot line revolves nearly around the mean of zero although it's drift from strict stationarity.

Figure 2. The Total Daily Confirmed Case (s) With COVID-19 at 1ˢᵗ Difference

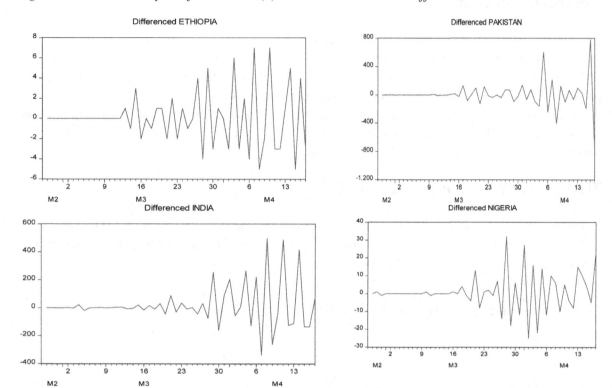

Testing Stationarity

The above time plot suggested that the series have non-stationary behavior at their levels and stationery at their first differences. The null hypotheses were rejected and stationarity was not satisfied at 5%. These results are presented in table 2 using Stata-14.2.

Table 2. Unit Roots Test Results [At levels]

Country	Level With Intercept		Level With Intercept and Trend	
	Test Statistic		Test Statistic	
	ADF	PP	ADF	PP
Ethiopia	0.007(0.954)	-1.644 (0.325)	-0.145 (0.721)	-0.0581 (0.921)
Pakistan	1.527(0.991)	-0.164 (0.825)	-1.209 (0.258)	-2.168 (0.453)
India	0.356 (0.978)	-0.176 (0.935)	-0.475 (0.981)	-2.883(0.176)
Nigeria	0.2671(0.974)	-1.697(0.427)	-1.827 (0.677)	-2.481(0.096)
5% Critical value	-2.923780	-2.919952	-3.500495	-3.500495
Decision	Non-stationary	Non-stationary	Non-stationary	Non-stationary

The above tests suggested that all are non-stationary since test critical values are greater than statistic values at 5% significance level. Also, p-values suggested in bracket are greater than 5%. As a result, the null hypothesis of a unit root is not rejected for all country's confirmed COVID-19 case (s).

After, taking 1[st]difference (period to period change), the series becomes stationary (have no unit root) at 5%. Thus, the total daily confirmed case (s) series under investigation are stationary after taking an appropriate difference of order-d.

Model Estimations

Table 3. The Selected MA (2) Result for Ethiopia COVID_19 Data

Type	Coefficient	S.E of Coefficient	T-statistic	p-Value
MA(1)	.8500	.001	850	.000
MA(2)	-.9039	.0262	-34.55	.000
Constant	-.00189	.01871	-.1010	.920

The MA (2) model with significant lags is written in equation (9) as

$$x_t = 0.850\varepsilon_{t-1} - 0.9039\varepsilon_{t-2} \tag{9}$$

From the above equation the coefficient for the second lag is negative, indicated that fraction of the shock two periods ago that is still felt in the current period (see table 3).

Table 4. The Selected ARIMA (1, 1, 1) Result for Pakistan COVID_19 Cases Data

Type	Coefficient	S.E of Coefficient	T-statistic	p-Value
AR(1)	-.3435	.1526	-2.25	.029
MA(1)	.9271	.0744	12.47	.000
Constant	11.314	1.763	6.42	.000

The Autoregressive Integrated Moving average with order one for all, ARIMA (1, 1, 1) of significant lags from the above result is (see table 4) given in equation (10) as:

$$x_t = 11.314 - 0.3435x_{t-1} + 0.9271\varepsilon_{t-1} \tag{10}$$

Table 5. The Selected ARIMA (2, 1, 1) Result for IndiaCOVID_19 Cases Data

Type	Coefficient	S.E of Coefficient	T-statistic	p-Value
AR(1)	.3463	.1874	1.85	.071
AR(2)	.5123	.1591	3.22	.002
MA(1)	.9457	.1647	5.74	.000
Constant	2.785	1.838	1.52	.136

The ARIMA (2, 1, 1) model with significant lags is written as (see table 5) given in equation (11) as:

$$x_t = 0.5123x_{t-2} + 0.9457\varepsilon_{t-1} \qquad (11)$$

Table 6. The Selected ARIMA (1, 1, 2) Result for NigeriaCOVID_19 Cases Data

Type	Coefficient	S.E of Coefficient	T-statistic	p-Value
AR(1)	.8050	.1347	5.98	.000
MA(1)	.5869	.020	29.34	.000
MA(2)	-.6102	79.50	-13.61	.000
Constant	.11757	.04515	2.60	.012

The ARIMA (1, 1, 2) model with significant lags is written as (see table 6) given in equation (12) as:

$$x_t = 0.1158 + 0.805x_{t-1} + 0.5869\varepsilon_{t-1} - 0.6102\varepsilon_{t-2} \qquad (12)$$

Figure 3. ACF&PACF Residual plot for Ethiopia

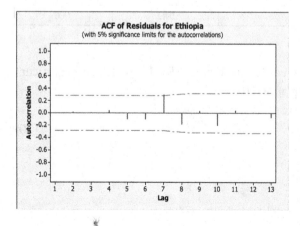 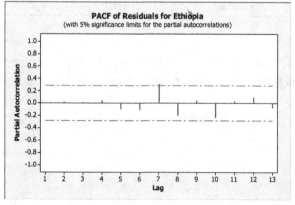

The above plots (see figure 3) showed that all strikes are in the confidence limit (red line), but very few touches the line, so it is implied that the assumption of no autocorrelation is not violated for Ethiopia COVID-19 case (s) data. Therefore, the residuals have no autocorrelation [28] problem (see figure 4).

Figure 4. Normal Probability plot (PP) and Histogram of Residual for Ethiopia

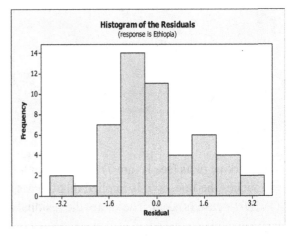

Therefore, it is indicated that no violation of normality assumption of Ethiopia COVID-19 case (s) data.

Figure 5. ACF & PACF Residual plot for Pakistan

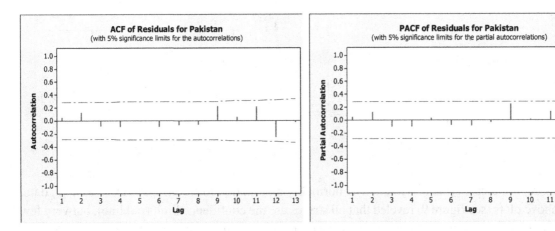

The above plots (see figure 5) showed that all strikes are in the confidence limit (red line), but very few touches the line, so it is implied that the assumption of no autocorrelation (Pramanik, S. and Bandyopadhyay, S. K. 2014) is not violated for the Pakistan COVID-19 case (s) data. Therefore, the residuals have no autocorrelation problem.

Therefore, it indicated that no violation of normality assumption for Pakistan COVID_19 case (s) data (see figure 6).

Figure 6. NPP and Histogram Residual for Pakistan

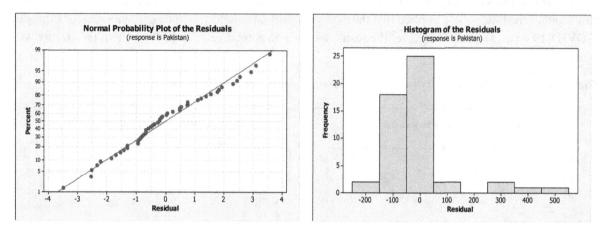

The above plots (see figure 7) showed that all strikes are in the confidence limit (red line), but very few touches the line, so it is implied that the assumption of no autocorrelation is not violated for India COVID-19 case (s) data. Therefore, the residuals have no autocorrelation problem (see figure 8).

Figure 7. ACF&PACF Residual plot for India

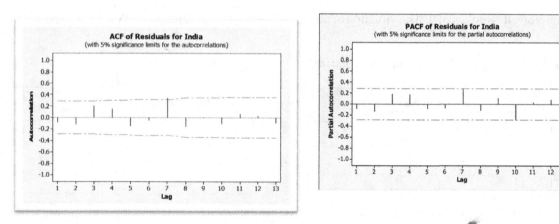

Therefore, it indicated that no violation of normality assumption for India COVID-19 case (s) data.

The above plots (see figure 9) reveled that all strikes are the confidence limit (red line), but very few touches the line, so it is implied that the assumption of no autocorrelation is not violated for Nigeria COVID-19 case (s) data. Therefore, the residuals have no autocorrelation problem (see figure 10).

Therefore, it indicated that no violation of normality assumption for Nigeria COVID-19 case (s) data. Overall, there is no strict violation of the underlined assumptions of the fitted time series ARIMA models for the four countries COVID-19 case data. Hence, it is feasible to be used those respective models for forecasting purpose, and the result is displayed below.

Figure 8. NPP and Histogram Residual for India

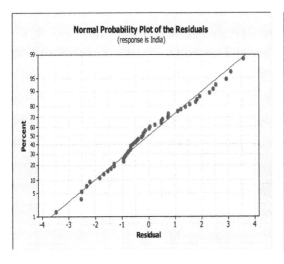
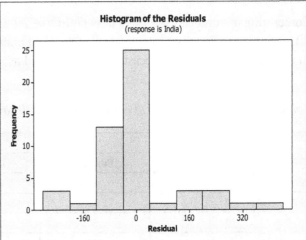

Figure 9. ACF & PACF Residual plot for Nigeria

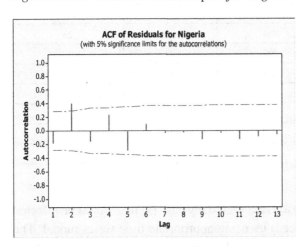
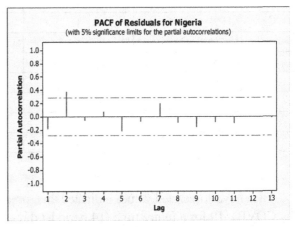

Figure 10. Normal Probability plot (PP) and Histogram of Residual for Nigeria

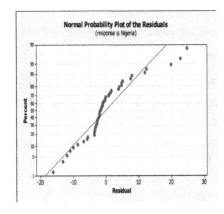
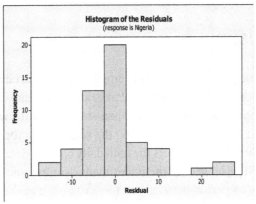

Forecasting From ARIMA Models

Forecasting is a crucial work especially in time series analysis (Pramanik, S. and Bandyopadhyay, S. K., 2014) to foresee (predict) future values of an event and/or series. Due to this, from the separate ARIMA model, two weeks ahead forecast value is obtained as shown in table 7. below.

Table 7. Two Weeks (14 days) forecasted value of Total case (s) COVID-19 for Countries

Period/Day	Forecasted Value for:			
	Ethiopia	Pakistan	India	Nigeria
51	*5*	*388*	*1,160*	*24*
52	*5*	*318*	*1,123*	*25*
53	5	353	1,156	25
54	5	352	1,151	26
55	4	364	1,169	26
56	4	371	1,175	27
57	4	380	1,190	27
58	4	388	1,201	28
59	5	397	1,215	28
60	4	405	1,228	29
61	5	413	1,243	29
62	4	422	1,257	31
63	4	430	1,273	31
64	6	439	1,288	31

The above-forecasted values indicated that, prediction of a total daily case (s) that would be affected by COVID_19 for a future time (14 periods/ days ahead) by using appropriate time series model. The very first two rows are all in sample estimate that used a sample of usable data to forecast values outside of the prediction time to equate it to equivalent known or real results, while the remaining values are now out of sample prediction.

CONCLUSION

The major objective of study was employed a univariate time series model to investigate the total daily confirmed case (s) for Ethiopia, Pakistan, India, and Nigeria. Over time considered, all three series remained non-stationary at level then stationary afterward first change. Consequently, entirely series stand united of order one. Based on the selected appropriate Autoregressive integrated moving average (ARIMA) time series models the total daily case (s) data of COVID-19 were fitted for Ethiopia, Pakistan, India and Nigeria; and all the underlined assumptions are tested and well satisfied. Following this, the two weeks (14 period/days) ahead forecasts were made for the whole countries.

- **Funding:** No funding is provided for the preparation of manuscript.
- **Conflict of interest:** All authors declare that they have no conflict of interest.
- **Ethical approval:** This article does not contain any studies with human participants or animals performed by any of the authors.
- **Authors' contributions**: This work was carried out in collaboration between authors. All authors read and approved the final manuscript.

REFERENCES

Ababulgu, A. N., & Wana, F. H. (2021). (2021). The Horrors of COVID-19 and the Recent Macroeconomy in Ethiopia. *Journal of the Knowledge Economy*. Advance online publication. doi:10.100713132-020-00713-6

Ashebir, Y. (2015). *Modelling and Forecasting the Balance of Trade in Ethiopia*. Academic Press.

Ayenew, B. (2020). Risk for Surge Maternal Mortality and Morbidity during the Ongoing Corona Virus Pandemic. *Med Life Clin.*, *2*(1), 1012.

Bhatnagar, V., Poonia, R. C., Nagar, P., Kumar, S., Singh, V., Raja, L., & Dass, P. (2020). Descriptive analysis of COVID-19 patients in the context of India. *Journal of Interdisciplinary Mathematics*, *24*(3), 489–504. doi:10.1080/09720502.2020.1761635

Birhanu, A. (2020). Challenges and opportunities to tackle COVID-19 spread in Ethiopia. *Journal of Peer Scientist, 2*(2), 1–6.

Bokam, Y., Guntupalli, C., Gudhanti, S., Kulandaivelu, U., Alavala, R., Alla, N., & Manne, R. (2021). Importance of Pharmacists as a Front Line Warrior in Improving Medication Compliance in Covid 19 Patients. *Indian Journal of Pharmaceutical Sciences*, *83*(2), 393–396.

Box, G. E. P., & Jenkins, G. M. (1976). *Time series analysis: forecasting and control*. Holden-Day.

Chatfield, C. (2004). *The analysis of time series: an introduction (6th ed.)*. CRC Press.

Daniel, B. (2021). The knowledge and practice towards COVID19 pandemic prevention among residents of Ethiopia. An online cross-sectional study. *PLoS One*.

Dong, E., Du, H., & Gardner, L. (2020). An interactive web-based dashboard to track covid-19 in real time. *The Lancet. Infectious Diseases*, *20*(5), 533–534. doi:10.1016/S1473-3099(20)30120-1 PMID:32087114

Emmanuel, S. (2020). *The Impact of COVID-19 in South Africa*. Africa Program Occasional Paper. Wilson Center.

Gujarati, D. N., & Sangeetha, S. (2007). *Basic Econometrics* (4th ed.). Tata McGraw - Hill Education.

Jembere, G. B., Cho, Y., & Jung, M. (2018). Decomposition of Ethiopian life expectancy by age and cause of mortality. *PLoS One*, *13*(10), 1990–2015. doi:10.1371/journal.pone.0204395 PMID:30281624

Khongsai, L. (2020). Combating the Spread of COVID-19 Through Community Participation. *Global Social Welfare: Research, Policy & Practice*. Advance online publication. doi:10.100740609-020-00174-4 PMID:32837833

Lee, H. (2020). COVID-19 perception, knowledge, and preventive practice: Comparison between South Korea, Ethiopia, and Democratic Republic of Congo. *African Journal of Reproductive Health*, *24*(2), 66–77.

Manne, R., & Kantheti, S. (2020). COVID-19 and Its Impact on Air Pollution. *International Journal for Research in Applied Science & Engineering Technology*, *8*(11), 344-346. doi:10.22214/ijraset.2020.32139

Mekonnen, H, Z. (2020). *COVID-19 in Ethiopia: Assessment of How the Ethiopian Government has Executed Administrative Actions and Managed Risk Communications and Community Engagement*. Academic Press.

Otom, R. (2020). *COVID-19 Awareness to Kenyans*. Academic Press.

Ozili, P, K. (2020). *Spillover of COVID-19 : impact on the Global Economy*. Academic Press.

Pandey, D. (2020). Infectivity, Preclusion, and Control (IPC) of Pandemic Novel COVID-19. *International Journal of Computer Engineering In Research Trends*, *7*(5), 1–8.

Pandey, D. (2021). *Covid-19: A Framework for Effective Delivering of Online Classes during Lockdown*. Human Arenas.

Parmeshwar, U. (2020). Global food security in the context of COVID-19: A scenario-based exploratory analysis. *Progress in Disaster Science*, 7.

Pramanik, S., & Bandyopadhyay, S. K. (2013). Application of Steganography in Symmetric Key Cryptography with Genetic Algorithm. *International Journal of Computers and Technology*, *10*(7), 1791–1799. doi:10.24297/ijct.v10i7.7027

Pramanik, S., & Bandyopadhyay, S. K. (2014). Hiding Secret Message in an Image. *International Journal of Innovative Science. Engineering and Technology*, *11*(3), 553–559.

Pramanik, S., & Bandyopadhyay, S. K. (2014). An Innovative Approach in Steganography. *Scholars Journal of Engineering and Technology*, *2*(2B), 276–280.

Pramanik, S., & Raja, S. S. (2020). A Secured Image Steganography using Genetic Algorithm. *Advances in Mathematics: Scientific Journal*, *9*(7), 4533–4541.

Pramanik, S., & Singh, R. P. (2017). Role of Steganography in Security Issues. *International Journal of Advance Research in Science and Engineering*, *6*(1), 119–1124.

Pramanik, S., & Bandyopadhyay, S. K. (2014). Image Steganography using Wavelet Transform and Genetic Algorithm. *International Journal of Innovative Research in Advanced Engineering, 1*.

Singh, R., & Adhikari, R. (2020). *Age-structured impact of social distancing on the COVID-19 epidemic in India*. Accessed from https://arxiv.org/pdf/2003.12055.pdf

Singh, V. (2020). Prediction of COVID-19 corona virus pandemic based on time series data using Support Vector Machine. *Journal of Discrete Mathematical Sciences & Cryptography.*

Sritha, Z. D. B. (2020). Acting tools of ICT to tackle Covid19. *OmniScience: A Multi-disciplinary Journal, 10*(1), 1–9.

Tanne, J. H., Hayasaki, E., Zastrow, M., Pulla, P., Smith, P., & Rada, A. G. (2020). Covid-19: How doctors and healthcare systems are tackling coronavirus worldwide. *BMJ (Clinical Research Ed.), 368*, 1–5. doi:10.1136/bmj.m1090 PMID:32188598

WHO. (2020). *COVID-19*. WHO.

Yakubu, L. (2021). Africa's low COVID-19 mortality rate: A paradox. *International Journal of Infectious Diseases, 102*, 18–122. PMID:33075535

Zhu, N., Zhang, D., Wang, W., Li, X., Yang, B., Song, J., Zhao, X., Huang, B., Shi, W., Lu, R., Niu, P., Zhan, F., Ma, X., Wang, D., Xu, W., Wu, G., Gao, G. F., & Tan, W. (2020). A novel coronavirus from patients with pneumonia in China, 2019. *The New England Journal of Medicine, 382*(8), 727–733. doi:10.1056/NEJMoa2001017 PMID:31978945

Chapter 10
Application of Machine Intelligence–Based Knowledge Graphs for Software Engineering

Raghavendra Rao Althar

First American India, India & Christ University (Deemed), India

Debabrata Samanta

ⓘ https://orcid.org/0000-0003-4118-2480

Christ University (Deemed), India

ABSTRACT

This chapter focuses on knowledge graphs application in software engineering. It starts with a general exploration of artificial intelligence for software engineering and then funnels down to the area where knowledge graphs can be a good fit. The focus is to put together work done in this area and call out key learning and future aspirations. The knowledge management system's architecture, specific application of the knowledge graph in software engineering like automation of test case creation and aspiring to build a continuous learning system are explored. Understanding the semantics of the knowledge, developing an intelligent development environment, defect prediction with network analysis, and clustering of the graph data are exciting explorations.

INTRODUCTION

The web has been expanding its wing to an extent never seen earlier; AI (Artificial Intelligence) has seen its fullest blossoming in the recent past; this combination has called for exciting methods of knowledge representation and extensive processing of the information. The emerging field of knowledge graph provides a good platform for variety of area like web applications and AI. An in-depth exploration of the web's knowledge has resulted in an extensive understanding of the ontology of the knowledge of

DOI: 10.4018/978-1-7998-7701-1.ch010

various domains. Google introduced the knowledge graph and put up a large-scale knowledge graph. The knowledge graphs' semantic networking capability has seen its application in areas like recommendations systems, question answering systems, and natural language processing. Deep learning with big data has put knowledge graphs in driving positions, leading to AI's rapid development. Even with the advent of large-scale knowledge graphs, some missing entities and relationships need further exploration and mining of data. Knowledge graphs can be traced back to the expert system that was developed decades ago.

The effort of a combination of the data and knowledge leads to a knowledge-based system. Various modalities of constructing knowledge graphs provide different categorization for them. Knowledge graphs can be domain-specific or generic ones. Based on the influence of time, static and dynamic knowledge graphs can be built. Knowledge representation, graph construction, and application are the core focus area of the research in knowledge graphs. These focus areas provide the base for information retrieval, data mining, extraction of information, and natural language processing. Explorations in the area of knowledge graphs offer opportunities in the area of development of software product capabilities. The knowledge hidden in a wide variety of sources across the landscape of software development processes can be meaningfully put in to use only if there is a method to put together all this information effectively. Since there is a continuous change of people involved in software engineering processes, knowledge graphs provide a base for capturing the knowledge systematically and avoiding the dependency on humans. This aspect builds towards facilitating autonomous and smart systems that can enhance software engineering systems' quality and productivity. Exploration of knowledge graphs in the area of intelligent development environment development finds its exciting applications. Automated test case generation has been another area of investigation to optimizing the testing process. As the exploration continues in the area of the knowledge graphs for software engineering, it will also enhance the possibilities of developing useful software engineering tools. Knowledge graphs exploration on the software source code finds its essential place in AI research in software engineering. Some of the challenges hauntings AI exploration around the data-hungry algorithms will ease with these knowledge systems being built to capture the process knowledge. Some of the challenges around generalizing the AI solutions also will find some direction with the knowledge graphs. This chapter explores some of the exciting work done in applications of the knowledge graphs in software engineering.

FOCUS OF THE CHAPTER

The focus of the work is to throw light on knowledge graphs and their possibilities in software engineering. The exploration covers various work done in this area as follows. It covers efforts towards proposing knowledge management architecture and leveraging the knowledge graph for inferencing the documents associated with software engineering. Continuous learning agents that build on their knowledge with the flow of new information is the aspiration. Exploration of leveraging on the words and phrases of natural language with knowledge graph representation. Intelligent Integrated Development Environment enablement with knowledge graphs that work on top of artifacts associated with software development. Network analysis being leveraged for predicting defects. Feature extraction with graph mining and graph learning. Pattern recognition methodologies that work with graph matching and related techniques. Vectoral and graphical representation studies that has text mining as its base. Models for vulnerabilities prediction to optimize the testing efforts associated with software development. With these areas of exploration, the work intends to bring out the key areas that can be focused on as future research through which it is possible to strengthen the domain of software engineering. As the exploration unfolds in this work, it also throws light on the methods used to validate these experiments and assess their relevance.

EXPLORATION OF KNOWLEDGE GRAPHS FOR SOFTWARE ENGINEERING

Knowledge graphs are the new thought process for the software engineering community. Some of the approaches that are explored in the field have shown good potential. But there is quite a bit of open questions specific to software engineering from the perspective of knowledge graphs. These challenges are seen in the fusion of knowledge, graph construction, representation of the learning, and knowledge reasoning. Some of the prospective areas that can help move in positive directions to answers these questions should investigate a wide variety of aspects of software engineering and its related areas. Natural Language Processing, data mining, and other formal methods need to be explored. Some of the focus areas of software engineering are knowledge graphs construction for software engineering artifacts, knowledge graph-based document retrieval and question answering, recommendations for software engineering tasks, and operations management with software knowledge graphs. Knowledge graphs have demonstrated great possibilities concerning tying together the fragmented knowledge, which is widely prevalent in software engineering practices. There is a need for collaboration among the efforts across various areas like knowledge engineering, big data analytics, statistics, machine learning, pattern recognition, knowledge visualization, to name a few. Further down in this section, we review the work done in various critical areas of knowledge graphs and their software engineering practices. Learnings from these works can help devise a comprehensive plan for progressing software engineering-related research and development.

In work (Sabou et al., 2018), authors explore enterprise knowledge graphs as a case study in software engineering. Work is positioned in the landscape of architecture knowledge management and exploratory search research. Software architects would need supporting information to explore knowledge hidden in software architecture; this knowledge would design decisions or design patterns. Look-up-based tasks are only executed in knowledge management tools, which are semantic-based, and this is a limitation. Explorations are not possible in this kind of setup. They are faceted search and do not explore these areas on web-scale knowledge graphs which lacks the enterprise-wide adaptation. The author explores the extent of usage of Enterprise-scale Knowledge Graphs (EKG) for the exploratory search for architectural knowledge. EKGs-based experimental search systems are proposed here for Siemens to evolve a STAR system used by around 300 architects.

Implicit organizational knowledge is made explicit with these systems, providing the details on metrics needed in support of exploration. EKG data will influence the performance of the metrics identified. For the selection of the right metrics, statistical and user-based inputs from their evaluations can be utilized. To make software architecture useful, there is a need for adequate architecture knowledge. This need is not enabled with good support for the architects in general. In the work author conduct exploration of EKG to gather architectural knowledge in an exploratory way. The exploratory approach used in this work applies to influence around 300 architects in Siemens. Ontology of EKG is used as base information to design-related metrics that underwent evaluation before they are implemented. Looking at ontology helped define a standard solution to a heterogeneous group of architects at Siemens. Assessment of these ontologies with the others related to the domain helped to derive more concepts. This ontology study helped to see common themes around design patterns of the domain, like qualities of the architecture, design-related decisions, which emerged as a recurring pattern. Semantic relations were made explicit in this work; unlike with earlier approaches, these were implicit when non-semantic solutions were used as bases. The method also helped explore the concepts from models specific to an organization that is influenced by the organizational standards. Authors intend to focus on future work to improve architectural knowledge quality with semi-automatic methods for acquiring the data and cleaning the same. Work here refers to the various ontologies defined in literature as part of work (Graaf et al., 2012; Tang et al., 2011). Some tools explore the ontologies with visualization making them intuitive (Kruchten et al., 2006). Some work uses the lighter background of ontologies with wiki-based systems to browse to manage service-oriented architecture documentation (Speer et al., 2017) and manage rationales of software design (López et al., 2012).

In work (Nayak et al., 2020), knowledge graphs are used as a base for automated test cases for software engineering. When the data houses the complex entities within it, a knowledge graph can be handy to store and retrieve this information efficiently. This representation will be convenient for software engineering as they store complex information in functions, classes, and modules. The method proposed in this work focuses on creating a knowledge graph from the documents associated with the software engineering for test case generation from requirements information, which are natural language. Test intent extraction is done using Constituency Parse Tree (CPT) pathfinding algorithm in a knowledge graph creation tool. Automatic feature extraction is done with Named Entity Recognition (NER) based on Conditional Random Field (CRF), and signal extraction is done with sentence vector embedding. An automotive domain software project is used as a base for experimentation. The methodology proposed in the paper works on the generation of knowledge graph from unstructured and sparse software documents. The knowledge graph will be extracting the test intent from the requirements document. This test intent will assist in querying the knowledge graph for extracting the associated test cases and derive the missing information.

Compared to the traditional method of automated test cases generation in this approach, knowledge graphs are formed from the large corpus of information in an automated fashion, which retains the domain knowledge. Tool for creating Knowledge graph is used to help conduct exhaustive testing. As the future step, there is potential to explore an alternate approach for NER with Transfer Learning as the approach, which uses a pre-trained neural network with a small amount of data that are domain labeled data. Work can also be extended to validate the requirements information with a knowledge graph. This cross-validation helps to assess the criteria information with that of the knowledge graph.

Figure 1. Automated test cases generation pipeline with knowledge graphs

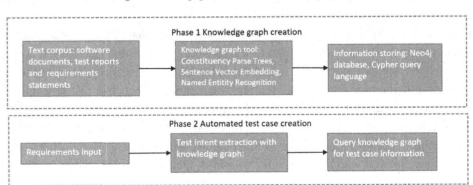

Figure 1 demonstrates the automated test cases generation with knowledge graphs. Stanford OpenID (Angeli et al., 2015) and ClausIE(Del et al., 2013) are widely used to create knowledge graphs related to requirements engineering. Some of the significant work in knowledge graphs are seen in Google knowledge graph (Singhal et al., 2012), ConceptNet(Speer et al., 2017), and Never-Ending language Learner (Carlson et al., 2010). These works have captured entities and relations in millions from a wide variety of datasets like Wikipedia. But these works have not been extensively extended for software engineering, where the data are unstructured and sparse.

Work (Carlson et al., 2010) focuses on building an architecture for never-ending language learning. The idea is to develop a computer agent that runs every day picking information from the web to learn by extracting and reading the story, to construct a growing knowledge base. This approach intends to improve the knowledge day after day by getting better compared to the previous day. Approach and design principles are proposed for the agent. Partial implementation of the system can be seen here, which learns from the knowledge base of 242,00 constructs with precision estimated to be 72% running for 67 days. Lessons learned in the preliminary attempt is used as a base for building a never-ending system. The shared knowledge base is used as a reference for the approach here, which grows continuously and is used by collecting subcomponents that learn the method by knowledge extraction approaches. Knowledge base kick starts with a starting point of few defined ontologies that defined categories and relations. The target of the work is to improve the knowledge base by reading and enhancing the reading continuously. This never-ending learner agent with a partial implementation where four sub-systems components are structured focuses on extracting the knowledge. Results provide confidence in the approach of using a variety of knowledge extraction methods for meaningful learning. This system also provides a database for storing the candidate's beliefs and information, confident enough. Some of the essential highlights are the ability to take a call on what is the next step. Human interaction daily is utilized as input for learning and discovering new predicates that can be learned and deriving additional information related to language. Entity level modeling and effective probabilistic modeling across implementation are the key highlights. Work derives its similarity with the (Erman et al., 1980) for the knowledge base developed in this work being like the "blackboard" in the speech recognition system. Frame-based implementation of the knowledge base is inspired by work (Mitchell et al., 1991), which focused on integrated learning and inference representation. Work (Thrun et al., 1995) focused on lifelong learning to bank on previously learned functions. In work (Banko et al., 2007), where the agent's lifelong learning objective explores to build theory for the domain where decision happens to decide the next step to tackle. Work in (Yarowsky, 1995) uses bootstrap learning for word sense disambiguation with classifiers being trained. Progress demonstrated in this work provides excellent confidence towards a never-ending learner system. New extraction rules being learned by the system is the building block for the system to keep exploring the new avenues of the knowledge base.

Work (Fabian et al., 2007) focus on the core semantic knowledge that unifies wordnet and Wikipedia. YAGO is the approach built in this work that is a lightweight, extensible ontology that ensures high coverage and quality. Entities and relations are the basis for YAGO; it encapsulates 5 million facts and 1 million entities. Non-taxonomic relations are built between entities. Facts from Wikipedia are automatically extracted and combined with WordNet, which utilizes a carefully designed combination including heuristics and rule-based combinations, covered in this work. A knowledge base created from this work contributes to getting WordNet to the next level in its quality where the knowledge about organizations, individuals, etc., are captured with their semantic relations. Quantity wise WordNet is taken to the next level in terms of its magnitude for the number of facts that are added on.

Fact correctness stands at 95% accuracy on the evaluation conducted on fact empirically. Extensibility, decidability, and compatibility with RDFS (Resource Description Framework Schema) is the YAGO's critical feature that provides a logically clean model. Work also points at the future possibility of extending YAGO towards information extraction capabilities. YAGO is a high quality, sizeable extendable ontology. Compared to publicly available ontology, YAGO has the rich set of ontologies. YAGO is accompanied by a clear set of semantics as presented by the proposed data model. For canonical storage of ontologies of YAGO small set of unique ontologies are ensured in the collection of equal ontologies. State of the art information extraction techniques is used for extracting information from web documents for the extensibility of YAGO.

As the fact's accumulation increased, it is demonstrated that the facts gathering improves. The author also set forth the hypothesis that this demonstration will help grow the knowledge base that will benefit the future. Various new challenges can be looked at with YAGO. Validation of the YAGO models with different other approaches from the perspectives of semantic aspects. The model can be further enriched by going beyond the arbitrary relation that is established now. This enrichment is possible by including information extracted from web pages and facts that come in with confidence from gazetteers. Work also intends to the possibility of gathering a positive loop of feedback for data. Semantic web vision will be the possibility with the availability of high-quality ontology. Various work focuses on creating general-purpose ontologies with the base representations. One of the categories of approaches focuses on extracting the knowledge structures from text corpora in an automated way, where pattern matching, statistical learning, and natural language parsing approaches of information extraction are used (Agichtein et al., 2000; Cafarella et al., 2005; Cunningham, 2002; Etzioni et al., 2004).

In work (Speer et al., 2017) open multilingual graph of general knowledge is explored. Machine learning done on the knowledge can get better when there is a supply of a combination of information from external sources and those specific to the target study. Work in this paper focuses on an open-source data resource called ConceptNet, which suits well with WordEmbeddings that are modern NLP (Natural Language Processing) approaches. In a graphical approach, labeled edges represent words and phrases of the natural language in the knowledge graphs. Information gathering is done from various sources like games with a purpose, information sources created by experts, and crowdsourcing. General knowledge involved in the understanding of the knowledge is represented in the approach designed here. This approach also enhances the natural language applications by exploring the meaning behind words that people put in to use. Word embeddings extracted from distributional semantics like that of word2vec, when combined with ConceptNet, will have the ability that cannot be generally seen in distributional semantics alone or from WordNet or DBPedia. This characteristic will end up being a resource for a narrower focus.

Intrinsic evaluation for the relatedness of the words helps improve the word vector applications with state of art. Word embeddings represent the distributional semantics. Another case where word embeddings represent relational knowledge and combination of both are experimented in variety and seen that the combination of both works well compared to individual ones. Distributional semantics include GloVe, Word2Vec, and LexVec, whereas ConceptNet represents relational knowledge. With word embeddings being the focus, ConceptNet will have its importance as the knowledge gathered in ConceptNet can be used by the word embeddings. ConceptNet also helps bring in correlations with human judgment and makes it robust; this is in-line with the state-of-the-art results ConceptNet achieves as part of its Numberbatch version. Human annotators are matched on evaluations conducted in multiple attempts.

Relational knowledge source needs to be accounted for any technique based on word embeddings. Another option is to start with pre-trained word embeddings, which has accounted for relational knowledge. ConceptNet intends to use this knowledge in a convenient form to be extended to multiple languages and many domains. (Singh et al., 2002) is used as a base to develop a knowledge graph version in ConceptNet. In (Lenat et al., 1989), predicate logic form for a decade of data is built which had common sense data as its ontology. The work of (Auer et al., 2007) creates DBPedia that gathered knowledge from Wikipedia to represent facts around named entities. Work in (Singhal, 2012) highlights that the Google knowledge graph is the largest but not publicly available. Work in (Faruqui et al., 2014) uses retrofitting, where the word embedding matrix is refined based on the knowledge graph.

In work (Lin et al., 2017) author talks about intelligent development environment and software knowledge graph. Making software development processes smart has been a prominent research area. In this work author has proposed an intelligent development environment (IntelliDE) and software knowledge graph for the first time. Software processes involve significant data aggregation and analysis which is done in the IntelliDE ecosystem, which provides smart assistance to software development processes. The system's architecture is discussed, and an explanation of the challenges associated with the research is provided. Software knowledge representation and management framework provided by software knowledge graph play an essential role in IntelliDE. Work here explores the construction of a system and ways to leverage the same. Intelligent ecosystem is the direction forward to move towards an intelligent development environment. Providing competent assistance, aggregating the data, and acquiring the knowledge are critical challenges. Software knowledge graph also acts as an infrastructure for knowledge acquisition of IntelliDE. Software knowledge graphs here cover primitive and derivative knowledge of the system. Primitive knowledge is extracted from the issue tracking system and source code, whereas derivative knowledge comes from API usage, traceability links, and latent topics. Semantic understanding and inferencing in IntelliDE are managed with software knowledge graph with software

Figure 2. IntelliDE architecture

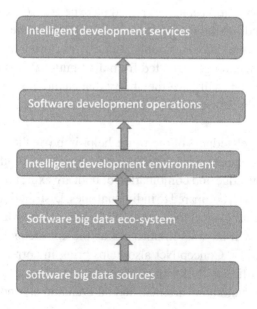

text semantic search engine. The authors also hope that this system paves the way for a smart question and answering system for the future.

Figure 2 represents the architecture of IntelliDE. Work in (Kaiser et al., 1998) explores building intelligent assistance on software development and maintenance, work (Robillard et al., 2009) focuses on recommendations systems for software engineering, and code completion for statistical language models are explored in work (Raychev et al., 2014). Ranking of relevant bug files is done based on the domain knowledge in work (Ye et al., 2014). Work (Trautsch et al., 2016) explores the addressing of problems with external validity of repository mining using smart data. Data aggregation is one of the critical challenges in IntelliDE, with its root in software big data being web-scale, multi-source, distributed, fast-growing, and dynamic. This aspect makes it essential for IntelliDE to have an up-to-date, large scale, and easy to access data repository to tackle the challenges with insufficiency of the data. Focus is needed on data collection, data fusion, and data update. For IntelliDE to handle the data by correctly sensing and accurately downloading adequate study for big data features in the software is essential. Data fusion throws up the problem of putting together the knowledge hidden across various parts of the software so that they can be put together to make sensible aggregated knowledge.

In work (Zimmermann et al., 2008), authors focus on predicting defects using network analysis on dependency graphs. Time and cost limitation of the resources are the critical aspects of software development concerning quality assurance. The experience of the managers plays a crucial role in allocating the resources for the software engineering work. Some of the metrics like code complexity may provide some backing for managers' decisions to be taken. Dependencies between various code parts make it complicated for the managers to devise their decision due to the lack of knowledge around it. These dependencies can be represented as the graph at a low level for the system knowledge representation. Dependency graphs are subjected to network analysis in this paper. This analysis helps managers figure out critical central units of software that are more likely to face defects. Authors also confirm that based on the evaluation conducted by then on windows server 2003, they see that recall performs better by 10% using network measure than complexity metrics. Critical binaries of 60% were identified by network measure, which was acknowledged by windows developers that would not be done with complexity metrics. Work demonstrates that dependency graphs subjected to network measures help predict the defects for windows sever 2003 binaries. This system enables managers' task allocation work for quality assurance. This ability helps to focus on the key areas where most of the defects exist. Authors present the result that complexity metrics fail to predict the binaries considered critical by developers; coverage is only about 30%. Network measures explore 60% of essential binaries. Indication and prediction of defects can be made possible on dependency graphs by network measures. Both network measures and complexity metrics have comparable precision, but recall stands higher for network measures. Though the author highlights that dependency data is essential and caution that it alone may not be the answer, it contributes as a crucial part to why software fails. Process metrics like code churn and code complexity metrics are other vital predictors. As a future work author want to focus on building more predictors and strengthen the framework. The human factor is the significant part as humans introduces the defects. Figuring out why failure is presented by the programmer and building tools to avoid such human shortcomings will be the critical part of future exploration. To enable a support system for the managers, various quality indicators are explored to build a prediction model that predicts software quality. One of the key indicators would be the complexity of the source code. McCabe's cyclomatic complexity has a higher correlation to defects as per multiple studies, but there is no standard model that applies universally for all types of projects (Basili et al., 1996; Nagappan et al., 2006; Subramanyam et al., 2003). Bugs have taken up a semantic nature in their databases. This aspect is prominently due to dynamic and static bug localization coming into the forefront (Auer et al., 2007). Network analysis is extensively used in software engineering to study open source development from a social network analysis perspective (Ghosh, 2003).

In work (Aggarwal et al., 2010), authors survey clustering algorithm for graph data. Exploration extends to clustering algorithm categories and clustering methods design in the recent past for graphical data. Node clustering type of algorithm focuses on edge behavior in the graph concerning the dense region. In the structural clustering algorithm, different graphs are clustered with structural behavior as the reference. Semi-structured data are focused upon in another type to explore their utility on graph mining algorithms. The wide application of clustering graphs in data management and data mining area creates a lot of interest in exploring this field. In the case of the node clustering algorithm, exploration covered the partition of graphs into a group of clusters; each of these cluster groups included densely connected nodes. A densely connected group of nodes depicts entities in the graph and their inter-connection, providing a significant amount of information. In graph clustering, algorithm clusters are determined from complete graph information by assessing the structural information in the graphs. In the real domains, XML data represents such a kind of graph clustering. Work here focused on establishing a trade-off between various clustering algorithms. Clustering large datasets remain the challenge in graph clustering. This limitation ends with the restriction of the extensive graph data only on disk. Examples for such a large graph are the ones with 107 nodes and 1013 edges. These scenarios even pose a challenge to store the graphs on disk as well. The behavior of the graphs on the disk should be accounted for in designing the algorithm for cases of disk stored graphs. This behavior poses a challenging situation as the graph data set's structural behavior interferes with edge processing capabilities in many scenarios. Another idea is to capture the graph's summary behavior in case they are large enough to be stored on the disk.

Further graph clustering can be conducted effectively with this stored information. Another challenging area is the clustering of the graph streams. In this type of graph, large graphs are processed as edge streams, which offers the capability to process the edge only once for computation purposes. Here the summary structures would be needed for clustering purposes and useful for large disk stored graphs. Traditional database management systems and clustering algorithms need to have the right design of algorithms for their practical interfacing. Useful query language and representations become critical in this situation. For purposes of further enhancement of the graph algorithms, this is one key area of research. Work in (Cook et al., 2006) provides a detailed study of the graph mining algorithms. A detailed discussion on clustering algorithms work (Jain et al., 1988) is a useful reference. Congestion detection, XML data integration, and facility location are scenarios where clustering algorithms find their prominent applications (Lee et al., 2002). One of the areas of exploration is the edges that exist between pair of nodes with higher probabilities. Closely associated with this scenario is the problem of identifying the shingles, which are sub-graphs with many standard associated links (Broder et al., 1997;Gibson et al., 2005).

In work (Foggia et al., 2014), pattern recognition is studied for understanding graph matching. Breaking up the components into sub-parts and exploring the relationship between them based on structural pattern recognition. A rich set of edges and nodes in the graphs provide a base of such systems. There is a limitation associated with the methods employed to study graphs, such as restriction on structure of graphs, working on planar graphs only as an example. In some cases, there is a restriction of the type of attributes that can work with, like, working on only a single real-valued attribute. Graph embeddings and graph kernels are explored in the application of graph for vector-based techniques, which are used in the case of learning theory and statistical approach for classification. Innovative scientific vision has gone into learning and classification; one such scenario matches the algorithms with their intrinsic complexity. Some of the new approaches are suggested by well-developed statistical pattern recognition fields, which provided a new paradigm for graph-based methods. One of the questions is to use vector-based operations in the case of graph matching and learning so that the statistical approach can be easily applied. Each approach has its pros and cons; in the case of graphs, there is an end to access to all the information but with some heavy algorithms. There is the possibility of losing the discriminating power as the graphs are converted to vectors; this must be counterbalanced with selecting suitable attributes from the statistical framework. Some approaches can combine both the format, like graph edit distance where nodes and edges of two graphs are matched, and distance information can be used to project the graph in metrics space. But graph edit distance will have its prominence as the information available in the sub-parts in the form of metrics can provide context without being changed over to a vectoral state.

This area's prospects lie in leveraging on the exact and inexact methods of graph matching those proposed by the researchers. These possibilities are explored in pattern recognition. Bioinformatics is one such area where more powerful calculators if made available, can help solve more real-time problems, which expands the capability of adequate representation of the graphs. Graph kernels theoretical and applicative points of view must be consolidated for prediction purposes. Classical graph matching methods have been the center of focus for the researchers. Graph embedding and its kernelization have enhanced the researchers with ready to use tools to tackle complex and new pattern recognition problems and avoid the need for technical understanding of the graphs. Image recognition from a genetic algorithm with inexact graph matching (Auwatanamongkol, 2007), structural classification and pattern recognition with first-order Gaussian graphs (Bagdanov et al., 2003), skeleton graph matching with path similarity (Bai et al., 2008), distribution algorithm estimation as a means to inexact graph matching (Bengoetxea, et al., 2002) and Self-organizing maps for multidimensional scaling of the graphs (Bonabeau et al., 2002) are some of the interesting works. Figure 3 represents graph embedding and vector space representation.

Figure 3. Graph embedding and vector space representation

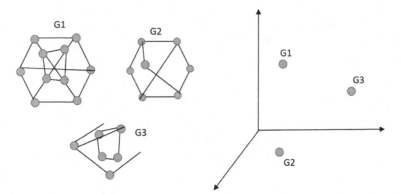

In work (Yamaguchi et al., 2014), code property graphs are explored for modeling and discovery of the vulnerabilities. Insecure code has been the leading cause for most security breaches. This situation calls for computer systems to be protected with the extensive process of identifying software system vulnerabilities, which is a complex process and an error-prone one. Just a single sneak in of the flaw would leave the system vulnerable in the hands of the attacker. This work focus on deriving information from the source code for vulnerabilities exploration. A unique approach of code property graphs is introduced in this work, which combines traditional program analysis methods like abstract syntax tree, control flow graphs, and program dependency graphs, putting it together with combined data structures. This representation method leads to graph traversals and common vulnerabilities templates being effectively modeled, which will help identify integer overflow, memory disclosures, and format string vulnerabilities. Eighteen previously unknown vulnerabilities of the Linux kernel source code are explored in this work with popular graph databases. Software systems security can be ensured by providing the software analysts with all support to discover the vulnerabilities. A large amount of code is mined for purposes of vulnerability exploration in this work. The code property graph is proposed as a unique approach for source code where graph traversal is done for representative modeling vulnerabilities. Concise and small traversals are proposed for well-known vulnerability types, like format string vulnerability, buffer overflow, and memory address leaks. Common types of vulnerabilities can be unearthed from graph traversals of code property graphs. An interesting fact is the possibility of tailoring code property graphs to identify vulnerabilities specific to the codebase. False-negative and false-positive rates related information is available with analysts based on a refinement of the traversals; this helps create vague traversals in the discovery phase of the analysis process, providing a specific traversal for the vulnerabilities that are correctly understood. Some of the applications of control flow graphs are seen in a security context where known malicious variants are explored (Gascon, et al., 2013) to guide fuzz testing tool (Sparks et al., 2007). Software testing concepts are used for vulnerabilities exploration applications with dynamic program analysis (Sutton, et al., 2007; Woo et al., 2013). Research around security explores approaches where analysts get enablement for their auditing purposes where static program analysis and expert knowledge are combined for vulnerabilities exploration (Vanegue et al., 2013).

Work in (Nguyen et al., 2010) proposes the prediction of vulnerabilities in the software components with dependency graphs. Prediction of vulnerabilities and security metrics has been on top of the list in the software community. A wide variety of software security metrics has been proposed in cohesion, coupling, and complexity metrics. Vulnerabilities prediction with the unique code metrics with the background of dependency graphs is explored in this work. The JavaScript engine of Firefox is used as a base for prediction modeling where efficiency validation is carried out. Good results shown in the study provides confidence that dependency graphs exhibit useful early vulnerabilities identification possibilities. New models proposed for vulnerabilities prediction helps to optimize testing efforts. Metrics generated from component dependency graphs of a software system are used to apply the machine learning technique. The quality of the prediction is demonstrated by experiments done on Mozilla Firefox 1.5 and 2.0. Good outcome on recall and accuracy shows that component dependency graphs are an excellent possibility in vulnerabilities prediction. The opportunity to build component dependency graphs at the design phase makes this approach identify the issues right at the beginning of the software system building.

Further exploration of this work is planned towards using nesting level metrics to make a better comparison with all the metrics. The author also plans to extend the experiments on various versions of the code base of Firefox. Extension of the metrics list for improving the quality of the graphs is another area of exploration. Typical issues with prediction modeling in this area are around the classification of the problems into vulnerabilities and defectives. The next area is the likelihood of vulnerabilities among the software components (Zimmermann et al., 2009). Different vulnerabilities prediction approaches extend from neural networks, case-based reasoning, genetic programming, statistic, and machine learning-based approaches (Catal et al., 2009). Earlier work in (Shin, 2008; Shin et al., 2008a; Shin et al., 2008b) on analysis of Java script engine has provided the base for authors to compare their work. The quality of the vulnerabilities database has a significant influence on the outcomes of this work. Bugs in the data collected were required to be eliminated with the proper sample checking approach.

FUTURE RESEARCH DIRECTIONS

There is a need to focus on the complex reasoning capabilities to be built into knowledge graphs. The need for continuous vector space to represent semantic associated with entities and relations will be critical. Reinforcement learning-based exploration will provide some light to solve the challenge. Interpretation of the AI solutions taking its priority, it holds equal importance for knowledge graph representation also. Large scale representation of the knowledge calls for the scalability of the models. Pre-trained models have enhanced knowledge aggregation capabilities, particularly the field of NLP (Natural Language Processing) has been benefiting from this. With the growing focus on the complexity of the knowledge graph, there is a need to focus on cutting down the manual intervention needed to build them.

Constraints around the knowledge lapse over time can be balanced with the dynamic knowledge graphs with the learning algorithm's capability that can capture the dynamics over time. Knowledge graphs will have a significant contribution in the area of explainable AI. Integration of the knowledge graphs needs to be focused on leveraging the ability to combine the knowledge sources. Knowledge graphs that capture semantic information form the center of innovations in the field of data-oriented products. Knowledge graphs have elevated online presence, and now it's time for extending them to real world applications. One of the software development industries' prominent expectations would be to leverage the vast amount of data hidden across the software development eco-system. Enhancing the customer experience and efficiency of the processes will be at the forefront as the software engineering experts explore knowledge graphs. There is a need for continuous information exchange and inferencing for constant visibility of the product in the market. This aspect will play a crucial role in deciding the proactiveness of the software development community. IoT (Internet of Things) presents great possibilities for improving the cognitive capabilities of software systems. Collaboration between business and technology partners is a focus area that can be enhanced with the knowledge's collaboration. Knowledge graph being the knowledge aggregator at its core, this technology business collaboration can be focused upon. Too much of the experts' dependency has not always helped the software engineering processes; knowledge graphs coming into rescue will ease these dependencies' load. Re-usability of the software components and knowledge has been an area of interest and can be served well with this focus. Optimization of the coding practices in software development can be assisted with knowledge graph capabilities. Code is the most natural and pattern-based entity beyond the human language. It presents the ability to make smart coding a reality cutting down on the manual effort involved, and possibilities of quality issues creeping into the system will be reduced.

CONCLUSION

Various areas explored in this domain of applying knowledge graphs in software engineering lead the software engineering industry to solve some of the long-standing issues. Consolidation of knowledge has been a primary aspiration of the industry, and knowledge graphs are the promising way ahead. Work reviewed here covers an exciting area of knowledge graphs. Exploring the ontologies inherent within the system and building a knowledge architecture is the right place surrounded. Validation of the new information coming into the system with the background of historical data will build confidence. For the organizations, fast-paced learning to put together a continuously learning system will help develop an active knowledge pool. Information extraction is further extended to pattern matching, statistical learning, and natural language parsing areas of applications.

Leveraging the knowledge systems' capabilities and the processes expert and balancing their capabilities is the right focus area. A large part of the software development community's aspiration can be addressed if the focus is to build a smart Integrated Development Environment that can leverage the knowledge extracted from the data that it handles. Network analysis thought process could be handy to integrate the knowledge scattered across various software development stakeholders like customers, technical team, industry peers, industry experts, and so on. The clustering algorithm enhances the takeaway from the knowledge graph by providing a capability to understand various characteristics within the knowledge graph area. Pattern recognition through graph matching is a critical area that brings together the useful features of both the area of knowledge graph and pattern recognition. There is a need to devise methods that help to tackle the complexity involved in the knowledge graph. This aspect can ease the domain's exploration without being constrained by the knowledge graphs' overwhelming complexity. Software testing and secure software development areas can explore some of the knowledge graph capabilities to optimize the current processes involved. Secured software systems are a significant concern for the software development community. However, a tremendous amount of knowledge lying around using the right information at the right point of time is a concern and needs attention. Knowledge data available on software systems' security is not a straightforward source to learn from; there is a need to build knowledge graph capabilities to validate these data sources with other alternate data sources and make them relevant for users. Software engineering and data science fields can work closer to support aspirations of each other and move towards a promising future that is technology-enabled. Knowledge being the core foundation of software engineering, knowledge graphs promises to be a hope to reduce the low-profile activities and help the community to focus on the more creative part of their work.

REFERENCES

Aggarwal, C. C., & Wang, H. (2010). A survey of clustering algorithms for graph data. In *Managing and mining graph data* (pp. 275–301). Springer. doi:10.1007/978-1-4419-6045-0_9

Agichtein, E., & Gravano, L. (2000, June). Snowball: Extracting relations from large plain-text collections. In *Proceedings of the fifth ACM conference on Digital libraries* (pp. 85-94). 10.1145/336597.336644

Angeli, G., Premkumar, M. J. J., & Manning, C. D. (2015, July). Leveraging linguistic structure for open domain information extraction. In *Proceedings of the 53rd Annual Meeting of the Association for Computational Linguistics and the 7th International Joint Conference on Natural Language Processing*:Vol. 1. *Long Papers* (pp. 344-354). 10.3115/v1/P15-1034

Auer, S., Bizer, C., Kobilarov, G., Lehmann, J., Cyganiak, R., & Ives, Z. (2007). Dbpedia: A nucleus for a web of open data. In *The semantic web* (pp. 722–735). Springer. doi:10.1007/978-3-540-76298-0_52

Auwatanamongkol, S. (2007). Inexact graph matching using a genetic algorithm for image recognition. *Pattern Recognition Letters, 28*(12), 1428–1437. doi:10.1016/j.patrec.2007.02.013

Bagdanov, A. D., & Worring, M. (2003). First order Gaussian graphs for efficient structure classification. *Pattern Recognition, 36*(6), 1311–1324. doi:10.1016/S0031-3203(02)00227-3

Bai, X., & Latecki, L. J. (2008). Path similarity skeleton graph matching. *IEEE Transactions on Pattern Analysis and Machine Intelligence, 30*(7), 1282–1292. doi:10.1109/TPAMI.2007.70769 PMID:18550909

Banko, M., & Etzioni, O. (2007, October). Strategies for lifelong knowledge extraction from the web. In *Proceedings of the 4th international conference on Knowledge capture* (pp. 95-102). 10.1145/1298406.1298425

Basili, V. R., Briand, L. C., & Melo, W. L. (1996). A validation of object-oriented design metrics as quality indicators. *IEEE Transactions on Software Engineering, 22*(10), 751–761. doi:10.1109/32.544352

Bengoetxea, E., Larranaga, P., Bloch, I., Perchant, A., & Boeres, C. (2002). Inexact graph matching by means of estimation of distribution algorithms. *Pattern Recognition, 35*(12), 2867–2880. doi:10.1016/S0031-3203(01)00232-1

Bonabeau, E. (2002). Graph multidimensional scaling with self-organizing maps. *Information Sciences, 143*(1-4), 159–180. doi:10.1016/S0020-0255(02)00191-3

Broder, A. Z., Glassman, S. C., Manasse, M. S., & Zweig, G. (1997). Syntactic clustering of the web. *Computer Networks and ISDN Systems, 29*(8-13), 1157–1166. doi:10.1016/S0169-7552(97)00031-7

Cafarella, M. J., Downey, D., Soderland, S., & Etzioni, O. (2005, October). Knowitnow: Fast, scalable information extraction from the web. In *Proceedings of Human Language Technology Conference and Conference on Empirical Methods in Natural Language Processing* (pp. 563-570). 10.3115/1220575.1220646

Carlson, A., Betteridge, J., Kisiel, B., Settles, B., Hruschka, E., & Mitchell, T. (2010, July). Toward an architecture for never-ending language learning. *Proceedings of the AAAI Conference on Artificial Intelligence, 24*(1), ●●●.

Catal, C., & Diri, B. (2009). A systematic review of software fault prediction studies. *Expert Systems with Applications, 36*(4), 7346–7354. doi:10.1016/j.eswa.2008.10.027

Cook, D. J., & Holder, L. B. (Eds.). (2006). *Mining graph data.* John Wiley & Sons. doi:10.1002/0470073047

Cunningham, H. (2002). GATE: A framework and graphical development environment for robust NLP tools and applications. In *Proc. 40th annual meeting of the association for computational linguistics (ACL 2002)* (pp. 168-175). Academic Press.

de Graaf, K. A., Tang, A., Liang, P., & van Vliet, H. (2012). Ontology-based software architecture documentation In *Proceedings of Joint Working Conference on Software Architecture & 6th European Conference on Software Architecture (WICSA/ECSA). WICSA 2012.* IEEE Computer Society. 10.1109/WICSA-ECSA.212.20

Del Corro, L., & Gemulla, R. (2013, May). Clausie: clause-based open information extraction. In *Proceedings of the 22nd international conference on World Wide Web* (pp. 355-366). 10.1145/2488388.2488420

Erman, L. D., Hayes-Roth, F., Lesser, V. R., & Reddy, D. R. (1980). The Hearsay-II speech-understanding system: Integrating knowledge to resolve uncertainty. *ACM Computing Surveys, 12*(2), 213–253. doi:10.1145/356810.356816

Etzioni, O., Cafarella, M., Downey, D., Kok, S., Popescu, A. M., Shaked, T., ... Yates, A. (2004, May). Web-scale information extraction in knowitall: (preliminary results). In *Proceedings of the 13th international conference on World Wide Web* (pp. 100-110). 10.1145/988672.988687

Fabian, M. S., Gjergji, K., & Gerhard, W. E. I. K. U. M. (2007). Yago: A core of semantic knowledge unifying wordnet and wikipedia. In *16th International World Wide Web Conference, WWW* (pp. 697-706). Academic Press.

Faruqui, M., Dodge, J., Jauhar, S. K., Dyer, C., Hovy, E., & Smith, N. A. (2014). *Retrofitting word vectors to semantic lexicons.* arXiv preprint arXiv:1411.4166.

Foggia, P., Percannella, G., & Vento, M. (2014). Graph matching and learning in pattern recognition in the last 10 years. *International Journal of Pattern Recognition and Artificial Intelligence, 28*(01), 1450001. doi:10.1142/S0218001414500013

Gascon, H., Yamaguchi, F., Arp, D., & Rieck, K. (2013, November). Structural detection of android malware using embedded call graphs. In *Proceedings of the 2013 ACM workshop on Artificial intelligence and security* (pp. 45-54). 10.1145/2517312.2517315

Ghosh, R. A. (2003). Clustering and dependencies in free/open source software development: Methodology and tools. *First Monday, 8*(4). Advance online publication. doi:10.5210/fm.v8i4.1041

Gibson, D., Kumar, R., & Tomkins, A. (2005, August). Discovering large dense subgraphs in massive graphs. In *Proceedings of the 31st international conference on Very large data bases* (pp. 721-732). Academic Press.

Jain, A. K., & Dubes, R. C. (1988). *Algorithms for clustering data.* Prentice-Hall, Inc.

Kaiser, G. E., Feiler, P. H., & Popovich, S. S. (1988). Intelligent assistance for software development and maintenance. *IEEE Software, 5*(3), 40–49. doi:10.1109/52.2023

Kruchten, P., Lago, P., & Van Vliet, H. (2006, June). Building up and reasoning about architectural knowledge. In *International conference on the quality of software architectures* (pp. 43-58). Springer. 10.1007/11921998_8

Lee, M. L., Yang, L. H., Hsu, W., & Yang, X. (2002, November). XClust: clustering XML schemas for effective integration. In *Proceedings of the eleventh international conference on Information and knowledge management* (pp. 292-299). 10.1145/584792.584841

Lenat, D. B., & Guha, R. V. (1989). *Building large knowledge-based systems; representation and inference in the Cyc project.* Addison-Wesley Longman Publishing Co., Inc.

Lin, Z. Q., Xie, B., Zou, Y. Z., Zhao, J. F., Li, X. D., Wei, J., Sun, H.-L., & Yin, G. (2017). Intelligent development environment and software knowledge graph. *Journal of Computer Science and Technology, 32*(2), 242–249. doi:10.100711390-017-1718-y

López, C., Codocedo, V., Astudillo, H., & Cysneiros, L. M. (2012). Bridging the gap between software architecture rationale formalisms and actual architecture documents: An ontology-driven approach. *Science of Computer Programming*, 77(1), 66–80. doi:10.1016/j.scico.2010.06.009

Mitchell, T. M., Allen, J., Chalasani, P., Cheng, J., Etzioni, O., Ringuette, M., &Schlimmer, J. C. (1991). Theo: A framework for self-improving systems. *Architectures for Intelligence*, 323-355.

Nagappan, N., Ball, T., & Zeller, A. (2006, May). Mining metrics to predict component failures. In *Proceedings of the 28th international conference on Software engineering* (pp. 452-461). 10.1145/1134285.1134349

Nayak, A., Kesri, V., & Dubey, R. K. (2020). Knowledge graph based automated generation of test cases in software engineering. In *Proceedings of the 7th ACM IKDD CoDS and 25th COMAD* (pp. 289-295). 10.1145/3371158.3371202

Nguyen, V. H., & Tran, L. M. S. (2010, September). Predicting vulnerable software components with dependency graphs. In *Proceedings of the 6th International Workshop on Security Measurements and Metrics* (pp. 1-8). 10.1145/1853919.1853923

Raychev, V., Vechev, M., & Yahav, E. (2014, June). Code completion with statistical language models. In *Proceedings of the 35th ACM SIGPLAN Conference on Programming Language Design and Implementation* (pp. 419-428). 10.1145/2594291.2594321

Robillard, M., Walker, R., & Zimmermann, T. (2009). Recommendation systems for software engineering. *IEEE Software*, 27(4), 80–86. doi:10.1109/MS.2009.161

Sabou, M., Ekaputra, F. J., Ionescu, T., Musil, J., Schall, D., Haller, K., ... Biffl, S. (2018, June). Exploring enterprise knowledge graphs: A use case in software engineering. In *European Semantic Web Conference* (pp. 560-575). Springer. 10.1007/978-3-319-93417-4_36

Shin, Y. (2008, October). Exploring complexity metrics as indicators of software vulnerability. *Proceedings of the 3rd International Doctoral Symposium on Empirical Software Engineering*.

Shin, Y., & Williams, L. (2008a, October). An empirical model to predict security vulnerabilities using code complexity metrics. In *Proceedings of the Second ACM-IEEE international symposium on Empirical software engineering and measurement* (pp. 315-317). 10.1145/1414004.1414065

Shin, Y., & Williams, L. (2008b, October). Is complexity really the enemy of software security? In *Proceedings of the 4th ACM workshop on Quality of protection* (pp. 47-50). 10.1145/1456362.1456372

Singh, P., Lin, T., Mueller, E. T., Lim, G., Perkins, T., & Zhu, W. L. (2002, October). Open mind common sense: Knowledge acquisition from the general public. In *OTM Confederated International Conferences On the Move to Meaningful Internet Systems* (pp. 1223-1237). Springer.

Singhal, A. (2012). *Introducing the knowledge graph: things, not strings*. Official Google Blog.

Sparks, S., Embleton, S., Cunningham, R., & Zou, C. (2007, December). Automated vulnerability analysis: Leveraging control flow for evolutionary input crafting. In *Twenty-Third Annual Computer Security Applications Conference (ACSAC 2007)* (pp. 477-486). IEEE. 10.1109/ACSAC.2007.27

Speer, R., Chin, J., & Havasi, C. (2017, February). Conceptnet 5.5: An open multilingual graph of general knowledge. *Proceedings of the AAAI Conference on Artificial Intelligence, 31*(1).

Subramanyam, R., & Krishnan, M. S. (2003). Empirical analysis of ck metrics for object-oriented design complexity: Implications for software defects. *IEEE Transactions on Software Engineering, 29*(4), 297–310. doi:10.1109/TSE.2003.1191795

Sutton, M., Greene, A., & Amini, P. (2007). *Fuzzing: brute force vulnerability discovery*. Pearson Education.

Tang, A., Liang, P., & Van Vliet, H. (2011, June). Software architecture documentation: The road ahead. In *2011 Ninth Working IEEE/IFIP Conference on Software Architecture* (pp. 252-255). IEEE. 10.1109/WICSA.2011.40

Thrun, S., & Mitchell, T. M. (1995). Lifelong robot learning. *Robotics and Autonomous Systems, 15*(1-2), 25–46. doi:10.1016/0921-8890(95)00004-Y

Trautsch, F., Herbold, S., Makedonski, P., & Grabowski, J. (2016, May). Adressing problems with external validity of repository mining studies through a smart data platform. In *Proceedings of the 13th International Conference on Mining Software Repositories* (pp. 97-108). 10.1145/2901739.2901753

Vanegue, J., & Lahiri, S. K. (2013, May). Towards practical reactive security audit using extended static checkers. In *2013 IEEE Symposium on Security and Privacy* (pp. 33-47). IEEE. 10.1109/SP.2013.12

Woo, M., Cha, S. K., Gottlieb, S., & Brumley, D. (2013, November). Scheduling black-box mutational fuzzing. In *Proceedings of the 2013 ACM SIGSAC conference on Computer & communications security* (pp. 511-522). 10.1145/2508859.2516736

Yamaguchi, F., Golde, N., Arp, D., & Rieck, K. (2014, May). Modeling and discovering vulnerabilities with code property graphs. In *2014 IEEE Symposium on Security and Privacy* (pp. 590-604). IEEE. 10.1109/SP.2014.44

Yarowsky, D. (1995, June). Unsupervised word sense disambiguation rivaling supervised methods. In *33rd annual meeting of the association for computational linguistics* (pp. 189-196). 10.3115/981658.981684

Ye, X., Bunescu, R., & Liu, C. (2014, November). Learning to rank relevant files for bug reports using domain knowledge. In *Proceedings of the 22nd ACM SIGSOFT International Symposium on Foundations of Software Engineering* (pp. 689-699). 10.1145/2635868.2635874

Zimmermann, T., & Nagappan, N. (2008, May). Predicting defects using network analysis on dependency graphs. In *Proceedings of the 30th international conference on Software engineering* (pp. 531-540). 10.1145/1368088.1368161

Zimmermann, T., & Nagappan, N. (2009, October). Predicting defects with program dependencies. In *2009 3rd international symposium on empirical software engineering and measurement* (pp. 435-438). IEEE. 10.1109/ESEM.2009.5316024

Chapter 11
A Comprehensive Study on Artificial Intelligence and Robotics for Machine Intelligence

Nagadevi Darapureddy
Chaitanya Bharathi Institute of Technology, Hyderabad, India

Muralidhar Kurni
 https://orcid.org/0000-0002-3324-893X
Anantha Lakshmi Institute of Technology and Sciences, Ananthapuramu, India

Saritha K.
Sri Venkateswara Degree and PG College, Ananthapuram, India

ABSTRACT

Artificial intelligence (AI) refers to science-generating devices with functions like reasoning, thinking, learning, and planning. A robot is an intelligent artificial machine capable of sensing and interacting with its environment utilizing integrated sensors or computer vision. In the present day, AI has become a more familiar presence in robotic resolutions, introducing flexibility and learning capabilities. A robot with AI provides new opportunities for industries to produce work safer, save valuable time, and increase productivity. Economic impact assessment and awareness of the social, legal, and ethical problems of robotics and AI are essential to optimize the advantages of these innovations while minimizing adverse effects. The impact of AI and robots affects healthcare, manufacturing, transport, and jobs in logistics, security, retail, agri-food, and construction. The chapter outlines the vision of AI, robot's timeline, highlighting robot's limitations, hence embedding AI to robotic real-world applications to get an optimized solution.

DOI: 10.4018/978-1-7998-7701-1.ch011

INTRODUCTION

Artificial Intelligence (AI), the field of science machinery development, can perform logical thinking, reasoning, learning, and planning. It means that machines are intelligent and conductive (Advani, 2021). A robot, an intelligent artificial machine, can sense and communicate with its environment using integrated sensors or computer vision (Perez et al., 2018). AI gives the robot a vision of the computer to explore, feel, and measure the reaction (Signorelli, 2018).

AI and Robotics are an enticing mix of automated activities (Wisskirchen et al., 2017). AI and Robotics affect healthcare, fabrication, transportation, and employment in logistics, protection, retail, agri-food, and construction (Smith & Anderson, 2014). Robots have penetrated our professional and personal lives variously, including manufacturing (assembly units), medicine (assisting surgeons), music (diligent orchestras), radiology (cancer detection), comedy (including bootstraps), restaurants (gourmet meals), military preparation, etc. (Schatsky& Ream, 2017). This section explains the Vision of AI as the timeline for robotics illustrates the shortcomings of robotics and then incorporates AI into real-world robotic applications for an integrated solution.

Origin

John McCarthy first invented the term artificial intelligence at the Dartmouth conference of 1956. However, the journey to understand if machines would truly think started a lot earlier. In 1945, Vannevar Bush published an article entitled "As we may think," proposing a framework that improves individuals' awareness and knowledge. A British mathematician, Alan Turing, wrote a paper in 1950 that machines can simulate like human beings and do intellectual things like chess (Benko & Sik Lányi, 2011).

Artificial intelligence (AI) is changing our planet. AI is an emerging area, with many AI technology evolving and improving with research and development, taking place in terms of a better technique for overcoming some of its technological constraints. It is a collection of problem-solving abilities to solve real problems by supplying intelligence through artificial technology to an entity created by human beings; not that happens naturally. It is an innovation in technology for the innovation of artificially produced machinery that can exhibit behavior like human beings without utilizing any living organism (OECD, 2019).

Types

AI can be categorized in many respects (see Figure 1). The two most common ways rely on their ability and functionality (Joshi, 2019), (Carroll, 2017).

1. Functionality-based:

    a. *Reactive machines*: The most ancient types of AI systems that do not have memory-based functionality. It means that they cannot learn. They are, therefore, reactive, responding only to present conditions. IBM Deep Blue is a reactive machine example that defeated Chess Grandmaster Kasparov.

    b. *Limited memory*: Limited memory machines can hold and use data to supplement their experience collection for a short time. Current forms of restricted memory are autonomous vehicles or self-driving cars. Self-driving cars can read and detect patterns or changes of external factors and, if necessary, modify their environment.

c. *Theory of mind*: Certain computers, such as voice assistants, have human skills but cannot conversing on principles such as human feelings, desires, memories, and mental models that affect their actions. Researchers of the Theory of Mind hope to construct human emotion-related computers and interpret human intelligence. Thus, the thinking process influences the "theory of mind" computers. Social interaction is an integral part of human interaction. To stabilize the theory of mind machines, the AI systems that regulate the now hypothesized machines should detect, recognize, retain, and remember emotional performance and actions while knowing how to respond. The robots Kismet and Sophia, developed in 2000 and 2016, are two remarkable examples.

d. *Self-awareness:* This is the final stage of AI production, which is only present today hypothetically. Self-awareness AI requires human-informed machines. This type of AI is not currently available but is considered man's most advanced artificial intelligence. Facets of AI include identifying and reproducing human behavior and thought the desired, and the perception of feelings. Self-aware AI is the promotion and extension of the theory of mind AI.

Figure 1. Types of AI

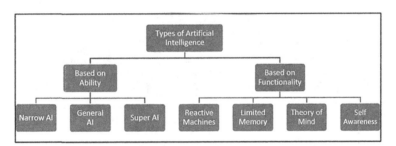

2. Based on Ability:

a. *Artificial Narrow Intelligence (ANI) / Weak AI / Narrow AI:* Narrow AI are the systems that can only carry out a particular mission with human capacity. These machines have very minimal or little power, so only as planned can they do. Narrow AI is programmed to perform specific tasks like facial recognition, speech acknowledgment, car driving assistants, or Internet search. Narrow AI can be either memory-restricted or reactive.

b. *Artificial General Intelligence (AGI) / Strong AI / Deep AI:* Machines that simulate human intelligence and general intelligence behaviors can learn and use their intellect to solve any problem. AI scientists and researchers have not yet produced solid AI. To succeed, they must find a way to understand machines and program different techniques of cognition. Fujitsu-built K is one of the fastest supercomputers and a major effort to achieve a strong AI, but only one second of a neural operation took 40 minutes to simulate, which is difficult to see whether or not strong AI has been achieved.

c. *Artificial superintelligence:* ASI is where machines are self-aware and transcend human intellect and capabilities. AI evolved to be so similar to human feelings and perceptions that the idea of artificial superintelligence does not understand them only; it evokes emotions, needs, values, and desires. ASI will be more memorable and more readily able to process and evaluate information and stimuli. Consequently, super-intelligent beings' decision-making and problem-solving capacities will be far superior to human beings. The potential for such powerful machines at our end may seem enticing, but there are many unknown implications for the idea itself. If super-intelligent self-conscious people were, they would be able to have ideas like self-preservation. Pure speculation is the effect and effects this would have on our way of life, humanity, and survival.

Subfields

Artificial intelligence can be used by implementing the following (see Figure 2) significant processes/ techniques to solve real-world problems (Seetharam et al., 2020):

Figure 2. Subfields of AI

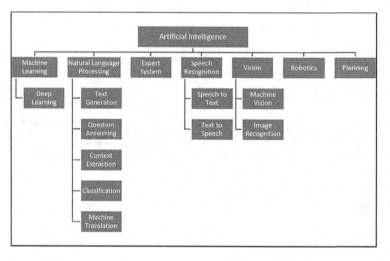

- *Machine learning:* The knowledge of computers interpreting, processing, and analyzing data to solve real-world problems.
- *Natural Language Processing:* Refers to the science of insight into the natural language of humans for communicating with computers and developing companies.
- *Expert systems:* An expert system is an AI computer system that learns and matches decision-making knowledge of human skills.
- *Speech recognition:* Spoken recognition is a technology that enables a machine to understand and translate the spoken language into a machine-readable format. It can also be said as automatic recognition of speech and computer speech. It is a way of talking to a machine and using it; a machine may perform a particular task based on a command.

- *Computer Vision:* It is a key aspect of artificial intelligence since it allows the computer to perceive, evaluate, and interpret visual data from images and visuals in the real world by collecting and intercepting them automatically.
- *Robotics:* It is an artificial intelligence branch focused on various branches and applications of robots. AI Robots are autonomous agents working in the real world to achieve results by accountable actions.
- *Planing:* Expert systems may also be used to plan and schedule those complex activities to accomplish the mission's purpose.

DEEP AI

AI uses data in a basic form, uses algorithms for data, and predicts performance. Data may be pictures, phrases, handwritten images, words, or numbers. A computer program is written by people who contain rules applied to predict data that allow machines to recognize handwriting. This form of AI is used in U.S. postal services to read emails automatically. This AI-based on rules works well for narrow applications but is difficult for complex applications or tasks. People cannot program them into algorithms if the rules cannot be described. Machine Learning in AI, machine construct either fully autonomously their rules or algorithms (unsupervised machine learning) or using human assistance (supervised or semi-supervised machine learning). In machine learning, we call it deep AI or deep learning (Marr & Ward, 2019) when we use multiple layers of artificial neural networks to learn from data. The two reasons why profound learning prospers today are,

1. We have data: data is the ingredient that fuels AI. It is gradually surrounded by intelligent machines that collect and distribute data due to our world's digitization.
2. We have computing power: Now, vast quantities of information can be stored and processed. Computing on devices such as mobile phones or other mobile-connected devices may now be done.

"Learning by doing" can now be replicated through reinforcement learning through machine learning algorithms. AI uses reinforcement learning algorithms to evaluate an optimal behavior based on environmental input that robots can autonomously walk, drive, or ride.

ROBOTICS

Robot an autonomous machine that substitutes for human effort (Moravec, 2021). It looks like human beings or performs tasks like human beings (Marr, 2020). Robotics is the engineering discipline involved in building, design, and service (Iborra et al., 2009). The term robotics was first published in the science fiction story of Isaac Asimov around 1942 (Wars et al., 2014).

CLASSIFICATION

Robots can be categorized (see Figure 3) based on their drive technology, kinematical structure, degrees of freedom, workshop geometry, and movement characteristics (enggmechanical.blogspot.com, 2010).

1.   Classification by Degrees of Freedom
2.   Classification by Kinematic Structure
3.   Classification by Drive Technology
4.   Classification by Workspace Geometry
5.   Classification by Motion Characteristics

Figure 3. Classification of robots

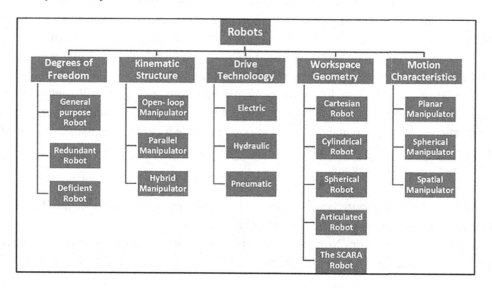

1.   ***Classification by Degrees Freedom:*** A robot is graded by degrees of freedom based on its freedom degrees. It should have 6 degrees of freedom in three-dimensional areas to freely manipulate an object. From this viewpoint, a robot might be a

   a.   *General-purpose robot*: These have 6 degrees of freedom.

   b.   *Redundant robot*: These have more than 6 degrees of freedom, capable of moving around obstacles. It works in an enclosed workplace.

   c.   *Deficient robot*: These have fewer than 6 degrees of freedom.

2.   ***Classification by kinematic structure:*** robots can be categorized as

   a.   *Open-loop Manipulator or serial robot*: These robots have their kinematic structure in the form of an open-loop chain. Example: Adept-One Robot.

   b.   *Parallel Manipulator*: These robots have a kinematic structure in the form of a closed-loop chain with greater rigidity, higher payload, higher stiffness, lower inertia. Example: Delta Robot.

    c.   *Hybrid Manipulator:* These robots are a combination of both serial and parallel manipulators. Example: S-900 W. Example. This type of robot is used in much industrial robotics.

3.   ***Classification by drive technology:*** manipulators are listed as drive technology

    a.   ***Electric:*** Electric AC servomotors or DC stepper motors are used in industrial manipulators.

    b.   ***Hydraulic:*** These manipulators are utilized for high speed and load transport. The risk of leaked oils is a significant drawback. Due to the bulk oil module, a hydraulic drive is inherently versatile.

    c.   ***Pneumatic:*** A pneumatic drive is clean and quick; while air is a compressible fluid, it is not easy to regulate it. They are used both for high-speed and high-load capacity transport.

4.   ***Classification by workspace geometry:*** The volume of space where the end of the effecter can reach is called the manipulator's workspace. There are two types of workspace:

    a.   Reachable working space is the space volume inside which each position can reach at least one orientation by the end.

    b.   A dexterous workspace is the space volume in which the final result can reach any orientation point. Geometry manipulators are classified according to the workplace as follows.

        i.   *Cartesian robot:* In a Cartesian robot, the robot arm's kinematic arrangement involves three prismatic joints that are mutually perpendicular. A cartesian robot's wrist position can be easily represented by three Cartesian coordinates associated with the three prismatic joints. The cartesian robot's local workspace is a rectangular box.

        ii.   *Cylindrical robot:* When the first or second joint of a cartesian robot is replaced with a revolute joint, it is considered a cylindrical robot. The cylindrical robot's center position is associated with three joint variable coordinate systems. Two concentrated cylinders of finite length comprise a workspace.

        iii.   *Spherical robot:* a robotic arm is referred to as a spherical robot if a spherical robot's wrist center position can be represented by a set of three joint variables of the spherical coordination system. Two concentrated spheres contain a cylindrical robot's working space.

        iv.   *Articulated robot:* In this robot, all three joints are revolute. The working area of this is very complicated, usually a cross-section with a crescent shape. Example: Puma robot.

        v.   *Robot SCARA (Selective Compliance Assembly Robot Arm):* A special robot consisting of two rotating joints and a prismatic joint. All joint axes are parallel and typically pointed towards gravity direction. The arm is free to a degree, so the whole robot has 4 degrees of freedom, useful for mounting components on a plane.

5.   ***Classification by Motion Characteristics:*** According to their design, manipulators categorized as

    a.   *Planar Manipulator:* It is named for its planer mechanism. They are used to manipulate an entity.

    b.   *Spherical Manipulator:* It is named for its spherical mechanism. The rigid body is in sphere movement if all body particles describe curves that lie in concentration. It is said that a mechanism is a spherical mechanism if all moving connections travel spherically around a common stationary point.

    c.   *Spatial Manipulator:* Named if at least one of the system's moving relations possesses a general spatial movement that is not described as a flat or spherical movement. In spatial mechanism instances, planer and spherical systems are considered exceptional.

PAST AND PRESENT

Some of the notable evolutions of AI & robots are as follows (Computer History Museum, n.d.)

- **1939** – The world's first robot, 'Elektro and Sparko', was introduced.
- **1941** – The term 'Robot' was used in the article 'Astounding Science Fiction' by Isaac Asimov, and three laws of robotics were introduced.
- **1948** – Norbert Wiener Cybernetics: for the first time, the term Cybernetic and possible integration of artificial intelligence and control systems were coined.
- **1950** – Grey Walter created biological robots 'Elsie and Elmer', the first two tortoises.
- **1950** – 'Imitation game' Alan Turing's Test to check machine's ability towards intelligence with a human.
- **1951** – 'Robot squee' squirrel-inspired robot that uses light sensors to senses the environment.
- **1959**– Computer-aided manufacturing (CAM) to automate manufacturing using the software was developed at MIT.
- **1961** – Unimate: General Motors began to run the first commercial mass production robot.
- **1963** – Ranco Arm's robot with six joints resembles a human arm, is the first computer-controlled robotic arm designed at the Californian hospital and later acquired by Stanford University.
- **1963** – The 'Orm Robot' was an unusual snake inspired robot developed by a team from Stanford University
- **1969** – Stanford Robot Arm with electric power is the first computer-controlled robot. The arm with 6 joints was fitted with optical and touch sensors.
- **1970** - Tentacle Arm by Marvin Minsky built an octopus-inspired tentacle arm robot with more than 12 tentacles assisted the robot in overcoming obstacles.
- **1970** – Robot Shakey's first mobile robot can navigate the AI environment, which has an inbuilt TV camera, laser range finders, and bumps sensors communicating wirelessly through a radio link antenna.
- **1974** – Silver arm robot can do small parts assembly by using touch and pressure sensors. This was designed by David silver, an MIT student.
- **1978**-Talk and Spell: Texas Instruments Inc., the first commercially available human artificial speech device. The machine was able to carry out the human-track mathematical model and produce human sounds.
- **1979** – Stanford Cart: A highly advanced mobile robot can travel through a chamber without hindrances. It was able to take images from various perspectives of the environment.
- **1981** – Direct Drive Arm: Takeo Kanade implemented direct drive arm technology for the first time. This helped in analyzing the friction and reverse effect of previous robots using chains and tendons.
- **1982** – FRED robot: Atari engineering created the first playful robot. It has never been on the market.
- **1984** – Hero Robot Kit: First human playful robot. It could navigate a sonar-led room and stay close to people by listening to their voices.
- **1985** – 'Sentry Robot,' first indoor security robot created by Denning, like security guard patrol for 14 hours, warns of unusual activities within a range of 150 feet.

- **1987** – 'Mitsubishi Movemaster' the first industrial machine robot with an arm gripper to do small items or to handle chemicals
- **1989** – Machine defeats human beings at chess: first time machines beat humans at chess.
- **1989**–MQ-1 Predator is an aircraft used by the U.S. Air Force with remote control. It was equipped with cameras for identification and could handle rockets.
- **1992** – Japan's multimillion-dollar initiative to develop the fifth generation's AI-based computer systems.
- **1997** – An international competition, 'First Robocup,' encourages AI and robotics for football and other dexterity games by robots. Around 40 teams participated in this event.
- **1997** – Deep blue beats Garry Kasparov: After a re-match in 2016, Garry Kasparov was defeated by deep blue by 2-1.
- **1999** – 'Aibo Robot' is a robotic dog trained to learn and communicate with its surroundings. It can respond to over 100 voice commands.
- **2000**-Asimo Robot: Honda's Asimo introduced the first humanoid robot, which could move like human beings, climb stairs, change direction, and recognize risks through a video camera.
- **2000** – DaVinci Surgical System: The FDA has approved a minimally invasive (keyhole) operating robot. A surgeon from a master console operates the robot.
- **2002** – 'Roomba' robot vacuum cleaner with sensors could detect and avoid obstacles. It can navigate in a house with onboard navigation.
- **2002** – Centibots of Darpa: the first mobile robot collaborative swarm that could track the region and construct a real-time map without human surveillance.
- **2004**– "Mars Exploration Rover" was twin robots, Spirit and Opportunity, landed on Mars. The mission was planned for 90 days, but they spent many years extending their lives and remaining operational until today for gathering information about Mars.
- **2005** – 'Big Dog' quadruped military Robot created Boston Dynamics. It could cross rough terrains with its four legs and carries heavy loads.
- **2007** – Checkers have been solved: the University of Alberta's program named Chinook has solved checkers and defeated humans in different competitions.
- **2010** – IBM Watson: IBM Watson's robot on the Jeopardy game show defeats human champions! By studying nature and seeking responses to questions faster and more reliably than their human competitors
- **2010** – Robotnaut 2: NASA has demonstrated a humanoid robot that can replace human astronauts using a wide range of sensors.
- **2010** – 3D Printing robots are made commercially available in the printing industry to create parts cost-effectively.
- **2014** – Pepper, Softbank launched the first robot for customer service in Japan. It has an emotional engine integrated to communicate with customers.
- **2014** – Robotic exoskeleton: A completely paralyzed person could go again using an Ekso Bionics-designed robotic exoskeleton.
- **2016** – 'Microfluidic robot' is a soft robot with microfluidic logic created by a Harvard University team.
- **2016** – 'Nanobots' are tiny robots developed that can deliver drugs to the disease-affected area without damaging the tissues and organs surrounding it.

- **2017** – Hanson Robotics, a Hong Kong based corporation, creates Sophia, a social humanoid robot. Sophia "became" a Saudi Arabia citizen in October 2017, the first robot to receive any country's citizenship.
- **2018** – The Walker can be activated by voice or touch screen and serves as the first bipedal to reach the commercial market as a robotic butler.
- **2020** – The 4-year-old AI robot will continue to play a robotic ambassador function to improve research on robotics and human-robot interactions developed by Hanson Robotics, based in Hong Kong.

HUMAN-ROBOT INTERACTION (HRI)

HRI is a field that involves understanding, constructing, and testing robotic systems for human use. There are two types of human-robotic contact and interaction (D & M.J., 2009)(Goodrich & Schultz, 2007).

- *Remote Interaction:* In this category, robots and humans are not placed in the same position and are divided spatially and temporally.
- *Proximate Interaction:* In this category, robots and humans are placed in the same position.

LIMITATIONS

In the past, robots existed without AI. These are the limitations.

- Not innovative or creative
- Cannot think alone
- Unable to take complicated decisions
- Unable to learn from errors
- They cannot easily adjust to changes in their climate.

ARE AI AND ROBOTICS THE SAME THING?

Is AI part of robotics? Is robotics part of AI? How are the two words different? AI and robotics are almost totally different fields. Venn diagram of the two will look like in figure 4 (Balajee, 2020)(Owen-Hill, 2017).

Often, people confuse the two due to their overlap: Artificially Intelligent Robots. Let us look at each word separately to understand how these three words relate to each other.

As learned from the sections above, robotics is a technology industry that manages robots. Robots are programmable devices, which can typically carry out many activities independently or semi-autonomously.

Experts can shockingly hardly agree precisely what a "robot" constitutes. Some people believe a robot needs to be able to "think" and decide. There is, however, no traditional "robotic thought" concept. The need for a robot to "think" indicates a certain degree of artificial intelligence. But you want to describe a

Figure 4. Venn diagram of robotics and AI

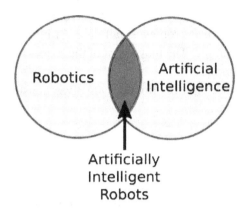

robot, and robotics includes physical robots design, construction, and programming. It includes artificial intelligence just a tiny portion (Balajee, 2020) (Owen-Hill, 2017).

Example of a robot: Basic cobot: A basic cobot is an example of an intelligent non-intelligent robot. For example, to pick an object up and put it somewhere, you can easily program a cobot. The cobot then keeps picking up and placing items in the same way before you turn them off. This is a self-sufficient feature because after being programmed, the robot doesn't need human input. The role requires no intellect because the cobot never changes what he does. Non-intelligent are most factory robots.

On the other hand, AI is a branch of computer science. It comprises the development of computer programs requiring understanding by the individual. AI algorithms can deal with learning, awareness, troubleshooting, logical reasoning. andlanguage understanding.

In today's world, AI is used in various ways like Google searches, the recommending Amazon engine, and the SatNavroute finders. Most AI programs are not part of robots' control.

While AI is used for powerful robots, including sensors, actuators, and AI programming (Balajee, 2020), AI algorithms only are part of the larger robotic system (Owen-Hill, 2017).

Example of a pure AI: AlphaGo: Games are among the most common examples of pure AI. Chess, where the AI Deep Blue beat Gary Kasparov in 1997, is the classic example. A more recent example is AlphaGo, an AI that in 2016 defeated Lee Sedol, the Go player world champion. AlphaGo did not have robotic parts. The play was moved by a person watching the robot move on the TV.

What are artificially intelligent robots? The bridge between Robotics and AI is an artificially intelligent robot (Balajee, 2020) (Owen-Hill, 2017). These are robots with AI controls. Many robots aren't smart artificially. All industrial robots can be programmed until recently only for repetitive movement, and repetitive movement does not require artificial intelligence.

In terms of functionality, non-smart robots are very constrained. To perform more complex tasks, AI algorithms are frequently required.

Example: Artificially Intelligent Cobot

You can increase a collaborative robot's capabilities by using AI (Balajee, 2020)(Owen-Hill, 2017). Imagine adding your cobot a camera. Robot vision is classified under the "perception" category, and normal AI algorithms are required. Tell that you wanted the cobot to detect the object it collected and put it elsewhere depending on the object type. This will require the preparation of a special vision program to identify the various object types. One way of doing this is to use a Template Matching AI algorithm.

The majority of artificially intelligent robots usually use AI in only one aspect. AI is used for the detection of objects only in our example. The movement of the robot is not AI since the object detector's performance affects their motions.

AI and robotics are two distinct entities. Robotics involves constructing robots, and AI involves intelligence programming.

AI TO ROBOTS

Artificial intelligence in robotics provides new market possibilities for improving efficiency and saving time (Raj & Seamans, 2019).

Where Robotics and AI Mix?

The distinction between robotic systems and artificial intelligence is robust, and people do not know the distinction between robotics and artificial intelligence is due to artificial intelligent robots (Khakurel et al., 2018). AI is, together, the brain, and robots are the body. You can program a simple robot so that an object can be picked up and placed somewhere until it has to stop. The robot can "see" an object, detect it, and determine where it should be placed with a camera and AI algorithm (Rubio et al., 2019). Artificially intelligent robots are very new. As research and development progress, we should expect artificially intelligent robots to resemble the humanoid characteristics we see in films.

Role of AI in Robotics

Important research is being conducted with AI to expand robot functionality. Commercially available robots provide the application of AI to (International Federation of Robotics, 2020):

- Allow robots to sense and react to their environment: This greatly increases the range of robotic functions.
- Robot performance enhancement, saving the company money.
- Allow robots to work in various environments, from public space, hospitals to retail outlets, and time-saving.

The key fields of AI research in robotics are (International Federation of Robotics, 2020):

- Expanded picking capability to work with objects that are not rigid or static.
- Extend robot mobility in non-standard conditions such as rugged terrain etc.
- Robotics control by verbal commands and gestures.
- Make programming robots easier. AI's application is ongoing research to enable robots to learn through video demonstrations and independent testing and error. Reducing robotic programming time and expense would make robot adoption by small and medium-sized businesses more efficient. Connected robots can also be studied together and operate as parallel computers effectively.

Intelligent Robot System

Perception is understood in robotics as a device for perceive, understand, and describe the environment by the robot (Premebida et al., 2016). In sensory information processing, environmental data representation and ML-based algorithms, as shown in Figure 5, are the basic components of a perception system since strong AI is still far from achieving in real-world robotics applications.

Main modules for a typical robotic perception system (Premebida et al., 2016):

- sensory data processing (visual and range perception focusing)
- data representatives related to tasks involved
- data analyzing and interpreting algorithms (AI / ML methods used)
- robot-environment-interaction behavior planning and implementation.

Figure 5. The essential components of a perception system

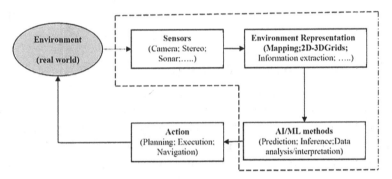

Robot perceptions are required to assess, plan and operate in real-world environments from occupancy grid mapping to object detection (Joo et al., 2020) through various functionalities and activities (Yan et al., 2020). Examples of robot perception subzones, including independent robot vehicles, are obstruction detection, object recognition, semantic position classification, 3D display, gesture and recognition of speech, activity classification, clustering of fields, path detection, detection of vehicles, pedestrian detection, tracking, and detection of human subjects (Garg et al., 2020). (Premebida et al., 2019).

Most systems use techniques from classical to an in-depth study of machine learning (ML). Machine learning for robotic perception may involve different approaches, including unsupervised learning, supervised classifications with handmade features, or trained neural networks, such as CNN.

Robotic Applications That Use AI

AI and robotics succeed in automating operations both inside and outside the factory. In the past few years, AI has been used in static applications for robotic solutions with mobility and learning skills. AI is still on the way in its progress, but many still feel; for several applications in the manufacturing sector.

There are a few key forms of AI deployment in today's global manufacturing market (Robotics Online Marketing Team, 2018).

1. Assembly: In robotic assembly applications, AI is a beneficial tool. AI can help real-time correction of courses, particularly useful in complex production areas such as aerospace in conjunction with advanced vision systems. AI can even enable a robot to find out which paths are the best for such operations.

2. Packaging: For faster, lower costs, and more precise packaging, robotic packaging also uses AI types. AI helps to maintain the motions that a robotic system produces by active refinement them, making it easier for others to mount and shift robotic systems.

3. Customer Service: Robots are also being used in customer care in hotels and retail stores worldwide. Most robots use AI to interact more humanely with customers. The higher the communication between these systems and people, the more they understand.

4. Open-Source Robotics: A few robotic systems with AI capability are available as open source. Therefore, users can teach their robots their duties, like small farming, according to their special applications. In the future, AI robots may be a significant trend in open-sources robotics and AI capability.

Robots are more intelligent, more precise, and more profitable when working together. AI also needs to get close to its full potential, but robotics can do so as it progresses.

Benefits of Robots and AI in Manufacturing

-   *Productivity optimization:* It is possible to program robotics equipment to run at a constant and optimum speed without delay. This means that automated machines generate more in a shorter time than human employees. Fast settings or issues can be resolved with remote administration. Automated devices are often highly scalable and can flexibly move between tasks, increasing the productivity of the output.

-   *Enhanced quality:* Intelligent machines delete human error and have almost 100% accuracy, improving production quality. This, in turn, increases consumer loyalty, as faulty goods are not accessible to end-users.

-   *Costs reduced:* After a robot can substitute hundreds of workers, factory owners saved on wages for employees. In two years, a fast ROI compensates the initial costs. Higher efficiency and speed are possible. The robots can operate in light-out conditions and do not require customer-saving climate control.

-   *Safety:*Many manufacturing workers entail elevated physical hazards and work in hazardous environments. Intelligent robots are replacing humans in dangerous conditions, mitigating accidents at the workplace, and reducing the harm to workers' health. High-risk industries such as mining and fertilizer manufacturing transform robots to prevent adverse accidents and guarantee their employees' protection.

LOOKING TO THE FUTURE

This section throws a lite on the robots of tomorrow (Schutten, 2014).

1. *Flying Cars:* The flying car's story is close to the history of space travel. In the beginning, people did not think about it, but experimental cars could drive. To fly, you need a pilot license and, above all, permission to access airspace so that air traffic cannot be disturbed or airplanes are threatened. Moreover, you cannot do anything with a Ford Flyer, Mercedes Boeing, Volkswagen copter, or Jet Jaguar. In fact, the flying car must function safely, efficiently, and ecologically on public roads and in the air. It also needs to fly without a professional pilot at the controls and competitive buying and operational costs for widespread adoption. There may be flying cars in future conceptions where air traffic is monitored safely by computers and robots. Cars are still going to undergo several improvements but in very different ways.

2. *Autopilot Cars or Robot Cars:* Why can't a car interfere at crucial moments? In this sense, the idea of autopilot vehicles was raised. The word 'autopilot' is used in the aircraft industry for decades, which helps pilots decrease cognitive stress by allowing a plane to fly practically during "cruising" portions of the journey. The same is true now that cars with autopilot are easily converted into autonomous cars, so drivers can allow cars to drive themselves along with some areas of the journey, such as highways. In 1995, a car was driving to Copenhagen in Denmark from Munich, Germany. It passed other cars and stopped at the crossroads for pedestrians.

3. Moreover, the human driver only had to take control in a few situations (5 percent of the whole journey). In 2010, about 8,000 miles from Italy to China were powered by four autopilot vans. The drivers had to take over a couple of times but arrived without a scratch in time. There are already buses that carry passengers without a driver to their destination. China tested the first automatic public buses on a highway between Zhengzhou and Kaifeng in Henan Province in 2015. At the end of 2019, Yandex announced that its self-driving cars in fully autonomous mode had spent 1 million miles in Russia, Israel, and the United States. Many major corporations and research organizations, including Mercedes-Benz, General Motor, Continental Automotive Systems, Bosch, Toyota, Nissan, Audi, Vislab, Volvo, University of Parma, Google, and Oxford University, have established independent operating vehicles. Since 2019, 29 states of the United States have passed legislation allowing autonomous cars. In Belgium, France, Italy, and the United Kingdom, Europe plans to operate driverless vehicles, while the robotic automotive testing was approved for Germany, the Netherlands, and Spain. The EU and Japan will also control autonomous cars in 2020.

4. *Robots in the Future:* The question now is: are we going to have a robot at home, but when and what are you going to do with it? It is going to be much fun. People had robots that do less for years, but they still loved them. When you can talk to your robot or play tennis with it, it will become fun. You should play with your robot if your dear ones do not want to play. These robots will not just be there as toys, however. A robot can be helpful for the elderly or disabled around the building. It will help you get out of bed, help you walk, and do all kinds of homework. It is not an excessive luxury. Right now, more people are working than older people, but that will change. It would be helpful for robots to take over the work of older people. Robots may function in the future at a hotel reception or in areas that require an information stand, such as tourist attractions and railway stations. Robots will do vital work such as leak nuclear power plants, firefighting, and skyscraper cleaning. They can also focus on human problems to get up in the atmosphere or deep in the ocean. They make cars, buses, and trains safer than humans. They will also do some "boring" jobs, such as night watch managing or safeguarding a hospital and other areas where nothing regulates people in museums.

5. *Nanobots:* Robots are not only becoming smarter and better; they are becoming much smaller. Some robots are currently so small that you cannot see with your naked eye. These "nanobots" do not look like in film and books, but they can do what other robots do. Scientists work on nanobots, which can search for and free from side effects for our bodies' cancer cells. The researchers are currently working with biotech companies to cheapen the nanobots that combat cancer. Nanobots are robots that perform a specific function and are about 50–100 nm wide. They are very efficiently used in disease-affected parts for medicines.

CASE STUDY: HOW COMPANIES USING AI IN ROBOTICS

For years it's a film and science fiction history with an intelligent fleet of serving human robots being too intelligent and overcoming the universe. The main story is also one that worries some people that artificial intelligence and robotics will be brought into reality thanks to technological advances. However, it all means promoting creativity for many others to use intelligence for robotics.

The following few companies do amazing things with AI in robotics, from producing intelligent consumer goods to developing the first human-like artificial brain (Schroer, 2020).

1. ***Boston Dynamics***: Boston Dynamics is a diverse, intelligent, and adaptive robotics engineering company. The company designs sensor-based controllers with nine different models to prepare robots for various environments and terrains. You saw ideas in movement from the Boston Dynamics if you're a fan of Netflixs' Black Mirror. The four-day episode of 'Metalhead' inspired Boston Dynamic's four-legged killer robot SpotMini.
2. ***Canvas Technology***: Canvas Technology constructs autonomous, intelligent and efficient warehouse technology. The essential product of the business, an autonomous industrial cart, operates in unstructured storage facilities and factory environments — interior or exterior. The cart is easily adaptable and installed designed to be quickly changed. The International Flavors and Fragrances based in New Jersey have found that only one autonomous canvas cart reduces 10 hours per week in its facility.
3. ***DroneSense***: For public safety purposes, DroneSense's drone program is used. Drones can be used autonomously or manually for emergency response teams, fire services, police, rescue and search operations. A drone is unique because it can capture and document a wide range of information and data—such as the source of a wildfire, the location or extent of catastrophic damage in a particular area, stolen assets, or persons missing.
4. ***iRobot***: It is a commercial robot company that manufactures robotics for home purification and lifestyle. The most famous product Roomba is a robot vacuum that cures dirt and small debris of floors and adapts to its environment. The most recent Roomba is incredibly clever compared to the original model and can choose room sizes, adjust to carpet or hardwood, pick suitable pathways and recall the location of a room. The robots improve progressively as they learn and adapt to their environment.
5. ***Miso Robotics***: For use in industrial kitchens, Miso Robotics develops independent robots. Its Flippy cooker offers a 3D and thermal view, which allows it to learn new skills from its surroundings. It has been stated that robotic use in kitchens reduces food waste and allows human labor to prepare

and assist customers. The Los Angeles Dodgers are using Flippy to prepare and serve chicken tenders in their stadium kitchens.

6. *Neurala*: "The Neurala Brain," AI software has been developed by Neurala that makes the collection of devices smarter. The technology is already embedded in over 9 million products, boosting the intelligence of automobiles, drones, phones,and cameras. This technology is used by leading enterprises, including DARPA, NASA, NVIDIA, andMotorola. Neural solutions are designed to make drones smarter. Early signs of corrosion in large machines, such as wind turbines, are found in the future, and elephant poaching is curbed by hunting and hunting.

7. **Rethink Robotics**: Rethink Robotics produces collaborative robots in the same environment as human beings. Simplified training involves rotating the arms of the robot to display its behavior. To create safer environments and expand the job capabilities, Rethink's robots automatically adapt the force required for a particular mission. A furniture manufacturer used Rethink's Sawyer model to meet increased production needs. Sawyer prepared assembly tools, finished inspections, and adapted to the changes in human workflow.

8. *Sea Machines*: For the marine and shipping industries, Sea Machines produces autonomous technology. The technology of the business links the equipment of the vessel to autonomous or remote navigation sensors. The machine works as a data recording device for remote tasks or the usual routing of workboats. On the Massachusetts coast, Sea Machines provides a retrofitting tool for ships and builds a sophisticated driver support system.

9. *Veo Robotics*: With AI, 3D-sensing, and computer vision capabilities, Veo Robotics creates industrial robots that increase manufacturing activities. Robots collaborate with humans to improve flexibility and efficiency in workplaces. Industrial robots are usually separated for safety purposes from personnel, but Veo robots use 3D sensors to detect and delay or prevent objects or people in the vicinity, where possible. The robots are used for heavy lifting in car assembly lines while people do more delicate tasks.

10. *Perceptive Automata*: This company uses machine learning to allow auto cars with one of the most important components of their job to anticipate human activities. Most autonomous vehicles drive stiffly, which means that they drive human actions conservatively and often in contrast. Perceptive Automata teaches autonomous vehicles constantly about human conduct and demonstrates the way vehicles function safely in the world of human drivers.

11. *Piaggio Fast Forward*: The AI-enabled robots of Piaggio Fast Forward work as an extra couple of hands. The Gita robot from Piaggio uses AI, which serves as a flexible, manually free carrier, to follow its owners nearly behind.

12. *UiPath*: The Robotic Process Automation program from UiPath introduces AI to robots that make it easier to do repeated work and to learn as they go. The unicorn AI software allows robots to speed up activities in a world that no longer requires human supervision, from the retail industry to manufacturing.

13. *Engineered Arts*: Engineered Arts creates semi-humanoid,humanoid robots and a software package to personalize and render robots attractive. The enterprise has many kinds of artificial intelligence in its hardware, including automatic speech recognition to recognize facial and objects and computer vision. In collaboration with researchers and experts from the Universities of Leeds and Oxford, Ai-Da, a robotic artist who uses AI algorithms and computer vision to produce drawings and images, has been created.

14. *Cruise*: Cruise blends AI and machine learning to build autonomous self-driving vehicles. They use AI in the whole plan, simulation, and infrastructure to ensure that the robot can see the world around them in real-time and respond safely.

15. **Hanson *Robotics***: It is an AI and robotics company that creates robots like human beings with a human appearance and eye contact and face recognition characteristics, voice, and natural conversation. The robots will generate high-quality expressions with a proprietary nanotechnology skin called Frubber. In many news segments and in panels and technical conferences, Hanson's best-known humanoid robot Sophia is featured.

16. *Brain Corp*: The proprietary technology of Brain Corp allows robots adaptable and scalable to handle unstructured environments such as floors and warehouses. The robots also can chart, route, detect anomalies of surfaces, prevent objects and collect cloud-based data. The EMMA (Enabling Mobile Machine Automation) robot Brain Corp was checked after hours for cleaning the floor in Walmart stores.

17. *Starship*: To generate products within a range of four-mile, Starship generates independent robots. These robots can manage food, order food, or even retail orders and supply them with AI on the way to customers.

18. *CloudMinds*: Due to the immense physical dimensions required by a humanoid robot's brain – which are far larger than that of the human brain – CloudMinds makes it possible for users to operate cloud robots using a mobile device operates in safe and secure network environments. CloudMinds offers robotic services such as airport patrol robots to produce single-task manufacturing robots (with view and browsing capabilities) and AI technology.

19. *Vicarious*: Vicarious creates intelligent architectural robots that train and adapt faster. The company's goal is to increase human intelligence using algorithms through abstract ideas through sensor motor experiences. The first breakthrough of Vicarious occurred when its AI passed the CAPTCHA response tests intended to assess and validate human/computer users.

CONCLUSION

A good blend of automation technology is artificial intelligence and robotics. In recent years AI has become more and more popular with robotics solutions, with mobility and learning skills incorporated into previously static applications. While AI is still in its early stages, it has changed applications in many areas. Robots in the next robotic wave will have a magical resemblance to humans with the help of AI. Artificial intelligence and robotics are not job-destroying, like other new technologies. The main goal of digital technology is not to replace people but to enhance the security and efficiency of all processes. It is a fruitful relationship between automated robotics and people, not a confrontation.

REFERENCES

Advani, V. (2021). *What is Artificial Intelligence? How does AI work, Types and Future of it?* Great Learning. https://www.mygreatlearning.com/blog/what-is-artificial-intelligence/

Balajee, N. (2020). *Robotics Vs. Artificial Intelligence: What is the difference?* Medium. https://medium.com/@nanduribalajee/robotics-vs-artificial-intelligence-what-is-the-difference-6adad236d997

Benko, A., & Sik Lányi, C. (2011). History of Artificial Intelligence. Encyclopedia of Information Science and Technology, 1759–1762. doi:10.4018/978-1-60566-026-4.ch276

Carroll, B. O. (2017). *What are the 3 types of AI? A guide to narrow, general, and super arti cial intelligence What is arti cial intelligence.* https://codebots.com/artificial-intelligence/the-3-types-of-ai-is-the-third-even-possible#:~:text=There

Computer History Museum. (n.d.). *Timeline of Computer History Elektro at the World's Fair The Three Laws.* https://www.computerhistory.org/timeline/ai-robotics/

D, F.-S., & M.J., M. (2009). Human Robot Interaction. In *Encyclopedia of Complexity and Systems Science* (pp. 4643–4659). Springer.

Garg, S., Sünderhauf, N., Dayoub, F., Morrison, D., Cosgun, A., Carneiro, G., Wu, Q., Chin, T.-J., Reid, I., Gould, S., Corke, P., & Milford, M. (2020). Semantics for Robotic Mapping, Perception and Interaction: A Survey. In Foundations and Trends® in Robotics (Vol. 8, Issues 1–2). doi:10.1561/2300000059

Goodrich, M. A., & Schultz, A. C. (2007). Human-robot interaction: A survey. *Foundations and Trends in Human-Computer Interaction*, *1*(3), 203–275. doi:10.1561/1100000005

Iborra, A., Caceres, D. A., Ortiz, F. J., Franco, J. P., Palma, P. S., & Alvarez, B. (2009). Design of service robots: Experiences using software engineering. *IEEE Robotics & Automation Magazine*, *16*(1), 24–33. doi:10.1109/MRA.2008.931635

International Federation of Robotics. (2020). *IFR presents world robotics report.* https://ifr.org/ifr-press-releases/news/record-2.7-million-robots-work-in-factories-around-the-globe

Joo, S. H., Manzoor, S., Rocha, Y. G., Bae, S. H., Lee, K. H., Kuc, T. Y., & Kim, M. (2020). Autonomous navigation framework for intelligent robots based on a semantic environment modeling. *Applied Sciences (Switzerland)*, *10*(9), 1–30. doi:10.3390/app10093219

Joshi, N. (2019). 7 Types of Artificial Intelligence. *Forbes*, 1–6. https://www.forbes.com/sites/cognitiveworld/2019/06/19/7-types-of-artificial-intelligence/#b69f9bb233ee

Khakurel, J., Penzenstadler, B., Porras, J., Knutas, A., & Zhang, W. (2018). The Rise of Artificial Intelligence under the Lens of Sustainability. In Technologies (Vol. 6, Issue 4, p. 100). doi:10.3390/technologies6040100

Marr, B. (2020). *Artificial Human Beings: The Amazing Examples Of Robotic Humanoids And Digital Humans.* Forbes. https://www.forbes.com/sites/bernardmarr/2020/02/17/artificial-human-beings-the-amazing-examples-of-robotic-humanoids-and-digital-humans/#7b1a2c9c5165

Marr, B., & Ward, M. (2019). *Artificial Intelligence in Practice.* Wiley.

Moravec, H. P. (2021). *Robot.* Britannica Online Encyclopedia. https://www.britannica.com/technology/robot-technology

OECD. (2019). *AI & Society.* https://www.oecd-ilibrary.org/content/publication/eedfee77-en

Owen-Hill, A. (2017). *What's the Difference Between Robotics and Artificial Intelligence?* Robotiq. https://blog.robotiq.com/whats-the-difference-between-robotics-and-artificial-intelligence

Perez, J. A., Deligianni, F., Ravi, D., & Yang, G.-Z. (2018). *Artificial Intelligence and Robotics.* https://www.mygreatlearning.com/blog/what-is-artificial-intelligence/

Premebida, C., & Ambrus, R. (2016). *Zoltan-Csaba.* Intelligent Robotic Perception Systems. In IntechOpen. https://www.intechopen.com/books/advanced-biometric-technologies/liveness-detection-in-biometrics

Premebida, C., Ambrus, R., & Marton, Z.-C. (2019). Intelligent Robotic Perception Systems. *Applications of Mobile Robots*, 111–127. doi:10.5772/intechopen.79742

Raj, M., & Seamans, R. (2019). Primer on artificial intelligence and robotics. In Journal of Organization Design (Vol. 8, Issue 1, pp. 1–14). doi:10.118641469-019-0050-0

Robotics Online Marketing Team. (2018). *How Artificial Intelligence is Used in Today ' s Robots.* Association for Advancing Automation.

Rubio, F., Valero, F., & Llopis-Albert, C. (2019). A review of mobile robots: Concepts, methods, theoretical framework, and applications. *International Journal of Asvanced Robotic Systems*, 1–22.

Schatsky, B. D., & Ream, J. (2017). Robots uncaged. *Deloitte Insights*, 1–8.

Schroer, A. (2020). *AI Robots: How 19 companies use artificial intelligence in robotics.* Builtin. https://builtin.com/artificial-intelligence/robotics-ai-companies

Schutten, J. P. (2014). *Hello from 2030.* Academic Press.

Seetharam, K., Raina, S., & Sengupta, P. P. (2020). The Role of Artificial Intelligence in Echocardiography. *Current Cardiology Reports*, 22(9), 99. Advance online publication. doi:10.100711886-020-01329-7 PMID:32728829

Signorelli, C. M. (2018). Can computers become conscious and overcome humans? *Frontiers in Robotics and AI*, 5(OCT), 1–20. doi:10.3389/frobt.2018.00121 PMID:33501000

Smith, A., & Anderson, J. (2014). *AI, Robotics, and the Future of Jobs.* Issue August.

Wars, S., Sojourner, R., Pathfinder, M., Robots, U., Capek, K., Asimov, I., One, L., Two, L., Law, F., & Three, L. (2014). *Robotics : A Brief History Early Conceptions of Robots.* Academic Press.

Wisskirchen, G., Thibault, B., Bormann, B. U., Muntz, A., Niehaus, G., Soler, G. J., & Von Brauchitsch, B. (2017). Artificial Intelligence and Robotics and Their Impact on the Workplace. *IBA Global Employment Institute*.

Yan, Z., Schreiberhuber, S., Halmetschlager, G., Duckett, T., Vincze, M., & Bellotto, N. (2020). Robot perception of static and dynamic objects with an autonomous floor scrubber. *Intelligent Service Robotics*, 13(3), 403–417. doi:10.100711370-020-00324-9

Chapter 12
Innovation and Creativity for Data Mining Using Computational Statistics

M. R. Sundara Kumar
ⓘ https://orcid.org/0000-0003-0941-8393
Sona College of Technology, Salem, India

S. Sankar
Sona College of Technology, Salem, India

Vinay Kumar Nassa
ⓘ https://orcid.org/0000-0002-9606-7570
South Point Group of Institutions, Sonepat, India

Digvijay Pandey
ⓘ https://orcid.org/0000-0003-0353-174X
Institute of Engineering and Technology, Lucknow, India

Binay Kumar Pandey
College of Technology, Govind Ballabh Pant University of Agriculture and Technology, India

Wegayehu Enbeyle
ⓘ https://orcid.org/0000-0002-0788-6137
Department of Statistics, Mizan-Tepi University, Ethiopia

ABSTRACT

In this digital world, a set of information about the real-world entities is collected and stored in a common place for extraction. When the information generated has no meaning, it will convert into meaningful information with a set of rules. Those data have to be converted from one form to another form based on the attributes where it was generated. Storing these data with huge volume in one place and retrieving from the repository reveals complications. To overcome the problem of extraction, a set of

DOI: 10.4018/978-1-7998-7701-1.ch012

rules and algorithms was framed by the standards and researchers. Mining the data from the repository by certain principles is called data mining. It has a lot of algorithms and rules for extraction from the data warehouses. But when the data is stored under a common structure on the repository, the values derived from that huge volume are complicated. Computing statistical data using data mining provides the exact information about the real-world applications like population, weather report, and probability of occurrences.

INTRODUCTION

Data Mining is the concept of extracting meaningful information from a huge repository with a set of rules and algorithms. Mainly pre-processing of the data has been done on the extraction phase for the removal of duplicated and unwanted data. To perform the retrieval of data from the warehouses their knowledge has been discovered by KDD for getting better results while processing. (Regin. et al (2021)) Data mining is the main technique used for data extraction, transformation, and loading (ETL) in all disciplines to perform data storage and data retrieval effectively. Various fields are used data mining principles for accessing their data from real-world scenarios and convert them into applications. A lot of traditional approaches are used to do data processing on the larger networks but the speed of the recovery was not at the expected time interval. So they were searching for better solutions to do this with minimal time and high speed. Mining is the only solution for all and it has given the exact match of the data from the user perspective view. Moreover, a lot of information is stored in a common place for easy recovery and retrieval but the data loss and leakage were not monitored and controlled by classical methods. Data pre-processing is the main concept used in data mining techniques to avoid data duplication and repetition during data transmission. Data cleaning is also used in data mining for removing extra noisy data on the databases to overcome the data corruption problem. Data loss and leakage can be managed in data mining techniques with several algorithms and rules from classical approaches. But the main challenges and limitations of all algorithms used in data mining react with less accuracy and high latency for all real-world problem applications. So if any user wants to retrieve the data from the huge repository they must wait for a long time to get output. When considering output from that method also not inaccurate about originality. So a lot of people are working on the data mining domain to overcome all the problems and provide their solutions with recent trends or algorithms. While handling huge datasets like big data and IoT time consumed for data extraction is more rather than normal databases. For rectifying this issue in data mining recent and modern tools are used to avoid latency among the network systems. Perhaps, new innovative approaches and algorithms are implemented in day-to-day field data mining has its hype among all. The capability of categorized the data on the network systems is the basic one for all researchers. In other words, Data Mining is the backbone of all research people and industry people to do the applications of real-world problems. Anyhow data processing on the larger network is controlled and monitored by the data mining approach with its effective data processing mechanism. Data mining principles (Graif. et al (2021)) are worked on the following areas for improving the performance of the data processing in order to accessing the data.

- **Dimensionality**

It is the main component of data mining techniques used to represent the information with the help of values as a variable. Their values are not constant and will be changed in every event. The analysis is made by dimensions as the values present with all data mining techniques. The values are changed every second due to the originality of the data. Because the data generated by both machines and humans are increased in size at every point of view. In this scenario, the normal standard procedures do not help with data processing whereas data mining provides much better results to this kind of problem neatly concerning time and accuracy.

- **Uncertainty**

The choices of the data can be selected from KDD's to perform data extraction as samples or values. But the output given by the system is not up to the level so their values not satisfied the requirements. Consistency-based data changes are noted at every time to avoid uncertainty over the networks. Samples have not given accurate results because of the size of the data from the repository. Sometimes wrong samples have been accessed for the mining process would lead to negative results. To overcome this problem lot of mathematical formulas and equations are used such as mean, median, average, and standard deviation (Pramanik, S. and Raja, S. S. 2020).

- **Scalability**

Data scalability plays a vital role in data mining concepts, for their complications of data generated at various sources. All data could collect on the databases and will be stored in a huge database as a warehouse for storage purposes. A rule-based algorithm is used to discover the exact match of the data among the system on a larger network. This issue can be solved with the help of storing it in different places. Storage in a commonplace or private place doesn't make any problem to the users but processing from the network has created a lot of issues like traffic, corruption, leakage, and loss. Data loss is the main cause for all the problems that happened in data mining principles. So while storing itself strong rules and structure should be followed for better results.

- **Sampling**

It is a statistical tool for maintaining scalability and take samples from huge datasets to make a decision based on that. If the sample data was not processed properly the entire system gets collapsed. The way samples have been treated on data mining techniques with lot theorems for processing. Multiply the input result with key values and get the output of the whole system is run in this method for data mining applications. Sampling theorems are used to increase the efficiency of the data accessing strategy by double the input to the output signals and their ratio should be strictly followed. The levels like 0 t0 3 are always maintained as a Signal to Noise Ratio (SNR) to monitor the amount of originality of the data on network systems.

- **Size**

In this method, data size is very huge means the entire data processing systems need storage concepts and feasibility reports of the storage data. In this report, all the functionalities of the data are represented clearly so that actions could be taken based on that. The number of volumes increased on the network will affect the speed of the system. Origin of the data generation is not the main role in data mining. All data must be generated from different sources over the network and some on the internet than by human-made machines. It is very difficult to store a huge volume of data at a common storage place with minimum security policies. Strong policies have to be written to provide security (Pramanik, S. and Singh, R. P. 2017) over the data transmission technique and will be monitored always for changes notification.

- **Visualization**

All output data must be explored properly to deliver exact results to the customers in figures or pictures. But the issues are final output will not be fixed as the values of the figures are keep changing their values. Data exploration is the main concept in data mining to display the processed data over the network. The entire process is called visualization and it leads data mining to the next level with perspective views. There are a lot of modern tools and software is available in the market for doing this visualization effectively. In General, classic approaches used for visualizations are working on less size of data exploration to the user whereas modern tools are handled big data over IoT platform even at real-world applications also.

- **Automated Analysis**

The analysis can make by the system itself by self-learning models to avoid conflicts between the users. At a time number of users accessing the same input data then the system speed will get slow down due to its size. The analysis is the concept of categorizing the input data into different forms and applying certain principles to them to segregate over new perspective views. Most of the automation doesn't focus on the optimization of the entire process. But in data mining principles lot of rules and relationships are worked together to get accurate results during data transmission from one form to another form. While doing this automation process has done by pre-processing procedure using classical methodologies.

- **Algorithms Used**

Various algorithms are used to perform data processing units with the help of rules and relationships such as association rules, apriori algorithm, and support vector machine. A lot of other methods are used to find the best and suitable solutions for data processing. Classical algorithms are used to perform data processing scenarios but they are working only for small datasets or minimum size of the data. If the structure of the data is very strong then easy retrieval is possible. When the unstructured digital data like pdf, text, audio, and video databases are in the warehouse the work is very tuff and the time consumed is very high in such cases. So which algorithms are used in data mining should reveal the importance of the data processing on a larger network. Scientific algorithms are used mathematical formulas and equations to solve complicated issues from the outside world. While handling mathematical equations algorithms are helping them to extract the original data from the databases. Otherwise, duplication or repetition will happen on the network, and hackers are stolen the data without their knowledge.

- **Relationships with Data Models**

Generalized the liner models of the statistics with demonstration, results of the data mining is to categorize by various models. Statistical tools are used to find the relationships with the data models and their values as variables. Machine learning techniques are also used to perform data processing with these models by various techniques like linear regression techniques, Logistic regressions, and generalized linear models. All these models have discussed the mathematical operations for extracting the original data during extraction. Several data models used in data mining decide the quality of the system or in other words, more data models used in data mining will provide more security to the entire data processing scheme. Relationships are giving bonds or binding the data and values to perform the operations at a particular time.

DATA MINING PRINCIPLES

Data mining deals with databases as a collection of files and attributes over the computer network than with information science for their analytics. Statistics is the concept of extracting the exact information from the databases or repository using mathematical representations and formulas for accuracy. Machine learning also used to perform data processing and retrieving for accessing a huge volume of data around the world. Now a day social media rules the world with its generated volume of data but the storage and retrieving is still complicated when it comes to the statistics field. Information science is a technique which is used to do analytics by its KDD based principles to get good results accurately at all time. To avoid fake calculations and duplicate data integration in all disciplines this information science helps to do the ETL process in the research field and industries. Data visualization is the concept of presenting the output to the people in a meaningful way with clear vision by various tools and technologies. All other disciplines like healthcare, education, military, media, and the internet are using the data mining concepts for displaying their applications to the real world. It reveals their business ideas insights into the future perspective point to improve them in all fields. Multiple domains are used data mining concepts to perform data processing for delivering their ideas and innovations to real-world people. Business Intelligence (Su, Y. S., & Wu, S. Y. (2021)) and the latest technologies are discussed about their insights as innovative principles and for performing ETL operations with the help of data mining concepts only. Data mining concepts are well known in research fields and other disciplines but still it has a lot of challenges and issues in the speed of the data extraction and latency. Those problems are addressed in this paper for getting the exact output from the huge repositories. The repository is placed in different places and all the data generated can be added as a network architecture system for maintaining the throughput of the entire system. The values and relationships between the data models are effectively run by the data mining algorithms to find an exact match of output on the network. The latest trends and technologies like machine learning concepts, artificial intelligence, deep learning, web applications, and neural networks are accessed data mining principles for deriving the data from the repository quickly with high speed. There are traffic problems that occurred in the network and will be addressed by creating data models with a combination of data mining algorithms. It helps to decide on the complicated applications with the easy method but have to work on statistical data (Saeed. et al (2021)). The following figure 1 represents the data mining concepts and techniques used areas.

Figure 1. Data mining concepts used in various fields

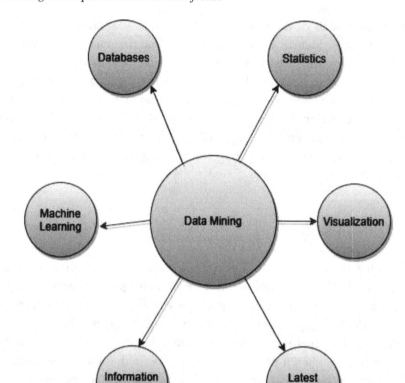

STATISTICS

Statistics is the process of manipulating meaningful data from different sources on the internet or generated by humans. Multiple steps are followed to do statistics for data processing into their insights. Initially, data has created from different places and modified into perspective insights according to their types. Modified data have to be merged or integrated with newly generated data on the network and will give new dimensions to its view. After pre-processing, all data concepts like cleaning, removing, updating, and merging computational calculations can make for results on real-world applications. Moreover, calculations are not only arithmetic and logical operations but also used for critical mathematical calculations like algebra, matrix, differentiations, calculus, probability, and equations (Wibawa. et al (2021)). Once calculations were made their results would be distributed to the classifiers or the models used for data processing. In this phase, all the modern tools and software are utilized by the data mining approach to delivering the results accurately. Unfortunately, the results have produced by those models are not checked and validated by a human are not reliable due to trial and error concepts. So the machines have to do this testing work with the help of previous data given by the models created during processing time. Finally, statistics procedures done on the databases with different perspective insights and results will be visualized using modern tools. Visualization can be done for the users to view the output in all views and make changes according to their requirements. The following figure 2 denotes the cyclic process of statistics process and its steps.

Figure 2. Statistics steps and their cycles

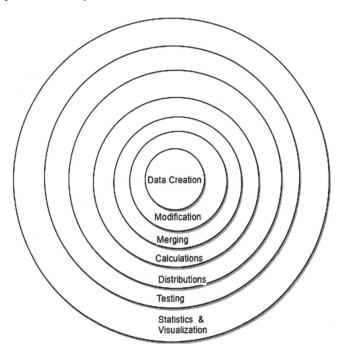

Data Creation
Modification
Merging
Calculations
Distributions
Testing
Statistics & Visualization

DIFFERENCES OF DATA MINING AND STATISTICS

Data Mining is the process of accessing the information on the larger network with certain rules and algorithms to perform the accurate result output extraction. Nevertheless, statistics is the process of getting results as numerical values and have predictive analysis for taking future decisions. All the concepts were analyzed and accessed with mathematical operations to deliver quality results with accuracy. Data mining features have differed from statistics in various points such as the roles, size of data, automation, and processing techniques. Finally, all the features of data mining and statistics are described the computational process is too quick and easy to get good results in less time. But the issues are created from its methodologies, instead of getting accurate results on time. The structure of the data is in data mining is not in a form whereas statistics have a good and well-defined structure for data processing. Moreover, major features of statistics are distinct from data mining especially the pre-processing techniques. Data pre-processing (Pramanik, S. and Bandyopadhyay, S. K. 2014) can be done before the data processing state and in statistics no need to do that. Other features are conveying the normal qualities of both techniques but all are doing the same work such as computational data from the huge databases on the larger network. The noise is affected in pre-processing data ratio and it differs in the range of 0 to 3 (Shousong,

C., & Jing, Z. (2021)). It will decide the time to remove the unwanted data quickly. The following table 1 summarizes the features are used and differences between data mining among statistics techniques.

Table 1. Differences of data mining and Statistics

Sl. No	Features	Data Mining	Statistics
1	Problem Type	Un structured and semi structured	Well defined structure
2	Role of inference	No explicit	Explicit
3	Objectives	Analysis, Modeling	Formulation, collection
4	Dataset size	Large and Heterogeneous	Small and homogeneous
5	Approach	Heuristic or inductive	Deductive
6	Noise ratio for signal	0 to 3	Greater than 3
7	Analysis type	Investigate	Authenticate
8	Variables used	Huge	less
9	Pre-Processing	Data Cleaning	No
10	Automation	Easy	Expert view needed
11	Data processing	Algorithms used	Mathematics used
12	Scope	Guidelines made	No guidelines
13	Techniques	Clusters, sensors, analysis, visualization	Descriptive and inferential analysis
14	Application	Theoretical oriented	Both theory and practical

STATISTICAL ANALYTICS AND TYPES

Statistical computation in data mining can be done in different types based on the availability of the data among the nodes in the larger network. Statistical analytics can be done with the concepts of data collection, Exploration, and organizing patterns with the recent trends used by the data mining concepts. Data exploration in a distinct way to real-world applications is the art to perform in statistical analytics. Patterns are the function points used to identify the layers of the data presented on the network and it will avoid confusion among the systems. To perform searching (Pramanik, S. and Bandyopadhyay, S. K. 2014) options technical algorithms and mathematical formulas are used in statistics with decision-making conditions (Viviane. et al (2021)). The hype of decision making is to deliver the results with accurate value and based on those applications modules have to be designed for real-world problems by the industries. Industry persons and researchers are confused with data mining concepts due to getting numerical values as a result when algorithms and rules were used. This issue will address by the types of statistical computations and the characteristics are controlled on a centric processed method (Pande. et al (2021)). The methods have identified the pattern of the data models created by data mining concepts and will be analyzed using powerful tools to get accurate results. Statistics is the term that is used in mathematical sciences and foundations for delivering the higher class output to the nodes on a network. The following figure 3 summarizes the types of statistical analysis methods in data mining.

Figure 3. Types of statistical analytics

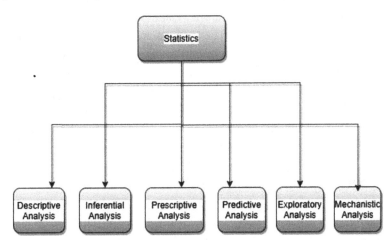

Descriptive Analysis

Descriptive analytics is the technique used to describe the data as a helping to suggest the way to use it. It will not be used for making conclusions during the data processing. Quantitative description of the data among the network nodes could be used for further processing like applying mathematical formulas, equations in constant methods. A simple example taken for this descriptive analysis is calculating the student's average mark from their exam results. The total sum value was taken and divided by the total number of subjects to provide the result of the average mark. But this result doesn't make any sense to find the conclusion of the problem. For larger data sets also the same problem is addressed and rectified with data mining algorithms. Because of the above-said example, one average value will not give a strong result of the student performance from the entire mark list. But from larger data set the same way is not followed to make decisions or conclusions (Mcandrew. et al (2021)). So there are two ways methods used to do this with the help of quantitative values.

- **Central Tenancy measure**: This method is used to find the mid-value or central point of the entire data set and it will be taken for the calculation. This value contains the information about the whole data set for deciding the results. The mathematical terms like mean, mid, average, and median are the example so this statistics. These measurements are used to get an exact match from the input data and will be used for the predictive analysis also. Prescriptive and predictive analysis can be done using mathematical representations with relationships among the data models. If the strong relationship between the model, and it has to be emphasized in data processing strategy successfully. The tenancy is the word that represents the occupancy of the data on the network to establish the connectivity among the relationships. If this tenancy principle will be centralized then all the controlling and monitoring will be maintained automatically with a systematic approach
- **Spread Measure**: The data processing using this method is finding the total average of the value i.e. if the average value of 150 students is 75 means then the average value less than 75 students are under the poor category from all students. This decision has been made due to this major value

used in the larger data set. This method is used to decide for finding the range, standard deviation, mean and median from the huge data sets. Data can be spread over the network along with the measurements taken from the input. The formats are different as input data from the users it will be insisted in to divide the entire system into small pieces for easy accessing. To enhance this method, measurements have taken from the tools or software can be spread to the entire system and the preferred module will get access among all systems particularly on large networks. So when compared with the previous data processing methods of tools and software's statistical data can be computed with this data mining concept with accuracy. So all the other techniques are accepted the data mining concepts are the backbone of data processing.

Inferential Analysis

This type of statistical analysis is used to make the decisions from the databases with the help of samples or references. The best example for this type is from the population analysis samples have taken and done the data processing. Results declared from those samples ie this year's population would be 78 percentage and the same way it will be continued means next year it would be 84 percent based on the samples taken from the databases. Even though the population sample value is changed in real-time but the result is not biased. Sampling errors occur during the data processing time because the population values samples taken from different places at various times. For the generalization of input data, two methods are used in inferential analytics.

- **Parameters estimation**: All the parameters used in statistics are estimated by their units to monitor those from the centralized controller. Measurements alone will not give the exact results on a network system. Their units are important to take decisions while dynamic processing of data from the repository. Multiple methods are used to deliver the results of data processing in real-time but the time and accuracy are not up to the mark. So this estimation of parameters has to be considered as an important factor in data mining techniques to deliver high accuracy results. Parameters are the attributes/ entities in the real world activities and handled them on a larger network is the major issue in data mining from the beginning itself. The research peoples are taken the estimation of the measurement from the generated data and convert them into meaningful information based on the units calculated. If several units are more than the average value has to be considered but it will not reflect the accuracy of the results.

- **Statistical Hypothesis Testing:** Tested the proposition of the statistical data on the larger network to avoid the errors in a result or fewer errors in a result. Here proposition helps to identify the relationship between the data models over the network. Their proposition is suitable then data processing can be very easy in small networks. But when the large network, the number of individual data generation is more so there are several possibilities to get wrong results. For that issue on a large network, this hypothesis testing has to be applied especially for statistical data. The statistical data are large in numbers but small in size. All because of the mathematical formulas and equations handled in that. The results of the testing are analyzed properly if any deviation is there then the proper technique will be used to rectify that in a short period. The results given from this testing play an important role in data mining techniques in a larger network.

Prescriptive Analysis

Asking a lot of questions to find the best solutions from the answer is called prescriptive analysis in the statistical method. For this method, the best solutions will be selected by the users from the huge dataset as answers. It gives the optimal recommendation for taking decisions from the results i.e. find the best among the possible choices of answers given. The best examples to study prescriptive analysis are graphs, simulations, algorithms, and machine learning concepts. In graphs and simulations, all the measurements and values used in the prescriptive analysis are telling only what was happened on the system. So it cannot provide accurate results whereas gives the approximate results in data mining. Assumption and approximations are the two types of results in this analysis, but the decision-making doesn't meet its requirements immediately. Generally, prescriptive analysis will give overall view about the problem, not at all an exact view of the problem.

Predictive Analysis

Future events decision will be measured or calculated from today's event is called predictive analysis. The historical and current event facts are the criteria for this kind of analysis. It is used to reduce the risks of taking negative decisions because from the current only all the data must process. The probability of providing fake or wrong results will avoid by this technique. All the financial sectors, online service providers, marketing companies, and insurance companies have used this method. Predictive analysis techniques examples are developed in data mining, Artificial Intelligence techniques, and modeling. In this analysis, existing system values are used to decide the future values. The existing values are changed automatically it will reflect in proposed values. It is used mainly in population survey calculation and helps to predict the future values from the existing data. Most of the real-world applications like weather reports, population survey, earth weight, share market, and other applications are predicting their results based on this analytics only. It may cause the reason for errors in the results not reflect in the total output like prescriptive analysis.

Casual Analysis

This analysis is made by collecting reasons from the peoples for the success and failures of the strategies used in companies. The root cause analysis was found and a decision will be taken based on that with remedies. It will implement for a futuristic purpose but have to use for both situations like present and future. From the root cause reason of the problems, it will analyze the main problem and then will be addressed for finding the solutions. Major IT companies and industries are tried to find the reasons for their failures after the recession period this type of analytics is used. The casual analysis doesn't have any timing for calculation. It will not do the analysis process frequently at the time of problem has come in the industries. But regular time intervals may be in a month or week it will be conducted. Their decision-making time refers to common failures and reasons for analytics to provide the results at a given time.

Exploratory Data Analysis

This type of statistics is used to identify the patterns and unknown relationships that can figure out at the end. To find the missing data and assumptions made on the decisions while unknown relationships occur in data transmission time. This type of statistics cannot be used alone and as its insights are giving a view of bird-eye. This can be taken while data pre-processing started in the initial level itself. Patterns are written in a well-structured format to be considered for analytics as an input of various sources. The values are as results might be varied in patterns in the form of missing data and assumptions. If missing data has been taken for consideration it may lead to calculate wrong results often. So while doing analytics check the possibilities of any missing data regularly.

Mechanistic Analysis

It is mostly used in the big industries because the exact changes of the situations are identified and rectified with solutions. It is not used as a common statistic method. It made interaction to all components in a system and it will affect the entire system if any small mistakes occur. The variables used in the analysis are depending on the other variables so that any changes occurred will affect the whole. External influences have not affected this system. Biological science-based companies and industries have used this strategy for making good decisions. Big industries are commonly affected by the failures of ideas and strategies. For identifying those reasons it would be very tuff before the tragedy has happened. It is not possible to take action immediately if something has happened in industries. After that happened we have to frame one policy and team members to do analytics step by step to rectify the problems. Later the entire team members and other peoples have accepted the faults and mistakes done in the situations where it was happened to be analyzed. The entire process takes a lot of time to get the results. It is a very slow and steady process in data mining concepts but not for use in small industries or immediate action required places. Multiple solutions will be given from this analysis but the suitable one which is accepted by all members of a team has to be the final output. A lot of suggestions and reasons told by the stakeholders or committee members all these points have been clarified and discussed before making the decision.

Statistical Data Analysis Techniques

Statistical data analysis can be done by the above types and some of the techniques are followed to do that effectively (Ley, J., & el Moctar, O. (2021)).

- Probability and Bayes theorem
- Random variables and Error Propagation
- Monte Carlo method & probability density functions.
- Multivariate methods & Correlation
- Statistical Tests with Parameter estimation
- Method of least squares
- Uncertainties

The entire data mining and statistical computational methods are explained in the following figure 4.

Figure 4. Computational Statistics in Data Mining Techniques

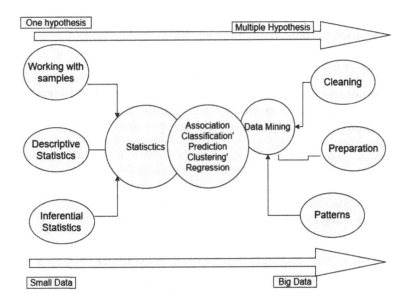

DATA SCIENTIST

Now a day's computational statistics in data mining can be done by the new innovative technique called data science and the person who did all work is called a data scientist. Data science is the recent technique used to perform complicated data analytics like statistical data with the help of mathematical problems and formulas. Their work is to collect the whole data from different sources and stored them in a single server than doing the analytics in either real-time or normal time. Real-time data can be calculated at every second so the immediate updates are captured and the status will store in a separate repository. So the values are generated by this data science method and all the other solutions for the complex problems will be addressed through this data science. The methods used in data science are handled by the person data scientist and their work has to be monitored the data processing continuously to avoid duplication or data cleaning in data mining applications (Wu, Y., &Xu, J. (2021 & (Zhang, C., Wang, X., & He, Z. (2021)). The following factors are considered for the challenges and solutions of the data scientist. The problems addressed in this paper will help future data scientists to handle huge computational statistical data.

• **Data Preparation**

Generally, 80% of the time is consumed by a data scientist for cleaning and quality improvement on the data which leads to the data accuracy and reliability on the larger network. Terabytes of unstructured data can be reviewed and prepared for the processes by the data scientists with different functions, solutions, and patterns. To overcome this problem Artificial Intelligence enabled technologies like augmented reality and analytics, auto feature engineering concepts are used. This will help the data scientist for preparing the data sets with high quality.

- **Multiple data sources**

Different sources are used as applications and tools were generated data continuously in social media or the internet with various formats. The decision taken by the data scientists from these various formats is typically complicated due to its size. This method requires physical entry and consumes a lot of time to complete the process in real-time. This will create errors, repetitions, and duplication of data while searching the data from the repository. An integrated centralized platform is used to perform data collection from multiple sources and aggregated in real-time. The time consumption is also reduced in this approach which leads to doing all the processes very effectively by the data scientists.

- **Data security**

When a huge amount of data can be stored in a cloud computing environment then security will be the most important factor in the network among the nodes. Data management is very complicated with security by the scientists. The major problems are vulnerable confidential data handling and cyber attacks. To provide the solutions for this problem machine learning-enabled security policies are encryption (Pramanik, S. and Bandyopadhyay, S. K. 2013) keys will be used. It will provide more security over network systems and data scientists are maintained their servers very safely.

- **Business problem understanding**

The business problem will be analyzed with a clear vision, mission with a mechanical approach and they thoroughly understand the business problem with stakeholders. Otherwise, their time and energy will be wasted on the entire process. Wok flow will be defined at the initial stage the analysis will be made with stakeholders as a communication process. They must maintain the checklist with all necessary items available in the data processing field to perform proper utilization of time.

- **Nontechnical stake holder's communication**

Client and stakeholders have good communication with each other to maintain effective time consumption and resource utilization with a proper understanding of the business problem. Their insights towards the goal of the company should be clearly understood by all stakeholders to avoid conflicts between them. They must concentrate on data storytelling, structured approach, powerful approach analysis (Pramanik, S. and Bandyopadhyay, S. K. 2014), and visualizations.

- **Collaboration with data engineers**

Misunderstanding between the data scientist end data engineer leads to the failure of the entire system due to their mismatched workflows, concepts, and thinking. To avoid such problems in the early stage take steps in enhancing the collaborations between them. Conduct more open communication meetings by appointing data science officers leads to avoid the misunderstanding.

- **KPI and Metrics**

There should be a well-defined metrics can be prepared and followed by data scientist among the data engineers and proper KPI's are used to perform analysis with their business impacts. KPI's are used to connect so many operations during program running time so that latency will be reduced literally and throughput is increased. The following figure 5 describes the data science process and its techniques. Key Performance Indicators (KPI) is the important factor using in data science process by data scientist. If this will be indentified during the analysis number of operations running would be more and more.

Figure 5. Data science and its techniques

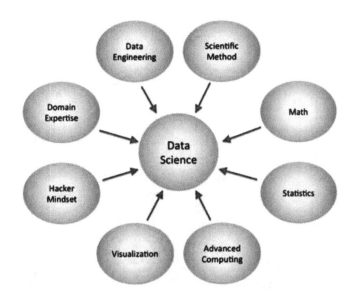

DATA VISUALIZATION

Finally, all the processing can be done in computational statistics with data mining concepts using various techniques all leads to exploring or visualized the output in graphical representations or a valid method for better understanding. Through this method, everyone understands the results given by the method, and immediate decision has been taken. Prescriptive and predictive analysis (Zhu, H., Shang, L., & Zhou, X. (2021))can be done in this visualization method using modern tools available in a market. The following are the basic or main methods to display the results as data visualization concepts (Malviya, M., Buswell, N. T., &Berdanier, C. G. (2021)).

- Column and stacked column Chart
- Bar and Stacked Bar Graph
- Area and Dual Axis Chart
- Line Graph
- Mekko Chart
- Pie and Waterfall Chart
- Bubble Chart
- Scatter Plot Chart

- Bullet Graph
- Funnel Chart
- Heat Map

The figure 6 shows some of the examples of data visualizations.

Figure 6. Data visualization examples

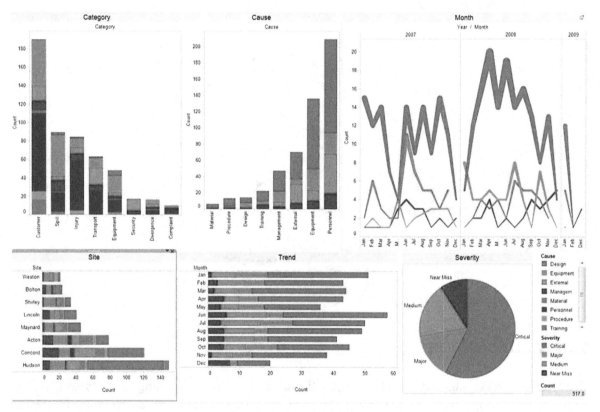

CONCLUSION

Data mining in computational statistics is a technique that is used to perform data processing has given high speed and accuracy results. The major disadvantages of data mining concepts are the very slow processing time of extraction from higher data sets and less accuracy in the result. The reliability level is low when the results given by the data mining concepts have noisy and duplicate. So the data mining systems are helped to retrieve the data from smaller data set is fine for latency and throughput factors. But it is not suitable for huge datasets. That problem can be solved by the latest techniques like machine learning and artificial intelligence as a new innovative approach. Statistics using the data mining approach leads to deal with the huge data sets and get the results with numerical values by applying mathematical formulas and equations. Statistical methods have challenges and issues it will be solved by using the latest techniques and algorithms for getting accurate results. Several types are used to perform data processing in statistical analysis it will help to select the appropriate method for doing analytics on the repositories. Each type has its limitations and solutions for the above-said problem. Data science with data engineers is the solution for computational statistical problems with advanced tools available for data processing on a larger network. Finally, all the data must be visualized with various representation formats by modern tools and software's. This chapter summarizes the entire challenges and solutions in data mining computational statistics and the methods to perform data extraction with less timing and high-speed accuracy. Finally data mining in computational statistics will give the accurate results on a larger network with less time.

REFERENCES

Graif, C., Freelin, B. N., Kuo, Y. H., Wang, H., Li, Z., & Kifer, D. (2021). Network spillovers and neighborhood crime: A computational statistics analysis of employment-based networks of neighborhoods. *Justice Quarterly*, *38*(2), 344–374. doi:10.1080/07418825.2019.1602160

Ley, J., & el Moctar, O. (2021). A Comparative Study of Computational Methods for Wave-Induced Motions and Loads. *Journal of Marine Science and Engineering*, *9*(1), 83. doi:10.3390/jmse9010083

Malviya, M., Buswell, N. T., & Berdanier, C. G. (2021). Visual and Statistical Methods to Calculate Intercoder Reliability for Time-Resolved Observational Research. *International Journal of Qualitative Methods*, *20*, 16094069211002418. doi:10.1177/16094069211002418

McAndrew, T., Wattanachit, N., Gibson, G. C., & Reich, N. G. (2021). Aggregating predictions from experts: A review of statistical methods, experiments, and applications. *Wiley Interdisciplinary Reviews: Computational Statistics*, *13*(2), e1514. doi:10.1002/wics.1514 PMID:33777310

Pande, A., Manchanda, M., Bhat, H. R., Bairy, P. S., Kumar, N., & Gahtori, P. (2021). Molecular insights into a mechanism of resveratrol action using hybrid computational docking/CoMFA and machine learning approach. *Journal of Biomolecular Structure & Dynamics*, 1–15. doi:10.1080/07391102.202 1.1910572 PMID:33829956

Pramanik, S., & Bandyopadhyay, S. K. (2013). Application of Steganography in Symmetric Key Cryptography with Genetic Algorithm. *International Journal of Computers and Technology*, *10*(7), 1791–1799. doi:10.24297/ijct.v10i7.7027

Pramanik, S., & Bandyopadhyay, S. K. (2014). Image Steganography using Wavelet Transform and Genetic Algorithm. *International Journal of Innovative Research in Advanced Engineering, 1.*

Pramanik, S., & Bandyopadhyay, S. K. (2014). An Innovative Approach in Steganography, Scholar. *Journal of Engineering Technology*, *2*(2B), 276–280.

Pramanik, S., & Bandyopadhyay, S. K. (2014). Hiding Secret Message in an Image, International Journal of Innovative Science. *Engineering and Technology*, *1*(3), 553–559.

Pramanik, S., & Raja, S. S. (2020). A Secured Image Steganography using Genetic Algorithm. *Advances in Mathematics: Scientific Journal*, *9*(7), 4533–4541.

Pramanik, S., & Singh, R. P. (2017). Role of Steganography in Security Issues. *International Journal of Advance Research in Science and Engineering*, *6*(1), 1119–1124.

Regin, R., Rajest, S. S., & Singh, B. (2021). Spatial Data Mining Methods Databases and Statistics Point of Views. *Innovations in Information and Communication Technology Series*, 103-109.

Saeed, S., Bagram, M. M. M., & Iqbal, M. M. (2021). An Intelligent Analysis of Crime Data using Data Mining Algorithms. *Technical Journal*, *26*(01), 102–115.

Shousong, C., & Jing, Z. (2021). Research on Healthcare Quality Evaluation Based on Data Mining. *Journal of Medical Imaging and Health Informatics*, *11*(4), 1117–1124.

Su, Y. S., & Wu, S. Y. (2021). Applying data mining techniques to explore user behaviors and watching video patterns in converged IT environments. *Journal of Ambient Intelligence and Humanized Computing*, 1–8. PMID:33425047

Viviani, E., Di Persio, L., & Ehrhardt, M. (2021). Energy Markets Forecasting. From Inferential Statistics to Machine Learning: The German Case. *Energies*, *14*(2), 364. doi:10.3390/en14020364

Wibawa, B., Siregar, J. S., Asrorie, D. A., & Syakdiyah, H. (2021, April). Learning analytic and educational data mining for learning science and technology. In AIP Conference Proceedings: Vol. 2331. *No. 1* (p. 060001). AIP Publishing LLC.

Wu, Y., &Xu, J. (2021). 13 Statistical Problems with Planted Structures: Information-Theoretical and Computational Limits. *Information-Theoretic Methods in Data Science*, 383.

Zhang, C., Wang, X., & He, Z. (2021). Efficient Importance Sampling in Quasi-Monte Carlo Methods for Computational Finance. *SIAM Journal on Scientific Computing*, *43*(1), B1–B29. doi:10.1137/19M1280065

Zhu, H., Shang, L., & Zhou, X. (2021). A Review of Statistical Methods for Identifying Trait-Relevant Tissues and Cell Types. *Frontiers in Genetics*, *11*, 1846. doi:10.3389/fgene.2020.587887 PMID:33584792

Compilation of References

Ababulgu, A. N., & Wana, F. H. (2021). (2021). The Horrors of COVID-19 and the Recent Macroeconomy in Ethiopia. *Journal of the Knowledge Economy*. Advance online publication. doi:10.100713132-020-00713-6

Adamopoulou, E., & Moussiades, L. (2020). Chatbots: History, technology, and applications. *Machine Learning with Applications*, *2*(November). doi:10.1016/j.mlwa.2020.100006

Advani, V. (2021). *What is Artificial Intelligence? How does AI work, Types and Future of it?* Great Learning. https://www.mygreatlearning.com/blog/what-is-artificial-intelligence/

Aggarwal, C. C., & Wang, H. (2010). A survey of clustering algorithms for graph data. In *Managing and mining graph data* (pp. 275–301). Springer. doi:10.1007/978-1-4419-6045-0_9

Agichtein, E., & Gravano, L. (2000, June). Snowball: Extracting relations from large plain-text collections. In *Proceedings of the fifth ACM conference on Digital libraries* (pp. 85-94). 10.1145/336597.336644

Ahmad Kuchy, S., Ahmed Khadri, S. K., Mukherjee, M., Samanta, D., & Le, D.-N. (2017). An Aggregation Approach Based on Elasticsearch. *Journal of Engineering and Applied Sciences (Asian Research Publishing Network)*, *12*, 9451–9454. doi:10.36478/jeasci.2017.9451.9454

Alcácer, V., & Cruz-Machado, V. (2019). Scanning the Industry 4.0: A Literature Review on Technologies for Manufacturing Systems. *Engineering Science and Technology, an International Journal, 22*(3), 899–919. doi:10.1016/j.jestch.2019.01.006

Aleya, K. F., & Samanta, D. (2013). *Automated damaged flower detection using image processing*. https://www.semanticscholar.org/paper/AUTOMATED-DAMAGED-FLOWER-DETECTION-USING-IMAGE-Aleya-Samanta/11f8ebd4082acef98b7329cecc81601b6ec20bc8

Allen, G. D. (1999). *A Brief History*. https://www.math.tamu.edu/~dallen/masters/alg_numtheory/pi.pdf

Althar, R. R., & Samanta, D. (2021). Building Intelligent Integrated Development Environment for IoT in the Context of Statistical Modeling for Software Source Code. In R. Kumar, R. Sharma, & P. K. Pattnaik (Eds.), Multimedia Technologies in the Internet of Things Environment. Studies in Big Data (Vol. 79). Springer. https://doi.org/10.1007/978-981-15-7965-3_7.

Althar, R. R., & Samanta, D. (2021). The realist approach for evaluation of computational intelligence in software engineering. *Innovations Syst Softw Eng*. doi:10.100711334-020-00383-2

Amoroso, E. G. (1994). *Fundamentals of Computer Security Technology*. Prentice-Hall.

Anderson, D. P. (2004). BOINC: A System for Public-Resource Computing and Storage. Grid, 4-10.

Anderson, J. A., Lorenz, C. D., & Travesset, A. (2008). General purpose molecular dynamics simulations fully implemented on graphics processing units. *Journal of Computational Physics, 227*(10), 5342 – 5359.

Anderson, A., & Semmelroth, D. (2015). *Statistics for Big Data For Dummies* (1st ed.). For Dummies.

Andieu, C., de Freitas, N., Doucet, A., & Jordan, M. (n.d.). *An introduction to MCMC for machine learning.* http://people.cs.ubc.ca/ nando/papers/mlintro.pdf

Angeli, G., Premkumar, M. J. J., & Manning, C. D. (2015, July). Leveraging linguistic structure for open domain information extraction. In *Proceedings of the 53rd Annual Meeting of the Association for Computational Linguistics and the 7th International Joint Conference on Natural Language Processing*:Vol. 1. *Long Papers* (pp. 344-354). 10.3115/v1/P15-1034

Ang, K. C. (2009). A simple stochastic model for an epidemic {numerical experiments with matlab. *The Electronic Journal of Mathematics & Technology, 1*(2), 117–128.

Annigeri, S. (2009). *Matrix Structural Analysis of Plane Frames using Scilab.* https://www.researchgate.net/publication/242759801

Ansari, F., & Taban, M. R. (2013, May). Implementation of sequential algorithm in batch processing for clutter and direct signal cancellation in passive bistatic radars. In *2013 21st Iranian Conference on Electrical Engineering (ICEE)* (pp. 1-6). IEEE.

Ansari, F., Taban, M. R., & Gazor, S. (2016). A novel sequential algorithm for clutter and direct signal cancellation in passive bistatic radars. *EURASIP Journal on Advances in Signal Processing, 2016*(1), 1–11. doi:10.118613634-016-0431-2

Antonov, I., & Saleev, V. (n.d.). *An economic method of computing lpt-sequences.* Academic Press.

Ashebir, Y. (2015). *Modelling and Forecasting the Balance of Trade in Ethiopia.* Academic Press.

Attalah, M. A., Laroussi, T., Aouane, A., & Mehanaoui, A. (2016, December). Adaptive filters for direct path and multipath interference cancellation: Application to FM-RTL-SDR based Passive Bistatic Radar. In *2016 7th International Conference on Sciences of Electronics, Technologies of Information and Telecommunications (SETIT)* (pp. 461-465). IEEE.

Attalah, M. A., Laroussi, T., Gini, F., & Greco, M. S. (2019). Range-Doppler fast block LMS algorithm for a DVB-T-based passive bistatic radar. *Signal, Image and Video Processing, 13*(1), 27–34. doi:10.100711760-018-1324-7

Auer, S., Bizer, C., Kobilarov, G., Lehmann, J., Cyganiak, R., & Ives, Z. (2007). Dbpedia: A nucleus for a web of open data. In *The semantic web* (pp. 722–735). Springer. doi:10.1007/978-3-540-76298-0_52

Auwatanamongkol, S. (2007). Inexact graph matching using a genetic algorithm for image recognition. *Pattern Recognition Letters, 28*(12), 1428–1437. doi:10.1016/j.patrec.2007.02.013

Ayenew, B. (2020). Risk for Surge Maternal Mortality and Morbidity during the Ongoing Corona Virus Pandemic. *Med Life Clin., 2*(1), 1012.

Badal, A., & Sempau, J. (2006). A package of Linux scripts for the parallelization of Monte Carlo simulations. *Computer Physics Communications, 175*(6), 440 – 450.

Bagdanov, A. D., & Worring, M. (2003). First order Gaussian graphs for efficient structure classification. *Pattern Recognition, 36*(6), 1311–1324. doi:10.1016/S0031-3203(02)00227-3

Bailey, R. (1994). Polar generation of random variates with the t-distribution. *Mathematics of Computation, 62*, 779–781.

Bai, X., & Latecki, L. J. (2008). Path similarity skeleton graph matching. *IEEE Transactions on Pattern Analysis and Machine Intelligence, 30*(7), 1282–1292. doi:10.1109/TPAMI.2007.70769 PMID:18550909

Balajee, N. (2020). *Robotics Vs. Artificial Intelligence: What is the difference?* Medium. https://medium.com/@nanduribalajee/robotics-vs-artificial-intelligence-what-is-the-difference-6adad236d997

Banko, M., & Etzioni, O. (2007, October). Strategies for lifelong knowledge extraction from the web. In *Proceedings of the 4th international conference on Knowledge capture* (pp. 95-102). 10.1145/1298406.1298425

Barndorff-Nielsen, O. E., & Shephard, N. (2001). Non-Gaussian Ornstein-Uhlenbeck based models and some of their uses in financial economics (with discussion). *Journal of the Royal Statistical Society. Series B. Methodological, 63*(2), 167–241. doi:10.1111/1467-9868.00282

Baruah, N. D., Berndt, B. C., & Chan, H. H. (2016). Ramanujan's Series for 1/π: A Survey. *The American Mathematical Monthly, 116*(7), 566–587. doi:10.4169/193009709X458555

Barzilay, R., & Lapata, M. (2008). Modeling local coherence: An entity-based approach. *Comput. Linguist., 34*(1), 1–34. doi:10.1162/coli.2008.34.1.1

Basili, V. R., Briand, L. C., & Melo, W. L. (1996). A validation of object-oriented design metrics as quality indicators. *IEEE Transactions on Software Engineering, 22*(10), 751–761. doi:10.1109/32.544352

Bengoetxea, E., Larranaga, P., Bloch, I., Perchant, A., & Boeres, C. (2002). Inexact graph matching by means of estimation of distribution algorithms. *Pattern Recognition, 35*(12), 2867–2880. doi:10.1016/S0031-3203(01)00232-1

Benko, A., & Sik Lányi, C. (2011). History of Artificial Intelligence. Encyclopedia of Information Science and Technology, 1759–1762. doi:10.4018/978-1-60566-026-4.ch276

Benth, F. E., Klüppelberg, C., Müller, G., & Vos, L. (2014). Futures pricing in electricity markets based on stable CARMA spot models. *Energy Economics, 44*, 392–406. doi:10.1016/j.eneco.2014.03.020

Bhatnagar, V., Poonia, R. C., Nagar, P., Kumar, S., Singh, V., Raja, L., & Dass, P. (2020). Descriptive analysis of COVID-19 patients in the context of India. *Journal of Interdisciplinary Mathematics, 24*(3), 489–504. doi:10.1080/09720502.2020.1761635

Bhowmick, A., & Hazarika, S. (2018). E-Mail Spam Filtering: A Review of Techniques and Trends. *Advances in Electronics, Communication and Computing, Lecture Notes in Electrical Engineering*, 583-590. doi:10.1007/978-981-10-4765-7_61

Birhanu, A. (2020). Challenges and opportunities to tackle COVID-19 spread in Ethiopia. *Journal of Peer Scientist, 2*(2), 1–6.

Biswal, A. K., Singh, D., Pattanayak, B. K., Samanta, D., & Yang, M.-H. (2021). IoT-Based Smart Alert System for Drowsy Driver Detection. *Wireless Communications and Mobile Computing.* doi:10.1155/2021/6627217

Black, F., & Scholes, M. S. (n.d.). The pricing of options and corporate liabilities. Journal of Political Economy, 637-654.

Bohouta, G., & Këpuska, V. (2018). Next-Generation of Virtual Personal Assistants (Microsoft Cortana, Apple Siri, Amazon Alexa and Google Home). *IEEE 8th Annual Computing and Communication Workshop and Conference (CCWC).* 10.1109/CCWC.2018.8301638

Bokam, Y., Guntupalli, C., Gudhanti, S., Kulandaivelu, U., Alavala, R., Alla, N., & Manne, R. (2021). Importance of Pharmacists as a Front Line Warrior in Improving Medication Compliance in Covid 19 Patients. *Indian Journal of Pharmaceutical Sciences, 83*(2), 393–396.

Bonabeau, E. (2002). Graph multidimensional scaling with self-organizing maps. *Information Sciences, 143*(1-4), 159–180. doi:10.1016/S0020-0255(02)00191-3

Bonomi, F., Milito, R., Zhu, J., & Addepalli, S. (2012).Fog computing and its role in the internet of things. In *Proc. of MCC*. ACM. 10.1145/2342509.2342513

Borkin, M. A., Bylinskii, Z., Kim, N. W., Bainbridge, C. M., Yeh, C. S., Borkin, D., Pfister, H., &Oliva, A. (2016). Beyond Memorability: Visualization Recognition and Recall. *IEEE Transactions on Visualization and Computer Graphics, 22*(1), 519–528. doi:10.1109/TVCG.2015.2467732

Box, G. E. P., & Jenkins, G. M. (1970). *Time series analysis: Forecasting and Control*. Holden-Day.

Box, G. E. P., & Jenkins, G. M. (1976). *Time series analysis: forecasting and control*. Holden-Day.

Box, G. E. P., Jenkins, G. M., & Reinsel, G. C. (1994). *Time Series Analysis: Forecasting and Control* (3rd ed.). Prentice Hall.

Box, G. E. P., & Muller, M. E. (1958). A note on the generation of random normal deviates. *Annals of Mathematical Statistics, 29*(2), 610–611.

Bradley, E. S., Toomey, M. P., Still, C. J., & Roberts, D. A. (2010, December). Multi-Scale Sensor Fusion With an Online Application: Integrating GOES, MODIS, and Webcam Imagery for Environmental Monitoring. *IEEE Journal of Selected Topics in Applied Earth Observations and Remote Sensing, 3*(4), 497–506. doi:10.1109/JSTARS.2010.2048419

Bratley, P., & Fox, B. L. (1988). Implementing Sobol's quasirandom sequence generator. *ACM Transactions on Mathematical Software, 14*(1), 88–100.

Brent, R. P. (2004). Note on Marsaglias xorshift random number generators. *Journal of Statistical Software, 11*(5), 1–4.

Brereton, P., Kitchenham, B. A., Budgen, D., Turner, M., & Khalil, M. (2007). Lessons from applying the systematic literature review process within the software engineering domain. *Journal of Systems and Software, 80*(4), 571-583.

Brereton, P., Kitchenham, B. A., Budgen, D., Turner, M., & Khalil, M. (2007). Lessons from applying the systematic literature review process within the software engineering domain. *Journal of Systems and Software, 80*(4), 571–583. doi:10.1016/j.jss.2006.07.009

Brin, S., & Page, L. (1998). The anatomy of a large-scale hypertextual web search engine. *Comput. Netw. ISDN Syst., 30*(1-7), 107–117. doi:10.1016/S0169-7552(98)00110-X

Brockwell, P. J. (2004). Representations of continuous-time ARMA processes. *Journal of Applied Probability, 41*(A), 375–382. doi:10.1239/jap/1082552212

Brockwell, P.J., &Lindner, A. (2014). Prediction of stationary Lévy-driven CARMA processes. *Journal of Econometrics, 189*(2), 263-271

Brockwell, P. J. (1994). On continuous-time threshold ARMA processes. *Journal of Statistical Planning and Inference, 39*(2), 291–303. doi:10.1016/0378-3758(94)90210-0

Brockwell, P. J., & Davis, R. A. (2001). Discussion of Levy-driven Ornstein–Uhlenbeck processes and some of their applications in financial economics. Shephard. *Journal of the Royal Statistical Society. Series B. Methodological, 63*, 218–219.

Brockwell, P. J., & Hannig, J. (2010). CARMA(p, q) generalized random processes. *Journal of Statistical Planning and Inference, 140*(12), 3613–3618. doi:10.1016/j.jspi.2010.04.028

Brockwell, P. J., & Hyndman, R. J. (1992). On continuous-time threshold autoregression. *International Journal of Forecasting, 18*(3), 439–454.

Brockwell, P. J., & Lindner, A. (2009). Existence and uniqueness of stationary Lévy-driven CARMA processes. *Stochastic Processes and Their Applications, 119*(8), 2660–2681. doi:10.1016/j.spa.2009.01.006

Brockwell, P. J., & Lindner, A. (2013). Integration of CARMA processes and spot volatility modelling. *Journal of Time Series Analysis, 34*(2), 156–167. doi:10.1111/jtsa.12011

Brockwell, P. J., & Marquardt, T. (2005). Lévy-driven and fractionally integrated ARMA processes with continuous time parameter. *Statistica Sinica, 15*, 477–494.

Broder, A. Z., Glassman, S. C., Manasse, M. S., & Zweig, G. (1997). Syntactic clustering of the web. *Computer Networks and ISDN Systems, 29*(8-13), 1157–1166. doi:10.1016/S0169-7552(97)00031-7

Cafarella, M. J., Downey, D., Soderland, S., & Etzioni, O. (2005, October). Knowitnow: Fast, scalable information extraction from the web. In *Proceedings of Human Language Technology Conference and Conference on Empirical Methods in Natural Language Processing* (pp. 563-570). 10.3115/1220575.1220646

Cardinali, R., Colone, F., Ferretti, C., & Lombardo, P. (2007, April). Comparison of clutter and multipath cancellation techniques for passive radar. In 2007 IEEE Radar Conference (pp. 469-474). IEEE.

Cardinali, R., Colone, F., Lombardo, P., Crognale, O., Cosmi, A., & Lauri, A. (2007). *Multipath cancellation on reference antenna for passive radar which exploits FM transmission*. Academic Press.

Carlson, A., Betteridge, J., Kisiel, B., Settles, B., Hruschka, E., & Mitchell, T. (2010, July). Toward an architecture for never-ending language learning. *Proceedings of the AAAI Conference on Artificial Intelligence, 24*(1), ●●●.

Carroll, B. O. (2017). *What are the 3 types of AI? A guide to narrow, general, and super arti cial intelligence What is arti cial intelligence*. https://codebots.com/artificial-intelligence/the-3-types-of-ai-is-the-third-even-possible#:~:text=There

Casamayor, A., Godoy, D., & Campo, M. (2010). Identification of non-functional requirements in textual specifications: A semi-supervised learning approach. *Information and Software Technology, 52*(4), 436-445.

Catal, C., & Diri, B. (2009). A systematic review of software fault prediction studies. *Expert Systems with Applications, 36*(4), 7346–7354. doi:10.1016/j.eswa.2008.10.027

Celik, N., Youn, H. S., Omaki, N., Lee, Y. L., Gagarin, R., & Iskander, M. F. (2011, July). Experimental evaluation of passive radar approach for homeland security applications. In *2011 IEEE International Symposium on Antennas and Propagation (APSURSI)* (pp. 224-227). IEEE.

Chakrabarti, S., & Samanta, D. (2016). Image Steganography Using Priority-Based Neural Network and Pyramid. In N. Shetty, N. Prasad, & N. Nalini (Eds.), Emerging Research in Computing, Information, Communication and Applications. Springer. https://doi.org/10.1007/978-981-10-0287-8_15.

Chambers, M. J., & Thornton, M. A. (2011). Discrete Time Representation of Continuous Time Arma Processes. *Econometric Theory, 28*(01), 219–238. doi:10.1017/S0266466611000181

Chapra, S. C., & Canale, R. P. (2010). *Numerical Methods for Engineers* (6th ed.). McGraw Hill.

Chatfield, C. (2004). *The analysis of time series: an introduction (6th ed.)*. CRC Press.

Chelva, M. S., Halse, S.V., & Ratha, B.K. (2016). Object Tracking In Real Time Embedded System Using Image Processing. *International conference on Signal Processing, Communication, Power and Embedded System (SCOPES)*.

Chen, F., Agüero, J. C., Gilson, M., Garnier, H., & Liu, T. (2017). EM-based identification of continuous-time ARMA Models from irregularly sampled data. *Automatica, 77*, 293–301. doi:10.1016/j.automatica.2016.11.020

Chioncel, C. P., Chioncel, P., Gillich, N., & Tirian, O. G. (2011). Wigner Ville Distribution in Signal Processing, using Scilab Environment. *Analele Universității "Eftimie Murgu" Reşiţa: Fascicola I, Inginerie, 18*(2), 101–106. http://anale-ing.uem.ro/

Chioncel, P., Gillich, N., Chioncel, C. P., & Elizabeta, S. (2016). *Digital Speed Cascade Control, using Scilab / Xcos Environment.* https://www.researchgate.net/publication/310124174

Chioncel, P., Silviu, D., & Chioncel, C. P. (2014). Calculation of Control Circuits in Time Domain using Scilab/ Xcos Environment. *Analele Universității "Eftimie Murgu" Reşiţa: Fascicola I, Inginerie, 21*(3). https://doaj.org/toc/1453-7397

Chong, J., Gonina, E., & Keutzer, K. (2010). Monte Carlo methods: a computational pattern for our pattern language. *Workshop on Parallel Programming Patterns.* 10.1145/1953611.1953626

Chopparapu, S. T., & Seventline, J. D. B. (Eds.). (2020). Object detection using Matlab, Scilab and Python. IAEME Publication. doi:10.34218/IJEET.11.6.2020.010

Chopparapu, S., & Seventline Dr, B. J. (2020). GUI for Object Detection using Voila Method in MATLAB. *International Journal of Electrical Engineering and Technology, 11*(4), 169–174.

Cioffi, R., Travaglioni, M., Piscitelli, G., Petrillo, A., & De Felice, F. (2020). Artificial intelligence and machine learning applications in smart production: Progress, trends, and directions. *Sustainability (Switzerland), 12*(2), 492. Advance online publication. doi:10.3390u12020492

Cleland-Huang, J., Settimi, R., Zou, X., & Solc, P. (2007). Automated classification of non-functional requirements. *Requirements Engineering, 12*(2), 103-120.

Coddington, P. D. (1996). *Random number generator for parallel computers. NHSE Review.*

Colone, F., Cardinali, R., & Lombardo, P. (2006, April). Cancellation of clutter and multipath in passive radar using a sequential approach. In *2006 IEEE Conference on Radar* (pp. 1-7). IEEE.

Colone, F., O'hagan, D. W., Lombardo, P., & Baker, C. J. (2009). A multistage processing algorithm for disturbance removal and target detection in passive bistatic radar. *IEEE Transactions on Aerospace and Electronic Systems, 45*(2), 698–722. doi:10.1109/TAES.2009.5089551

Colone, F., Palmarini, C., Martelli, T., & Tilli, E. (2016). Sliding extensive cancellation algorithm for disturbance removal in passive radar. *IEEE Transactions on Aerospace and Electronic Systems, 52*(3), 1309–1326. doi:10.1109/TAES.2016.150477

Computer History Museum. (n.d.). *Timeline of Computer History Elektro at the World ' s Fair The Three Laws.* https://www.computerhistory.org/timeline/ai-robotics/

Cook, D. J., & Holder, L. B. (Eds.). (2006). *Mining graph data.* John Wiley & Sons. doi:10.1002/0470073047

Cunningham, H. (2002). GATE: A framework and graphical development environment for robust NLP tools and applications. In *Proc. 40th annual meeting of the association for computational linguistics (ACL 2002)* (pp. 168-175). Academic Press.

D, F.-S., & M.J., M. (2009). Human Robot Interaction. In *Encyclopedia of Complexity and Systems Science* (pp. 4643–4659). Springer.

Daniel, B. (2021). The knowledge and practice towards COVID19 pandemic prevention among residents of Ethiopia. An online cross-sectional study. *PLoS One.*

Dastjerdi, A. V., & Buyya, R. (2016). Fog Computing: Helping the Internet of Things Realize Its Potential. *Computer*, *49*(8), 112–116. doi:10.1109/MC.2016.245

David, B. T., Lee, H., & Wayne, L. (2009). A comparison of CPUs, GPUs, FPGAs, and massively parallel processor arrays for random number generation. FPGA, 63–72.

de Graaf, K. A., Tang, A., Liang, P., & van Vliet, H. (2012). Ontology-based software architecture documentation In *Proceedings of Joint Working Conference on Software Architecture & 6th European Conference on Software Architecture (WICSA/ECSA). WICSA 2012*. IEEE Computer Society. 10.1109/WICSA-ECSA.212.20

Del Corro, L., & Gemulla, R. (2013, May). Clausie: clause-based open information extraction. In *Proceedings of the 22nd international conference on World Wide Web* (pp. 355-366). 10.1145/2488388.2488420

Di Blas, N., Mazuran, M., Paolini, P., Quintarelli, E., & Tanca, L. (2017). Exploratory computing: A comprehensive approach to data sensemaking. *International Journal of Data Science and Analytics*, *3*(1), 61–77. doi:10.100741060-016-0039-5

Di Lallo, A., Fulcoli, R., & Timmoneri, L. (2008, September). Adaptive spatial processing applied to a prototype passive covert radar: Test with real data. In *2008 International Conference on Radar* (pp. 139-143). IEEE. 10.1109/RADAR.2008.4653906

Dixon, M., Chong, J., & Keutzer, K. (2009). Acceleration of market value-at-risk estimation. *Workshop on High Performance Computing in Finance at Super Computing 2009*.

Dixon, M. F. (2012). Monte Carlo Based Financial Market Value-at-Risk Estimation on GPUs. In T. Bradley, J. Chong, & K. Keutzer (Eds.), *GPU Computing Gems Jade Edition* (pp. 337–353). Morgan Kaufmann. doi:10.1016/B978-0-12-385963-1.00025-3

Dong, E., Du, H., & Gardner, L. (2020). An interactive web-based dashboard to track covid-19 in real time. *The Lancet. Infectious Diseases*, *20*(5), 533–534. doi:10.1016/S1473-3099(20)30120-1 PMID:32087114

Doob, J. L. (1944). The elementary Gaussian processes. *Annals of Mathematical Statistics*, *25*(3), 229–282. doi:10.1214/aoms/1177731234

Dörr, J., Kerkow, D., Von Knethen, A., & Paech, B. (2003, June). Eliciting efficiency requirements with use cases. In Ninth international workshop on requirements engineering: foundation for software quality. In *Conjunction with CAiSE* (Vol. 3, pp. 22-23). Academic Press.

Duffy, F. H., Jones, K., Bartels, P., Albert, M., McAnulty, G. B., & Als, H. (1990). Quantified Neurophysiology with mapping: Statistical inference, Exploratory and Confirmatory data analysis. *Brain Topography*, *3*(1), 3–12. doi:10.1007/BF01128856 PMID:2094310

Dunde & Koteswara Rao. (2019). Weight matrix- based representation of sub-optimum disturbance cancellation filters. *International Journal of Intelligent Systems and Applications*, *11*(10), 15-24.

Dutta & Ghatak. (2017). A.D.M.G.S.D.: Feature selection based cluster- ing on micro-blogging data. *International Conference on Computational Intelligence in Data Mining (ICCIDM-2017)*.

Dutta & Ghatak. (2017). S.G.A.K.D.: A genetic algorithm based tweet clustering technique. *2017 International Conference on Computer Communication and Informatics (ICCCI)*, 1–6. 10.1109/ICCCI.2017.8117721

Dutta, S., Ghatak, S., Roy, M., Ghosh, S., & Das, A. K. (2015). A graph based clustering technique for tweet summarization. In *Reliability, Infocom Technologies and Optimization (ICRITO)(Trends and Future Directions), 2015 4th International Conference on* (pp. 1–6). IEEE. 10.1109/ICRITO.2015.7359276

Eleftheriou, M., Moreira, J., & Ryu, K. (Eds.). (2009). *WHPCF 2009: Proceedings of the 2nd Workshop on High Performance Computational Finance.* ACM.

El-Hadary, H., & El-Kassas, S. (2014). Capturing security requirements for software systems. *Journal of Advanced Research, 5*(4), 463-472.

Emmanuel, S. (2020). *The Impact of COVID-19 in South Africa.* Africa Program Occasional Paper. Wilson Center.

Erbacher, R. F. (2007). Panel Position Statement: The Future of CMV. *Fifth International Conference on Coordinated and Multiple Views in Exploratory Visualization (CMV 2007),* 75-75, 10.1109/CMV.2007.17

Erickson, B. H., & Nosanchuk, T. A. (1992). *Understanding data* (2nd ed.). University of Toronto Press.

Erkan, G., & Radev, D.R. (2004). *LexRank: Graph-based lexical centrality as salience in text summarization.* Academic Press.

Erman, L. D., Hayes-Roth, F., Lesser, V. R., & Reddy, D. R. (1980). The Hearsay-II speech-understanding system: Integrating knowledge to resolve uncertainty. *ACM Computing Surveys, 12*(2), 213–253. doi:10.1145/356810.356816

Etzioni, O., Cafarella, M., Downey, D., Kok, S., Popescu, A. M., Shaked, T., ... Yates, A. (2004, May). Web-scale information extraction in knowitall: (preliminary results). In *Proceedings of the 13th international conference on World Wide Web* (pp. 100-110). 10.1145/988672.988687

F.Y., O., J.E.T., A., O., A., O., H. J., O., O., & J., A. (2017). Supervised Machine Learning Algorithms: Classification and Comparison. *International Journal of Computer Trends and Technology, 48*(3), 128–138. doi:10.14445/22312803/IJCTT-V48P126

Fabbri, R. (n.d.). *Scilab & SIP for Image Processing.* Institute of Mathematical and Computer Sciences, University of Sao Paulo, Brazil.

Fabian, M. S., Gjergji, K., & Gerhard, W. E. I. K. U. M. (2007). Yago: A core of semantic knowledge unifying wordnet and wikipedia. In *16th International World Wide Web Conference, WWW* (pp. 697-706). Academic Press.

Faruqui, M., Dodge, J., Jauhar, S. K., Dyer, C., Hovy, E., & Smith, N. A. (2014). *Retrofitting word vectors to semantic lexicons.* arXiv preprint arXiv:1411.4166.

Favila, A., & Shivam, P. (2018). *Systems and methods for online fraud detection.* U.S. Patent Application No. 15/236,077.

Fernandes, R., & Cowie, A. (2004). *Capturing informal requirements as formal models* (Doctoral dissertation). Deakin University.

Figueiredo Filho, D. B., Rocha, E. C., Batista, M., Paranhos, R., & Silva, J. A. Jr. (2014). Reply on the Comments on When is Statistical Significance not Significant? *Brazilian Political Science Review, 8*(3), 141–150. doi:10.1590/1981-38212014000100024

Firesmith, D. G. (2005, August). A taxonomy of security-related requirements. In *International Workshop on High Assurance Systems (RHAS'05)* (pp. 29-30).Firesmith.

Fisher, J. (2007). *Owasp Application Security Requirements.* Available: https://www.owasp.org/index.php/File:OWASP_Application_Security_Requirements_-_Identification_and_Authorisation_v0.1_ (DRAFT).doc

Foggia, P., Percannella, G., & Vento, M. (2014). Graph matching and learning in pattern recognition in the last 10 years. *International Journal of Pattern Recognition and Artificial Intelligence, 28*(01), 1450001. doi:10.1142/S0218001414500013

Fractal Analysis of geomorphologic processes in the Linux environment using SCILAB. (2011). https://www.academia.edu/12089182/

Furmanova, K., Gratzl, S., Stitz, H., Zichner, T., Jaresova, M., Ennemoser, M., . . . Streit, M. (2017). *Taggle: Scalable visualization of tabular data through aggregation.* arXiv preprint arXiv:1712.05944.

Gao, H., Nie, H., & Li, K. (2019). Visualisation of Pareto Front Approximation: A Short Survey and Empirical Comparisons. *2019 IEEE Congress on Evolutionary Computation (CEC)*, 1750-1757. 10.1109/CEC.2019.8790298

Garcia, I., Klüppelberg, C., & Müller, G. (2010). Estimation of stable CARMA models withan application to electricity spot prices. *Statistical Modelling, 11*(5), 447–470. doi:10.1177/1471082X1001100504

Garg, S., Sünderhauf, N., Dayoub, F., Morrison, D., Cosgun, A., Carneiro, G., Wu, Q., Chin, T.-J., Reid, I., Gould, S., Corke, P., & Milford, M. (2020). Semantics for Robotic Mapping, Perception and Interaction: A Survey. In Foundations and Trends® in Robotics (Vol. 8, Issues 1–2). doi:10.1561/2300000059

Garland, M., Grand, S. L., Nickolls, J., Anderson, J. A., Hardwick, J., Morton, S., Phillips, E., Zhang, Y., & Volkov, V. (2008) Parallel Computing Experiences with CUDA. *Micro, IEEE, 28*(4), 13–27.

Garry, J. L., Smith, G. E., & Baker, C. J. (2015, June). Direct signal suppression schemes for passive radar. In *2015 Signal Processing Symposium (SPSympo)* (pp. 1-5). IEEE. 10.1109/SPS.2015.7168278

Gartner. Inc. (n.d.). *Gartner Customer Choice Awards - Analytics and Business Intelligence Platform.* https://www.gartner.com/reviews/customer-choice-awards/analytics-business-intelligence-platforms//

Gascon, H., Yamaguchi, F., Arp, D., & Rieck, K. (2013, November). Structural detection of android malware using embedded call graphs. In *Proceedings of the 2013 ACM workshop on Artificial intelligence and security* (pp. 45-54). 10.1145/2517312.2517315

Ghosh, R. A. (2003). Clustering and dependencies in free/open source software development: Methodology and tools. *First Monday, 8*(4). Advance online publication. doi:10.5210/fm.v8i4.1041

Gibson, D., Kumar, R., & Tomkins, A. (2005, August). Discovering large dense subgraphs in massive graphs. In *Proceedings of the 31st international conference on Very large data bases* (pp. 721-732). Academic Press.

Gibson, D., & de Freitas, S. (2016). Exploratory Analysis in Learning Analytics. *Tech Know Learn, 21*(1), 5–19. doi:10.100710758-015-9249-5

Gilberto, E. (2001). Probability Distributions with SCILAB. Academic Press.

Giles, M. B., Kuo, F. Y., Sloan, I. H., & Waterhouse, B. J. (2008). Quasi-Monte Carlo for finance applications. *The ANZIAM Journal, 50*, 308–323.

Gillberg, J., & Ljung, L. (2005). Frequency-Domain Identification of Continuous-Time Arma Models From Sampled Data. *IFAC Proceedings Volumes, 38*(1), 225–230. 10.3182/20050703-6-CZ-1902.00038

Gillich, G. R., & Chioncel, C. P. (2005). Simulation of dynamical systems with linear and non-linear behavior in SCICOS environment. *Annals of „Dunărea de Jos" University of Galati, Fascicle XIV. Mechanical Engineering*, 55–60.

Glasserman, P. (2003). *Monte Carlo Methods in Financial Engineering. Appl. of Math., 53.*

Godfrey, P., Gryz, J., & Lasek, P. (2016). Interactive Visualization of Large Data Sets. *IEEE Transactions on Knowledge and Data Engineering, 28*(8), 2142–2157. doi:10.1109/TKDE.2016.2557324

Gomathy, V., Padhy, N., & Samanta, D. (2020). Malicious node detection using heterogeneous cluster based secure routing protocol (HCBS) in wireless adhoc sensor networks. *Journal of Ambient Intelligence and Humanized Computing*, *11*, 4995–5001. https://doi.org/10.1007/s12652-020-01797-3

Gonzalez, R. G. (2002). *Digital Image Processing* (2nd ed.). Prentice Hall.

Goodrich, M. A., & Schultz, A. C. (2007). Human-robot interaction: A survey. *Foundations and Trends in Human-Computer Interaction*, *1*(3), 203–275. doi:10.1561/1100000005

Graif, C., Freelin, B. N., Kuo, Y. H., Wang, H., Li, Z., & Kifer, D. (2021). Network spillovers and neighborhood crime: A computational statistics analysis of employment-based networks of neighborhoods. *Justice Quarterly*, *38*(2), 344–374. doi:10.1080/07418825.2019.1602160

Gratzl, S., Gehlenborg, N., Lex, A., Pfister, H., & Streit, M. (2014). Domino: Extracting, comparing, and manipulating subsets across multiple tabular datasets. *IEEE Transactions on Visualization and Computer Graphics*, *20*(1), 2023–2032. Advance online publication. doi:10.1109/TVCG.2014.2346260 PMID:26356916

Gratzl, S., Lex, A., Gehlenborg, N., Pfister, H., & Streit, M. (2013). LineUp: Visual Analysis of Multi-Attribute Rankings. *IEEE Transactions on Visualization and Computer Graphics*, *19*(12), 2277–2286. doi:10.1109/TVCG.2013.173 PMID:24051794

Guha, A., & Samanta, D. (2020). Real-Time Application of Document Classification Based on Machine Learning. In L. Jain, S. L. Peng, B. Alhadidi, & S. Pal (Eds.), Intelligent Computing Paradigm and Cutting-edge Technologies. ICICCT 2019. Learning and Analytics in Intelligent Systems (Vol. 9). Springer. https://doi.org/10.1007/978-3-030-38501-9_37.

Guha, A., & Samanta, D. (2021). Hybrid Approach to Document Anomaly Detection: An Application to Facilitate RPA in Title Insurance. *Int. J. Autom. Comput.*, *18*, 55–72. doi:10.100711633-020-1247-y

Gujarati, D. N., & Sangeetha, S. (2007). *Basic Econometrics* (4th ed.). Tata McGraw - Hill Education.

Gujre, V. S., & Anand, R. (2020). Machine learning algorithms for failure prediction and yield improvement during electric resistance welded tube manufacturing. *Journal of Experimental & Theoretical Artificial Intelligence*, *32*(4), 601–622. doi:10.1080/0952813X.2019.1653995

Guo, Y., Yin, C., Li, M., Ren, X., & Liu, P. (2018). Mobile e-Commerce Recommendation System Based on Multi-Source Information Fusion for Sustainable e-Business. *Sustainability*, *10*(2), 147. Advance online publication. doi:10.3390u10010147

Haddara, M., & Elragal, A. (2015). The Readiness of ERP Systems for the Factory of the Future. *Procedia Computer Science*, *64*, 721–728. doi:10.1016/j.procs.2015.08.598

Haley, C., Laney, R., Moffett, J., & Nuseibeh, B. (2008). Security requirements engineering: A framework for representation and analysis. *IEEE Transactions on Software Engineering*, *34*(1), 133-153.

Hall, A., & Kabaila, A. P. (1977). *Basic Concepts of Structural Analysis*. Pitman Publishing.

Hallak, J. A., & Azar, D. T. (2020). The AI revolution and how to prepare for it. *Translational Vision Science & Technology*, *9*(2), 1–3. doi:10.1167/tvst.9.2.16 PMID:32818078

Hassel, M. (2007). *Universitets service, T., Ab, U.: Resource lean and portable automatic text summarization*. Tech. rep.

Hatebur, D., Heisel, M., & Schmidt, H. (2007, April). A pattern system for security requirements engineering. In *The Second International Conference on Availability, Reliability and Security (ARES'07)* (pp. 356-365). IEEE. 10.1109/ARES.2007.12

Haussmann, A. (2020). *Polynomial Regression: The Only Introduction You'll Need*. Towardsdatascience. https://towardsdatascience.com/polynomial-regression-the-only-introduction-youll-need-49a6fb2b86de

Haykin, S. (2002). *Adaptive filter theory*. Prentice Hall.

Heer, J., & Shneiderman, B. (2012). Interactive dynamics for visual analysis. *Queue, 10*(2), 30. doi:10.1145/2133416.2146416

High, R. (2012). *The era of Cognitive Systems: An Inside look at IBM Watson and how it Works*. IBM Corporation.

Hill, D. R. C. (1997). Object-oriented pattern for distributed simulation of large scale ecosystems. *SCS Summer Computer Simulation Conference*, 945-950.

Hissoiny, S., Després, P., & Ozell, B. (2011). *Using graphics processing units to generate random numbers*. Academic Press.

Hitaj, A., Mercuri, L., & Rroji, E. (2019). Lévy CARMA models for shocks in mortality. *Decisions Econ Finan, 42*(1), 205–227. doi:10.100710203-019-00248-9

Hoch, R. (1994). Using ir techniques for text classification in document analysis. In *Proceedings of the 17th Annual International ACM SIGIR Conference on Research and Development in Information Retrieval, SIGIR '94* (pp. 31–40). Springer-Verlag. https://dl.acm.org/citation.cfm?id=188490.188498

Howes, L., & Thomas, D. (2007). Efficient random number generation and application using CUDA. In H. Nguyen (Ed.), *GPU Gems 3, NVIDIA*. Addison Wesley.

Howland, P. E., Maksimiuk, D., & Reitsma, G. (2005). FM radio based bistatic radar. *IEE Proceedings. Radar, Sonar and Navigation, 152*(3), 107–115. doi:10.1049/ip-rsn:20045077

Huang, Ward, & Rundensteiner. (2005). Exploration of dimensionality reduction for text visualization. *Coordinated and Multiple Views in Exploratory Visualization (CMV'05)*, 63-74. . doi:10.1109/CMV.2005.8

Huba, M., Bisták, P., Fikar, M., & Kamenský, M. (2006). Blended Learning Course 'Constrained PID Control'. *7th IFAC Symposium on Advances in Control Education ACE'06*, Madrid, Spain.

Hubbard, D. (2007). *How to Measure Anything: Finding the Value of Intangibles in Business*. John Wiley & Sons.

Hu, M., Sun, A., & Lim, E. P. (2007). Comments-oriented blog summarization by sentence extraction. In *Proceedings of the Sixteenth ACM Conference on Conference on Information and Knowledge Man- agement, CIKM '07* (pp. 901–904). ACM. doi:10.1145/1321440.1321571

Hyderabad blasts. (2013). https://en.wikipedia.org/wiki/2013Hyderabadblasts

Hyndman, R. J. (1992). *Continuous-time threshold autoregressive modelling*. Unpublished doctoral dissertation]. University of Melbourne. https://robjhyndman.com/papers/PhDThesis.pdf

Iacus, S. M., Mercuri, L., & Rroji, E. (2017). COGARCH(p,q): Simulation and Inference with the yuima Package. *Journal of Statistical Software, 80*(4), 1–49. doi:10.18637/jss.v080.i04 PMID:30220889

Iacus, S. M., & Yoshida, N. (2018). *Simulation and Inference for Stochastic Processes with YUIMA: A Comprehensive R Framework for SDEs and other Stochastic Processes*. Springer. doi:10.1007/978-3-319-55569-0

Iborra, A., Caceres, D. A., Ortiz, F. J., Franco, J. P., Palma, P. S., & Alvarez, B. (2009). Design of service robots: Experiences using software engineering. *IEEE Robotics & Automation Magazine, 16*(1), 24–33. doi:10.1109/MRA.2008.931635

Idreos, S., Papaemmanouil, O., & Chaudhuri, S. (2015). Overview of data exploration techniques. In *Proceedings of the 2015 ACM SIGMOD International Conference on Management of Data*. ACM. 10.1145/2723372.2731084

Ikhankar, R., Kuthe, V., Ulabhaje, S., Balpande, S., & Dhadwe, M. (2015). Pibot:The Raspberry Pi Controlled MultiEn-vironment Robot For Surveillance & Live Streaming. In *2015 International Conference on Industrial Instrumentation and Control (ICIC)*. College of Engineering Pune,.

Im, J.-F., Villegas, F. G., & McGuffin, M. J. (2013). VisReduce: Fast and responsive incremental information visualization of large datasets. *IEEE International Conference on Big Data*, 25-32.

Indriasari, E., Soeparno, H., Gaol, F. L., & Matsuo, T. (2019). Application of Predictive Analytics at Financial Institutions: A Systematic Literature Review. *Proceedings - 2019 8th International Congress on Advanced Applied Informatics, IIAI-AAI 2019, February 2020*, 877–883. 10.1109/IIAI-AAI.2019.00178

Information Visualization, Second Edition: Perception for Design (Interactive Technologies) by Colin Ware (2004–04-21). (1783). Morgan Kaufmann.

International Federation of Robotics. (2020). *IFR presents world robotics report*. https://ifr.org/ifr-press-releases/news/record-2.7-million-robots-work-in-factories-around-the-globe

Isaacs, E., Damico, K., Ahern, S., Bart, E., & Singhal, M. (2014). Footprints: A visual search tool that supports discovery and coverage tracking. *IEEE Transactions on Visualization and Computer Graphics*, *20*(12), 1793–1802. doi:10.1109/TVCG.2014.2346743 PMID:26356893

ISO. (2009). *Evaluation, ISO/IEC 15408: Information technology - Security techniques*. Available: https://www.iso.org/iso/home/store/catalogue_ics/catalogue_detail_ics.htm

ISO. (2015). *ISO/IEC 27001 - Information security management*. Available: https://www.iso.org/iso/home/standards/management-standards/iso27001.htm

Iyer, G., DuttaDuwarah, S., & Sharma, A. (2012). DataScope: Interactive visual exploratory dashboards for large multidimensional data. *IEEE Workshop on Visual Analytics in Healthcare (VAHC)*, 17-23.

Jain, A. K., & Dubes, R. C. (1988). *Algorithms for clustering data*. Prentice-Hall, Inc.

Jakab, F., Andoga, V., Kapova, L., & Nagy, M. (2006). Virtual Laboratory:Component Based Architecture Implementation Experience. Electronic Computer and Informatics.

James, F. (1990). A review of pseudorandom number generators. *Computer Physics Communications*, *60*, 329–344.

Janorkar, D. S. (2018). True Value of Pi (π) Now is 3.141592653 we Call This as Goba Constant we Symbolic it as This Goba, This Letter. *International Journal of Mathematics Trends and Technology*, *59*(1), 27–34. doi:10.14445/22315373/IJMTT-V59P505

Janowczyk, A., Chandran, S., & Aluru, S. (2008). Fast, Processor-Cardinality Agnostic PRNG with a Tracking Application. In *Computer Vision, Graphics and Image Processing, 2008. ICVGIP08. Sixth Indian Conference on*, (pp. 171–178). Academic Press.

Jan, S. (2004). GATE: A simulation toolkit for PET and SPECT. *Physics in Medicine and Biology*, *49*, 4543–4561.

Jarrah, A. A., & Jamali, M. M. (2016). A parallel implementation of extensive cancellation algorithm (ECA) for passive bistatic radar (PBR) on a GPU. *Journal of Signal Processing Systems for Signal, Image, and Video Technology*, *85*(2), 201–209. doi:10.100711265-015-1066-5

Javed, W., & Elmqvistm N. (2013). ExPlates: spatializing interactive analysis to scaffold visual exploration. *Computer Graphics Forum*, *32*, 441-450. doi:10.1111/cgf.12131

Jembere, G. B., Cho, Y., & Jung, M. (2018). Decomposition of Ethiopian life expectancy by age and cause of mortality. *PLoS One*, *13*(10), 1990–2015. doi:10.1371/journal.pone.0204395 PMID:30281624

Jhon, W. T. (1961). The future of Data Analysis. Research Sponsored by the Army Research Office. Princeton University.

Ji, C., Yang, L., Zhu, W., Liu, Y., & Deng, K. (2018). On. Bayesian Inference for Continuous-time Autoregressive Models without Likelihood. *21st International Conference on Information Fusion*, 2137-214. 10.23919/ICIF.2018.8455660

Jichuan, L., Yaodong, Z., Yongke, Z., & Xiaode, L. (2013, August). Direct path wave purification for passive radar with normalized least mean square algorithm. In *2013 IEEE International Conference on Signal Processing, Communication and Computing (ICSPCC 2013)* (pp. 1-4). IEEE. 10.1109/ICSPCC.2013.6663987

Jin, X., Wah, B. W., Cheng, X., & Wang, Y. (2015). Significance and Challenges of Big Data Research. *Big Data Research*, *2*(2), 59–64. doi:10.1016/j.bdr.2015.01.006

Joe, S., & Kuo, F. (2003). Remark on algorithm 659: Implementing Sobol's quasi-random sequence generator. *ACM Transactions on Mathematical Software*, *29*(1), 49–57.

Johnstone, I. M., & Titterington, D. M. (2009). Statistical challenges of high-dimensional data. *Philos. Trans. R. Soc.*, *367*(1906), 4237–4253. doi:10.1098/rsta.2009.0159

Jondeau, E., Poon, S., & Rockinger, M. (2007). *Financial Modeling Under Non-Gaussian Distributions*. Springer Finance.

Jónsdóttir, G. M., Hayes, B., & Milano, F. (2018).Continuous-Time ARMA Models for Data-Based Wind Speed Models. *Power Systems Computation Conference (PSCC)*, 1-7. 10.23919/PSCC.2018.8442659

Joo, S. H., Manzoor, S., Rocha, Y. G., Bae, S. H., Lee, K. H., Kuc, T. Y., & Kim, M. (2020). Autonomous navigation framework for intelligent robots based on a semantic environment modeling. *Applied Sciences (Switzerland)*, *10*(9), 1–30. doi:10.3390/app10093219

Jöreskog, K. G., & Sörbom, D. (1993). *LISREL 8: Structural equation modeling with the SIMPLIS command language. Scientific Software International*. Lawrence Erlbaum Associates, Inc.

Jorion, P. (2007). *Value-at-Risk: The New Benchmark for Managing Financial Risk* (3rd ed.). McGraw-Hill.

Joshi, N. (2019). 7 Types of Artificial Intelligence. *Forbes*, 1–6. https://www.forbes.com/sites/cognitiveworld/2019/06/19/7-types-of-artificial-intelligence/#b69f9bb233ee

Kaiser, G. E., Feiler, P. H., & Popovich, S. S. (1988). Intelligent assistance for software development and maintenance. *IEEE Software*, *5*(3), 40–49. doi:10.1109/52.2023

Kaisler, S., Armour, F., Espinosa, J. A., & Money, W. (2013). Big data: Issues and challenges moving forward. *46th Hawaii International Conference on System Sciences (HICSS)*, 995-1004. 10.1109/HICSS.2013.645

Kaiya, H., Osada, A., & Kaijiri, K. (2004, September). Identifying stakeholders and their preferences about NFR by comparing use case diagrams of several existing systems. In *Proceedings. 12th IEEE International Requirements Engineering Conference*, 2004 (pp. 112-121). IEEE. 10.1109/ICRE.2004.1335669

Kalyanam, J. (2017). *Machine Learning and Applications on Social Media Data*. UC San Diego. Retrieved from https://escholarship.org/uc/item/6545w71z

Kalyani, A., Premalatha, B., & Ravi Kiran, K. (2018). Real Time Emotion Recognition from Facial Images using Raspberry Pi. IJATIR, 10(1), 13-16.

Kamat, N., & Nandi, A. (2014). *InfiniViz: Interactive Visual Exploration using Progressive Bin Refinement.* arXiv preprint arXiv:1710.01854.

Karmshu. (2012). *Probabilistic Simulation and Monte Carlo Method.* INSPIRE Science Camp, ISM Dhanbad.

Karthikeyan, M. P., Samanta, D., Banerjee, A., Roy, A., & Inokawa, H. (2021). Design and Development of Terahertz Medical Screening Devices. In M. Chakraborty, R. K. Jha, V. E. Balas, S. N. Sur, & D. Kandar (Eds.), Trends in Wireless Communication and Information Security. Lecture Notes in Electrical Engineering (Vol. 740). Springer. https://doi.org/10.1007/978-981-33-6393-9_40.

Kassimali, A. (1999). *Matrix Analysis of Structures.* Brooks/Cole Publishing Company.

Kaur, J., Bedi, R., & Gupta, S. K. (2018). Product Recommendation Systems a Comprehensive Review. *International Journal on Computer Science and Engineering, 6*(6), 1192–1195. doi:10.26438/ijcse/v6i6.11921195

Kedzie, C., McKeown, K., & Diaz, F. (2015). Predicting Salient Updates for Disaster Summarization. *Proc. ACL.*

Keim, D. A. (2002). Information visualization and visual data mining. *IEEE Transactions on Visualization and Computer Graphics, 8*(1), 1–8. doi:10.1109/2945.981847

Kelly, J. E. (2015). Computing, cognition and the future of knowing. Whitepaper IBM Res.

Keutzer, K., & Mattson, T. (2009). *Our pattern language (opl).* Academic Press.

Khadri, S. K. A. (2014). Approach of Message Communication Using Fibonacci Series. Cryptology. Lecture Notes on Information Theory. doi:10.12720/lnit.2.2.168-171

Khakurel, J., Penzenstadler, B., Porras, J., Knutas, A., & Zhang, W. (2018). The Rise of Artificial Intelligence under the Lens of Sustainability. In Technologies (Vol. 6, Issue 4, p. 100). doi:10.3390/technologies6040100

Khamparia, A., Singh, P. K., Rani, P., Samanta, D., Khanna, A., & Bhushan, B. (2020). An internet of health things-driven deep learning framework for detection and classification of skin cancer using transfer learning. *Trans Emerging Tel Tech.* doi:10.1002/ett.3963

Khongsai, L. (2020). Combating the Spread of COVID-19 Through Community Participation. *Global Social Welfare: Research, Policy & Practice.* Advance online publication. doi:10.100740609-020-00174-4 PMID:32837833

Kim, S., & Hur, J. (2020). A Probabilistic Modeling Based on Monte Carlo Simulation of Wind Powered EV Charging Stations for Steady-States Security Analysis. *MDPI.* doi:10.3390/en13205260

Kitapci, H., & Boehm, B. W. (2006, September). Using a hybrid method for formalizing informal stakeholder requirements inputs. In *Fourth International Workshop on Comparative Evaluation in Requirements Engineering (CERE'06-RE'06 Workshop)* (pp. 48-59). IEEE. 10.1109/CERE.2006.8

Kitchenham, B. (2004). Procedures for perfoming systematic review, software engineering group. Department of Computer Science, Keele University, United Kingdom and Empirical Software Engineering, National ICT Australia Ltd., TR/SE-0401.

Kiyavitskaya, N., Zeni, N., Breaux, T. D., Antón, A. I., Cordy, J. R., Mich, L., & Mylopoulos, J. (2008, October). Automating the extraction of rights and obligations for regulatory compliance. In *International Conference on Conceptual Modeling* (pp. 154-168). Springer. 10.1007/978-3-540-87877-3_13

Knauss, E., Houmb, S., Schneider, K., Islam, S., & Jürjens, J. (2011, March). Supporting requirements engineers in recognising security issues. In *International Working Conference on Requirements Engineering: Foundation for Software Quality* (pp. 4-18). Springer. 10.1007/978-3-642-19858-8_2

Kof, L. (2005). *Text analysis for requirements engineering* (Doctoral dissertation). Technische Universität München.

Kolawole, E. B. (2001). *Tests and Measurement, AdoEkiti.* Yemi Printing Services.

Komorowski, M., Marshall, D. C., Salciccioli, J. D., & Crutain, Y. (2016). Exploratory Data Analysis. In *Secondary Analysis of Electronic Health Records.* Springer. doi:10.1007/978-3-319-43742-2_15

Konopka, B. M., Lwow, F., Owczarz, M., & Łaczmański, Ł. (2018). Exploratory data analysis of a clinical study group: Development of a procedure for exploring multidimensional data. *PLoS One, 13*(8), e0201950. doi:10.1371/journal.pone.0201950 PMID:30138442

Krekhov, A., Cmentowski, S., Waschk, A., & Krüger, J. (2020, January). Deadeye Visualization Revisited: Investigation of Preattentiveness and Applicability in Virtual Environments. *IEEE Transactions on Visualization and Computer Graphics, 26*(1), 547–557. doi:10.1109/TVCG.2019.2934370 PMID:31425106

Kruchten, P., Lago, P., & Van Vliet, H. (2006, June). Building up and reasoning about architectural knowledge. In *International conference on the quality of software architectures* (pp. 43-58). Springer. 10.1007/11921998_8

Kulpa, K. (2008, September). The CLEAN type algorithms for radar signal processing. In *2008 Microwaves, Radar and Remote Sensing Symposium* (pp. 152-157). IEEE. 10.1109/MRRS.2008.4669567

Kumar, V., & Garg, M. L. (2018). Predictive Analytics: A Review of Trends and Techniques. *International Journal of Computers and Applications, 182*(1), 31–37. doi:10.5120/ijca2018917434

Kupiec, J., Pedersen, J., & Chen, F. (1995). A trainable document summarizer. In *Proceedings of the 18th Annual International ACM SIGIR Conference on Research and Development in Information Retrieval, SIGIR '95* (pp. 68–73). ACM. doi:10.1145/215206.215333

Kureethara, V., Biswas, J., & Debabrata Samanta, N. G. (n.d.). Balanced Constrained Partitioning of Distinct Objects. *International Journal of Innovative Technology and Exploring Engineering.* Doi:10.35940/ijitee.K1023.09811S19

L'ecuyer, P. (1988). Efficient and portable combined random number generators. *Commun. ACM, 31*(6), 742–751.

L'Ecuyer, P. (1990). Random numbers for simulation. Communications of the ACM, 85-98.

L'ecuyer, P. (1999). Tables of linear congruential generators of different sizes and good lattice structure. *Math. Comput., 68*(225), 249–260.

L'Ecuyer, P., & Simard, R. (2003). TESTU01: a software library in ANSI C for empirical testing of random number generators. Department d'Informatique et de Recherche Operationnelle, University of Montreal.

L'ecuyer, P., & Simard, R. (2007). TestU01: A C library for empirical testing of random number generators. *ACM Trans. Math. Softw., 33*(4), 22.

Lami, G., Gnesi, S., Fabbrini, F., Fusani, M., & Trentanni, G. (2004). *An automatic tool for the analysis of natural language requirements. Informe técnico, CNR Information Science and Technology Institute.*

Langdon, B. (2009). A Fast High Quality Pseudo Random Number Generator for nVidia CUDA. *GECCO 2009 Workshop, Tutorial and Competition on Computational Intelligence on Consumer Games and Graphics Hardware CIGPU.*

Larsson, E. K., Mossberg, M., & Soderstrom, T. (2006). An Overview of Important Practical Aspects of Continuous-Time ARMA System Identification. *Circuits, Systems, and Signal Processing, 25*(1), 17–46. doi:10.100700034-004-0423-6

Lash & Zhao. (2016). Early Predictions of Movie Success: the Who, What, and When of Profitability Artificial Intelligence (cs.AI). *Social and Information Networks.*

Lauri, A., Colone, F., Cardinali, R., Bongioanni, C., & Lombardo, P. (2007, March). *Analysis and emulation of FM radio signals for passive radar. In 2007 IEEE Aerospace Conference.* IEEE.

Lazaro, D., Breton, V., & Buvat, I. (2004). Feasibility and value of fully 3D Monte-Carlo reconstruction in single photon emission computed tomography. *Nuclear Instruments & Methods in Physics Research. Section A, Accelerators, Spectrometers, Detectors and Associated Equipment, 527,* 195–200.

Lazaro, D., El Bitar, Z., Breton, V., Hill, D. R. C., & Buvat, I. (2005). Fully 3D Monte Carlo reconstruction in SPECT: A feasibility study. *Physics in Medicine and Biology, 50,* 3739–3754.

Leão, C. P., & Rodrigues, A. E. (2004). Transient and steady-state models for simulated moving bed processes: Numerical solutions. *Computers & Chemical Engineering, 28*(9), 1725–1741.

Lee, P. S., & Howe, B. (2015). Dismantling composite visualizations in the scientific literature. In *International Conference on Pattern Recognition Applications and Methods.* ICPRAM. 10.5220/0005213100790091

Lee, H. (2020). COVID-19 perception, knowledge, and preventive practice: Comparison between South Korea, Ethiopia, and Democratic Republic of Congo. *African Journal of Reproductive Health, 24*(2), 66–77.

Lee, M. L., Yang, L. H., Hsu, W., & Yang, X. (2002, November). XClust: clustering XML schemas for effective integration. In *Proceedings of the eleventh international conference on Information and knowledge management* (pp. 292-299). 10.1145/584792.584841

Lee, P. S., West, J. D., & Howe, B. (2018). Viziometrics: Analyzing Visual Information in the Scientific Literature. *IEEE Transactions on Big Data, 4*(1), 117–129. doi:10.1109/TBDATA.2017.2689038

Lee, S. W., Muthurajan, D., Gandhi, R. A., Yavagal, D., & Ahn, G. J. (2006). Building decision support problem domain ontology from natural language requirements for software assurance. *International Journal of Software Engineering and Knowledge Engineering, 16*(06), 851–884. doi:10.1142/S0218194006003051

Lenat, D. B., & Guha, R. V. (1989). *Building large knowledge-based systems; representation and inference in the Cyc project.* Addison-Wesley Longman Publishing Co., Inc.

Lewis, J. M., Hull, P. M., Weinberger, K. Q., & Saul, L. K. (2008). Mapping Uncharted Waters: Exploratory Analysis, Visualization, and Clustering of Oceanographic Data. *2008 Seventh International Conference on Machine Learning and Applications,* 388-395, 10.1109/ICMLA.2008.125

Ley, J., & el Moctar, O. (2021). A Comparative Study of Computational Methods for Wave-Induced Motions and Loads. *Journal of Marine Science and Engineering, 9*(1), 83. doi:10.3390/jmse9010083

Li, T. (2017, December). Identifying security requirements based on linguistic analysis and machine learning. In *2017 24th Asia-Pacific Software Engineering Conference (APSEC)* (pp. 388-397). IEEE. 10.1109/APSEC.2017.45

Li, Y., & Mascagni, M. (2003). Improving Performance via Computational Replication on a Large-Scale Computational Grid. *CCGRID, 3rd International Symposium on Cluster Computing and the Grid,* 442-446.

Liguš, J., Ligušová, J., & Zolotová, I. (2005). *Distributed Remote Laboratories in Automation Education.* 16th EAEEIE Annual Conf. on Innovation in Education for Electr. and Information Eng., Lappeenranta, Finland.

Lin, C.-Y. (2004). ROUGE: A package for automatic evaluation of summaries. In *Proc. Workshop on Text Summarization Branches Out.* ACL.

Lin, L., Nuseibeh, B., Ince, D., Jackson, M., & Moffett, J. (2003). Introducing abuse frames for analyzing security requirements. *Proceedings of the 11th IEEE international requirements engineering conference (RE'03)*, 371–2. 10.1109/ICRE.2003.1232791

Lin, Z. Q., Xie, B., Zou, Y. Z., Zhao, J. F., Li, X. D., Wei, J., Sun, H.-L., & Yin, G. (2017). Intelligent development environment and software knowledge graph. *Journal of Computer Science and Technology*, *32*(2), 242–249. doi:10.100711390-017-1718-y

Liu, Y., & Man, H. (2005, March). Network vulnerability assessment using Bayesian networks. In *Data mining, intrusion detection, information assurance, and data networks security 2005* (Vol. 5812, pp. 61–71). International Society for Optics and Photonics. doi:10.1117/12.604240

Lönnblad, L. (1994). CLHEP – a project for designing a C++ class library for high energy physics. *Computer Physics Communications*, *84*, 307–316.

López, C., Codocedo, V., Astudillo, H., & Cysneiros, L. M. (2012). Bridging the gap between software architecture rationale formalisms and actual architecture documents: An ontology-driven approach. *Science of Computer Programming*, *77*(1), 66–80. doi:10.1016/j.scico.2010.06.009

Luhn, H.P. (1958). The automatic creation of literature abstracts. *IBM J. Res. Dev.*, *2*(2), 159–165. doi:10.1147/rd.22.0159

Lynch, C. (2008). Big data: How do your data grow? *Nature*, *455*(7209), 28–29. doi:10.1038/455028a PMID:18769419

Macgillivray, H. T., & Dodd, R. J. (2004). Monte-Carlo simulations of galaxy systems. Academic Press.

Mackie, S., McCreadie, R., Macdonald, C., & Ounis, I. (2014). Comparing Algorithms for Microblog Summarisation. *Proc. CLEF*.

Magesh, S., & Krishnan, N. (2016). *A survey on machine learning approaches to social media analytics*. Academic Press.

Mahesh, K.(1997). *Hypertext Summary Extraction for Fast Document Browsing*. Academic Press.

Maheswari, & Geetha, Kumar, Karuppiah, Samanta, & Park. (n.d.). PEVRM: Probabilistic Evolution based Version Recommendation Model for Mobile Applications. *IEEE Access: Practical Innovations, Open Solutions*. Advance online publication. doi:10.1109/ACCESS.2021.3053583

Mahmood, N.H., & Mansor, M.A. (2012). Red Blood Cells Estimation Using Hough Transform Technique. *SIPIJ*, *3*(2).

Maigne, L., Hill, D. R. C., Calvat, P., Breton, V., Reuillon, R., Lazaro, D., Legre, Y., & Donnarieix, D. (2004). Parallelization of Monte Carlo simulations and submission to a grid environment. *Parallel Processing Letters*, *14*, 177–196.

Maini, R., & Aggarwal, H. (n.d.). Study and Computational of Various Image Edge Detection Techniques. *International Journal of Image Processing*, *3*(1).

Maitra, M., Gupta, R.K., & Mukherjee, M. (2012). Detection and Counting of Red Blood Cells in Blood Cell Images using Hough Transform. *International Journal of Computer Application, 53*(16).

Malhotra, R., Chug, A., Hayrapetian, A., & Raje, R. (2016, February). Analyzing and evaluating security features in software requirements. In *2016 International Conference on Innovation and Challenges in Cyber Security (ICICCS-INBUSH)* (pp. 26-30). IEEE. 10.1109/ICICCS.2016.7542334

Malviya, M., Buswell, N. T., & Berdanier, C. G. (2021). Visual and Statistical Methods to Calculate Intercoder Reliability for Time-Resolved Observational Research. *International Journal of Qualitative Methods*, *20*, 16094069211002418. doi:10.1177/16094069211002418

Manne, R., & Kantheti, S. (2020). COVID-19 and Its Impact on Air Pollution. *International Journal for Research in Applied Science & Engineering Technology*, 8(11), 344-346. doi:10.22214/ijraset.2020.32139

Mao, J., & Jain, A. K. (1995, March). Artificial neural networks for feature extraction and multivariate data projection. *IEEE Transactions on Neural Networks*, 6(2), 296–317. doi:10.1109/72.363467 PMID:18263314

Marr, B. (2020). *Artificial Human Beings: The Amazing Examples Of Robotic Humanoids And Digital Humans*. Forbes. https://www.forbes.com/sites/bernardmarr/2020/02/17/artificial-human-beings-the-amazing-examples-of-robotic-humanoids-and-digital-humans/#7b1a2c9c5165

Marr, B., & Ward, M. (2019). *Artificial Intelligence in Practice*. Wiley.

Marsaglia, G. (1995). *Diehard, a battery of tests for random number generators*. Academic Press.

Marsaglia, G. (1997). *A random number generator for C*. Sci. Math. Num-analysis news group.

Marsaglia, G., & Zaman, A. (1987). *Toward a Universal Random Number Generator*. Florida State University.

Marsaglia, G., & Zaman, A. (1991). A New Class of Random Number Generators. *The Annals of Applied Probability*, 1(3), 462-480.

Marsaglia, G. (2003). Random number generation. In *Encyclopedia of Computer Science* (pp. 1499–1503). John Wiley and Sons Ltd.

Marsaglia, G. (2003). Xorshift RNGs. *Journal of Statistical Software*, 8(14), 2003.

Mascagni, M., Ceperley, D., & Srinivasan, A. (2000). SPRNG: A scalable library for pseudorandom number generation. *ACM Transactions on Mathematical Software*, 26, 618–619.

Mascagni, M., & Chi, H. (2004). Parallel linear congruential generators with Sophie-Germain moduli. *Parallel Computing*, 30, 1217–1231.

Mascagni, M., & Srinivasan, A. (2004). Parameterizing parallel multiplicative lagged-Fibonacci generators. *Parallel Computing*, 30, 899–916.

Matsumoto, M., & Nishimura, T. (1998). Mersenne twister: a 623-dimensionally equidistributed uniform pseudo-random number generator. *ACM Trans. Model. Comput. Simul.*, 8(1), 3–30.

Matsumoto, M., & Nishimura, T. (1997). Mersenne Twister: A 623-dimensionally equidistributed uniform pseudorandom number generator. *Proceedings of the 29th conference on Winter simulation*, 127-134.

Matsumoto, M., & Nishimura, T. (2000). *Dynamic creation of pseudorandom number generators* (Vol. 1998). Monte Carlo and Quasi- Monte Carlo Methods.

Ma, Y., Shan, T., Zhang, Y. D., Amin, M. G., Tao, R., & Feng, Y. (2016). A novel two-dimensional sparse-weight NLMS filtering scheme for passive bistatic radar. *IEEE Geoscience and Remote Sensing Letters*, 13(5), 676–680.

McAndrew, T., Wattanachit, N., Gibson, G. C., & Reich, N. G. (2021). Aggregating predictions from experts: A review of statistical methods, experiments, and applications. *Wiley Interdisciplinary Reviews: Computational Statistics*, 13(2), e1514. doi:10.1002/wics.1514 PMID:33777310

McDonald. (2015). *Parallel and Iterative Processing for Machine Learning Recommendations with Spark*. https://www.mapr.com/blog/parallel-and-iterative-processing-machine-learning-recommendations-spark

Mei, S., Montanari, A., & Nguyen, P. M. (2018). A mean field view of the landscape of two-layer neural networks. *Proceedings of the National Academy of Sciences of the United States of America, 115*(33), E7665–E7671. doi:10.1073/pnas.1806579115 PMID:30054315

Mekonnen, H, Z. (2020). *COVID-19 in Ethiopia: Assessment of How the Ethiopian Government has Executed Administrative Actions and Managed Risk Communications and Community Engagement.* Academic Press.

Mellado, D., Blanco, C., Sánchez, L. E., & Fernández-Medina, E. (2010). A systematic review of security requirements engineering. *Computer Standards & Interfaces, 32*(4), 153–165. doi:10.1016/j.csi.2010.01.006

Mellado, D., Fernández-Medina, E., & Piattini, M. (2007). A common criteria based security requirements engineering process for the development of secure information systems. *Computer Standards & Interfaces, 29*(2), 244–253. doi:10.1016/j.csi.2006.04.002

Metropolis, N. (1987). The Beginning of the Monte Carlo Method. *Los Alamos Science*, 125–130.

Meyer-Baese, A., Wismueller, A., & Lange, O. (2004, September). Comparison of two exploratory data analysis methods for fMRI: Unsupervised clustering versus independent component analysis. *IEEE Transactions on Information Technology in Biomedicine, 8*(3), 387–398. doi:10.1109/TITB.2004.834406 PMID:15484444

Michael, L. O. (2004). Numerical Computing with IEEE Floating Point Arithmetic. SIAM Publications.

Mitchell, T. M., Allen, J., Chalasani, P., Cheng, J., Etzioni, O., Ringuette, M., &Schlimmer, J. C. (1991). Theo: A framework for self-improving systems. *Architectures for Intelligence*, 323-355.

Moffett, J. D., & Nuseibeh, B. A. (2003). *A framework for security requirements engineering.* Report-University of York Department of Computer Science YCS.

Moffett, J. D., Haley, C. B., & Nuseibeh, B. (2004). Core security requirements artefacts. Department of Computing, The Open University, Milton Keynes, UK, Technical Report, 23.

Moon, N., Hsu, Y.-W., & Singh, R. (2006). A Multiple-Perspective, Interactive Approach for Web Information Extraction and Exploration. *22nd International Conference on Data Engineering Workshops (ICDEW'06)*, 41-41. 10.1109/ICDEW.2006.11

Moravec, H. P. (2021). *Robot.* Britannica Online Encyclopedia. https://www.britannica.com/technology/robot-technology

Moro, B. (1995). The full monte. *Risk Mag., 8*(2), 57–58.

Moulick, R. (2019). Calculating the value of Pi (π): A Monte Carlo Scheme in Scilab. *International Journal of Emerging Technologies and Innovative Research, 6*(1), 600-603. www.jetir.org

Mukherjee, M., & Samanta, D. (2014, June). Fibonacci Based Text Hiding Using Image Cryptography. *Lecture Notes on Information Theory, 2*(2), 172–176. doi:10.12720/lnit.2.2.172-176

Mulaik, S. A. (1984). Empiricism and exploratory statistics. *Philosophy of Science, 52*, 410–430. doi:10.1086/289258

Müller, G., & Seibert, A. (2019). Bayesian estimation of stable CARMA spot models for electricity prices. *Energy Economics, 78*, 267–277. doi:10.1016/j.eneco.2018.10.016

Munzner, T., Johnson, C., Moorhead, R., Pfister, H., Rheingans, P., & Yoo, T. (2006). NIH-NSF visualization research challenges report summary. *IEEE Computer Graphics and Applications, 26*(2), 20–24. doi:10.1109/MCG.2006.44 PMID:16548457

Murshed, S. Z., Dutta, A., Chatterjee, S., Mondal, I., Saha, A., Saha, S., Kundu, D., Ghosh, S., & Das Gupta, S. (2016). Controlling an Embedded Robot through Image Processing based Object Tracking using MATLAB. *10th International Conference on Intelligent Systems and Control (ISCO)*. 10.1109/ISCO.2016.7726922

Nagappan, N., Ball, T., & Zeller, A. (2006, May). Mining metrics to predict component failures. In *Proceedings of the 28th international conference on Software engineering* (pp. 452-461). 10.1145/1134285.1134349

Natrella, M. (2010). *NIST/SEMATECHe-Handbook of Statistical Methods*. NIST/SEMATECH.

Nayak, A., Kesri, V., & Dubey, R. K. (2020). Knowledge graph based automated generation of test cases in software engineering. In *Proceedings of the 7th ACM IKDD CoDS and 25th COMAD* (pp. 289-295). 10.1145/3371158.3371202

Nepal earthquake. (2015). https://en.wikipedia.org/wiki/2015Nepalearthquake

Nguyen, N.T., Duong, A.D., & Vu, H.Q. (2011). Cell Splitting with High Degree of Overlapping in Peripheral Blood Smear. *International Journal of Computer Theory and Engineering, 3*(3).

Nguyen, M. T., Kitamoto, A., & Nguyen, T. T. (2015). Tsum4act: A framework for retrieving and summarizing actionable tweets during a disaster for reaction. *Proc. PAKDD*. 10.1007/978-3-319-18032-8_6

Nguyen, V. H., & Tran, L. M. S. (2010, September). Predicting vulnerable software components with dependency graphs. In *Proceedings of the 6th International Workshop on Security Measurements and Metrics* (pp. 1-8). 10.1145/1853919.1853923

North India floods. (2013). https://en.wikipedia.org/wiki/2013NorthIndiafloods

NVIDIA Corporation Inc. (2009). *NVIDIA CUDA Compute Unified Device Architecture Programming Guide version 2.3*. Author.

Nyland, L., Harris, M., & Prins, J. (2007). Fast N-Body Simulation with CUDA. In H. Nguyen (Ed.), *GPU Gems 3*. Addison Wesley Professional.

OECD. (2019). *AI & Society*. https://www.oecd-ilibrary.org/content/publication/eedfee77-en

Ojha, S., & Sakhare, S. (2015). Image Processing Techniques for Object Tracking in Video Surveillance- A Survey. *2015 International Conference on Pervasive Computing (ICPC)*.

Olariu, A. (2014). Efficient online summarization of microblogging streams. *Proc. EACL(short paper), 236–240*. 10.3115/v1/E14-4046

Oliveira, S., & Stewart, D. (2006). *Writing Scientific Software: A Guide to Good Style*. Cambridge University Press., doi:10.1017/CBO9780511617973

Otom, R. (2020). *COVID-19 Awareness to Kenyans*. Academic Press.

OWASP. (2015). *Category: OWASP Application Security Verification Standard Project*. Available: https://www.owasp.org/index.php/Main_Page

Owen-Hill, A. (2017). *What's the Difference Between Robotics and Artificial Intelligence?* Robotiq. https://blog.robotiq.com/whats-the-difference-between-robotics-and-artificial-intelligence

Ozili, P, K. (2020). *Spillover of COVID-19 : impact on the Global Economy*. Academic Press.

Page, L., Brin, S., Motwani, R., & Winograd, T. (1999). *The pagerank citation ranking: Bringing order to the web*. Academic Press.

Palmer, J. E., & Searle, S. J. (2012, May). Evaluation of adaptive filter algorithms for clutter cancellation in passive bistatic radar. In *2012 IEEE Radar Conference* (pp. 493-498). IEEE.

Panciatici, P., & Chieh, A. S. (2011). Equation-based hybrid modeling of power systems for time-domain simulation. *Power and Energy Society General Meeting. IEEE*, 1-9. 10.1109/PES.2011.6039155

Pande, A., Manchanda, M., Bhat, H. R., Bairy, P. S., Kumar, N., & Gahtori, P. (2021). Molecular insights into a mechanism of resveratrol action using hybrid computational docking/CoMFA and machine learning approach. *Journal of Biomolecular Structure & Dynamics*, 1–15. doi:10.1080/07391102.2021.1910572 PMID:33829956

Pandey, D. (2020). Infectivity, Preclusion, and Control (IPC) of Pandemic Novel COVID-19. *International Journal of Computer Engineering In Research Trends*, 7(5), 1–8.

Pandey, D. (2021). *Covid-19: A Framework for Effective Delivering of Online Classes during Lockdown*. Human Arenas.

Panneton, F., & L'ecuyer, P. (2005). On the xorshift random number generators. *ACM Trans. Model. Comput. Simul.*, 15(4), 346–361.

Park, S. K., & Miller, K. W. (1988). Random number generators: good ones are hard to find. *Commun. ACM, 31*(10), 1192–1201.

Parmeshwar, U. (2020). Global food security in the context of COVID-19: A scenario-based exploratory analysis. *Progress in Disaster Science*, 7.

Pawlikowski, K. (2003). Towards credible and fast quantitative stochastic simulation. *Proceedings of International SCS Conference on Design, Analysis and Simulation of Distributed Systems, DASD'03*.

Pearson, R. K. (2018). *Exploratory Data Analysis using R*. CRC Press. doi:10.1201/9781315382111

Peng, R. (2016). *Exploratory Data Analysis with R*. lulu.com.

Perez, J. A., Deligianni, F., Ravi, D., & Yang, G.-Z. (2018). *Artificial Intelligence and Robotics*. https://www.mygreatlearning.com/blog/what-is-artificial-intelligence/

Perin, C., Vuillemot, R., & Fekete, J. (2013, December). SoccerStories: A Kick-off for Visual Soccer Analysis. *IEEE Transactions on Visualization and Computer Graphics*, 19(12), 2506–2515. doi:10.1109/TVCG.2013.192 PMID:24051817

Phillips, C., & Swiler, L. P. (1998, January). A graph-based system for network-vulnerability analysis. In *Proceedings of the 1998 workshop on New security paradigms* (pp. 71-79). 10.1145/310889.310919

Piringer, H., Berger, W., & Hauser, H. (2008). Quantifying and Comparing Features in High-Dimensional Datasets. *2008 12th International Conference Information Visualisation*, 240-245. 10.1109/IV.2008.17

Podlozhnyuk, V. (2007). *Parallel Mersenne Twister*. NVIDIA Corporation Inc.

Pradhan, Siddappa, Kavitha, & Samanta. (2019). Analysis & Improvement of Wireless Network Security Based on Biometrics. In *Proceedings of International Conference on Sustainable Computing in Science, Technology and Management (SUSCOM)*. Amity University Rajasthan. https://ssrn.com/abstract=3356360

Pramanik, S., & Bandyopadhyay, S. K. (2014). Image Steganography using Wavelet Transform and Genetic Algorithm. *International Journal of Innovative Research in Advanced Engineering, 1*.

Pramanik, S., & Bandyopadhyay, S. K. (2013). Application of Steganography in Symmetric Key Cryptography with Genetic Algorithm. *International Journal of Computers and Technology*, 10(7), 1791–1799. doi:10.24297/ijct.v10i7.7027

Pramanik, S., & Bandyopadhyay, S. K. (2014). An Innovative Approach in Steganography, Scholar. *Journal of Engineering Technology*, 2(2B), 276–280.

Pramanik, S., & Bandyopadhyay, S. K. (2014). An Innovative Approach in Steganography. *Scholars Journal of Engineering and Technology*, 2(2B), 276–280.

Pramanik, S., & Bandyopadhyay, S. K. (2014). Hiding Secret Message in an Image, International Journal of Innovative Science. *Engineering and Technology*, 1(3), 553–559.

Pramanik, S., & Bandyopadhyay, S. K. (2014). Hiding Secret Message in an Image. *International Journal of Innovative Science. Engineering and Technology*, 11(3), 553–559.

Pramanik, S., & Raja, S. S. (2020). A Secured Image Steganography using Genetic Algorithm. *Advances in Mathematics: Scientific Journal*, 9(7), 4533–4541.

Pramanik, S., & Singh, R. P. (2017). Role of Steganography in Security Issues. *International Journal of Advance Research in Science and Engineering*, 6(1), 1119–1124.

Pramanik, S., & Singh, R. P. (2017). Role of Steganography in Security Issues. *International Journal of Advance Research in Science and Engineering*, 6(1), 119–1124.

Praveen, B., Samanta, D., Prasad, G., Ranjith Kumar, C., & Prasad, M. L. M. (2020). Protecting Medical Research Data Using Next Gen Steganography Approach. In L. Jain, S. L. Peng, B. Alhadidi, & S. Pal (Eds.), Intelligent Computing Paradigm and Cutting-edge Technologies. ICICCT 2019. Learning and Analytics in Intelligent Systems (Vol. 9). Springer. https://doi.org/10.1007/978-3-030-38501-9_34.

Premebida, C., Ambrus, R., & Marton, Z.-C. (2019). Intelligent Robotic Perception Systems. *Applications of Mobile Robots*, 111–127. doi:10.5772/intechopen.79742

Premebida, C., & Ambrus, R. (2016). *Zoltan-Csaba*. Intelligent Robotic Perception Systems. In IntechOpen. https://www.intechopen.com/books/advanced-biometric-technologies/liveness-detection-in-biometrics

Puolamäki, K., Oikarinen, E., Kang, B., Lijffijt, J., & De Bie, T. (2020). Interactive visual data exploration with subjective feedback: An information-theoretic approach. *Data Mining and Knowledge Discovery*, 34(1), 21–49. doi:10.100710618-019-00655-x

QiangS.BayatiM. (2016). Dynamic Pricing with Demand Covariates. doi:10.2139srn.2765257

Radev, D. R., Allison, T., Blair-Goldensohn, S., Blitzer, J. C., Elebi, A., Dimitrov, S., Dr'abek, E., Hakim, A., Lam, W., Liu, D., Otterbacher, J., Qi, H., Saggion, H., Teufel, S., Topper, M., Winkel, A., & Zhang, Z. (2004). MEAD - A platform for multidocument multilingual text summarization. *Proceedings of the Fourth International Conference on Language Resources and Evaluation, LREC 2004*. http://www.lrec-conf. org/proceedings/lrec2004/pdf/757.pdf

Raghavan, P., & Gayar, N. (2019). Fraud Detection using Machine Learning and Deep Learning. *International Conference on Computational Intelligence and Knowledge Economy (ICCIKE)*, 334-339. 10.1109/ICCIKE47802.2019.9004231

Raj, M., & Seamans, R. (2019). Primer on artificial intelligence and robotics. In Journal of Organization Design (Vol. 8, Issue 1, pp. 1–14). doi:10.118641469-019-0050-0

Randhawa, K., Loo, C. K., Seera, M., Lim, C. P., & Nandi, A. K. (2018). Credit Card Fraud Detection Using Ada-Boost and Majority Voting. *IEEE Access: Practical Innovations, Open Solutions*, 6, 14277–14284. doi:10.1109/ACCESS.2018.2806420

Rao, S. (2002). *Applied Numerical Methods for Engineers and Scientist* (3rd ed.). Pearson Prentice Hall Education.

Rau, L. F., Jacobs, P. S., & Zernik, U. (1989). Information extraction and text summarization using linguistic knowledge acquisition. *Information Processing & Management, 25*(4), 419–428. doi:10.1016/0306-4573(89)90069-1

Raychev, V., Vechev, M., & Yahav, E. (2014, June). Code completion with statistical language models. In *Proceedings of the 35th ACM SIGPLAN Conference on Programming Language Design and Implementation* (pp. 419-428). 10.1145/2594291.2594321

Regin, R., Rajest, S. S., & Singh, B. (2021). Spatial Data Mining Methods Databases and Statistics Point of Views. *Innovations in Information and Communication Technology Series*, 103-109.

Reimer, U., & Hahn, U. (1988). Text condensation as knowledge base abstraction. *Artificial Intelligence Applications, 1988., Proceedings of the Fourth Conference on*, 338–344. 10.1109/CAIA.1988.196128

REST API Resources. (n.d.). *Twitter Developers*. https://dev.twitter.com/docs/api

Restivo, M. T., Mendes, J., Lopes, A. M., Silva, C. M., & Chouzal, F. (2009). A Remote Lab in Engineering Measurement. *IEEE Transactions on Industrial Electronics, 56*.

Reuillon, R., Hill, D. R. C., Gouinaud, C., Bitar, Z. E., Breton, V., & Buvat, I. (2008). Monte Carlo simulation with the GATE software using grid computing. *8th International Conference on New Technologies in Distributed Systems*.

Reuillon, R., Hill, D.R.C, & Bitar, Z. (2008). Rigorous Distribution of Stochastic Simulations Using the Dist Me Toolkit. *IEEE Transactions on Nuclear Science*.

Riaz, M., King, J., Slankas, J., & Williams, L. (2014, August). Hidden in plain sight: Automatically identifying security requirements from natural language artifacts. In *2014 IEEE 22nd international requirements engineering conference (RE)* (pp. 183-192). IEEE.

Risk Management Systems in the Aftermath of the Financial Crisis Flaws, Fixes and Future Plans. (2010). A GARP report prepared in association with SYBASE.

Robillard, M., Walker, R., & Zimmermann, T. (2009). Recommendation systems for software engineering. *IEEE Software, 27*(4), 80–86. doi:10.1109/MS.2009.161

Robotics Online Marketing Team. (2018). *How Artificial Intelligence is Used in Today's Robots*. Association for Advancing Automation.

Rodríguez-Ibáñez, M., Muñoz-Romero, S., Soguero-Ruiz, C., Gimeno-Blanes, F., & Rojo-Álvarez, J. L. (2019). Towards Organization Management Using Exploratory Screening and Big Data Tests: A Case Study of the Spanish Red Cross. *IEEE Access: Practical Innovations, Open Solutions, 7*, 80661–80674. doi:10.1109/ACCESS.2019.2923533

Roger, D. (2016). *Exploratory Data Analysis with R*. Leanpub.

Rubio, F., Valero, F., & Llopis-Albert, C. (2019). A review of mobile robots: Concepts, methods, theoretical framework, and applications. *International Journal of Asvanced Robotic Systems, 1–22*.

Rudra, K., Ghosh, S., Goyal, P., Ganguly, N., & Ghosh, S. (2015). Extracting situational information from microblogs during disaster events: A classification-summarization approach. *Proc. ACM CIKM*. 10.1145/2806416.2806485

Ryan, A. J., & Ulrich, M. J. (2020). *quantmod: Quantitative Financial Modelling Framework. R package version 0.4.17*. https://CRAN.R-project.org/package=quantmod

Sabou, M., Ekaputra, F. J., Ionescu, T., Musil, J., Schall, D., Haller, K., ... Biffl, S. (2018, June). Exploring enterprise knowledge graphs: A use case in software engineering. In *European Semantic Web Conference* (pp. 560-575). Springer. 10.1007/978-3-319-93417-4_36

Saeed, S., Bagram, M. M. M., & Iqbal, M. M. (2021). An Intelligent Analysis of Crime Data using Data Mining Algorithms. *Technical Journal, 26*(01), 102–115.

Salian, I. (2018). *SuperVize Me: What's the Difference Between Supervised, Unsupervised, Semi- Supervised and Reinforcement Learning?* NVIDIA. https://blogs.nvidia.com/blog/2018/08/02/supervised-unsupervised-learning/

Salini, P., & Kanmani, S. (2012). Survey and analysis on security requirements engineering. *Computers & Electrical Engineering, 38*(6), 1785–1797. doi:10.1016/j.compeleceng.2012.08.008

Salleh, Z. (2011). *Fundamental of Numerical Methods for Scientists and Engineers*. Lambert Academic Publishing.

Salleh, Z., & Yusop, M. Y. M. (2012). Basic of numerical computational using Scilab programming. *2nd International Conference on Mathematical Applications in Engineering (ICMAE2012)*.

Samanta, D. (2020). Distributed Feedback Laser (DFB) for Signal Power Amplitude Level Improvement in Long Spectral Band. *Journal of Optical Communications*. www.degruyter.com

Samanta, D., & Sanyal, G. (2012). Novel Shannon's Entropy Based Segmentation Technique for SAR Images. In K. R. Venugopal & L. M. Patnaik (Eds.), Wireless Networks and Computational Intelligence. ICIP 2012. Communications in Computer and Information Science (Vol. 292). Springer. https://doi.org/10.1007/978-3-642-31686-9_22

Samanta, D., Sivaram, M., Rashed, A., Boopathi, C. S., Sadegh Amiri, I., & Yupapin, P. (2020). Distributed Feedback Laser (DFB) for Signal Power Amplitude Level Improvement in Long Spectral Band. *Journal of Optical Communications*. . doi:10.1515/joc-2019-0252

Samanta, Mousumi, Khutubuddin, & Khadri. (2013). Message Communication Using Phase Shifting Method (PSM). *International Journal of Advanced Research in Computer Science, 4*(11), 9–11. doi:10.26483/ijarcs.v4i11.1936

Sandy Hook Elementary School shooting. (2012). https://en.wikipedia.org/wiki/SandyHookElementarySchoolsshooting

Schatsky, B. D., & Ream, J. (2017). Robots uncaged. *Deloitte Insights*, 1–8.

Schroer, A. (2020). *AI Robots: How 19 companies use artificial intelligence in robotics*. Builtin. https://builtin.com/artificial-intelligence/robotics-ai-companies

Schutten, J. P. (2014). *Hello from 2030*. Academic Press.

Scott, D., & Dong, C.Y. (2018). *VarianceGamma: The Variance Gamma Distribution. R package version 0.4-0*. CRAN.R-project.org/package=VarianceGamma

Seetharam, K., Raina, S., & Sengupta, P. P. (2020). The Role of Artificial Intelligence in Echocardiography. *Current Cardiology Reports, 22*(9), 99. Advance online publication. doi:10.100711886-020-01329-7 PMID:32728829

Shan, T., Ma, Y., Tao, R., & Liu, S. (2014). Multi-channel NLMS-based sea clutter cancellation in passive bistatic radar. *IEICE Electronics Express, 11*(20), 11–20140872. doi:10.1587/elex.11.20140872

Sharif, J. M., Miswan, M. F., Ngadi, M. A., & Salam Md, S. H. (2012). Red Blood Cell Segmentation Using Masking And Watershed Algorithm: A Preliminary Study. *International Conference On Biomedical Engineering*, 27-28.

Shetty, M. (2004). *Geometric Estimation of Value of Pi*. http://www.ijoart.org/docs/Geometric-Estimation-of-Value-of-Pi.pdf

Shin, Y. (2008, October). Exploring complexity metrics as indicators of software vulnerability. *Proceedings of the 3rd International Doctoral Symposium on Empirical Software Engineering*.

Shin, Y., & Williams, L. (2008a, October). An empirical model to predict security vulnerabilities using code complexity metrics. In *Proceedings of the Second ACM-IEEE international symposium on Empirical software engineering and measurement* (pp. 315-317). 10.1145/1414004.1414065

Shin, Y., & Williams, L. (2008b, October). Is complexity really the enemy of software security? In *Proceedings of the 4th ACM workshop on Quality of protection* (pp. 47-50). 10.1145/1456362.1456372

Shlomo, S. A. W. I. L. O. W. S. K. Y. (2003). Deconstructing Arguments From The Case Against Hypothesis Testing. *Journal of Modern Applied Statistical Methods; JMASM*, *2*(2), 467–474. doi:10.22237/jmasm/1067645940

Shou, L., Wang, Z., Chen, K., & Chen, G. (2013). Sumblr: Continuous summarization of evolving tweet streams. *Proc. ACM SIGIR*. 10.1145/2484028.2484045

Shousong, C., & Jing, Z. (2021). Research on Healthcare Quality Evaluation Based on Data Mining. *Journal of Medical Imaging and Health Informatics*, *11*(4), 1117–1124.

Siddique, S., & Sayyed, R. (2016, March). Automated RBCs Segmentation & Counting using SCILAB. *International Journal of Engineering Research & Technology*, *5*(3). Advance online publication. doi:10.17577/IJERTV5IS030872

Signorelli, C. M. (2018). Can computers become conscious and overcome humans? *Frontiers in Robotics and AI*, *5*(OCT), 1–20. doi:10.3389/frobt.2018.00121 PMID:33501000

Silhavy, R., Silhavy, P., & Prokopova, Z. (2017). Analysis and selection of a regression model for the Use Case Points method using a stepwise approach. *Journal of Systems and Software*, *125*, 1–14. doi:10.1016/j.jss.2016.11.029

Sindre, G., & Opdahl, A. L. (2005). Eliciting security requirements with misuse cases. *Requirements Engineering*, *10*(1), 34–44. doi:10.100700766-004-0194-4

Singh, P., Lin, T., Mueller, E. T., Lim, G., Perkins, T., & Zhu, W. L. (2002, October). Open mind common sense: Knowledge acquisition from the general public. In *OTM Confederated International Conferences On the Move to Meaningful Internet Systems* (pp. 1223-1237). Springer.

Singh, R., & Adhikari, R. (2020). *Age-structured impact of social distancing on the COVID-19 epidemic in India*. Accessed from https://arxiv.org/pdf/2003.12055.pdf

Singh, U. (2013). *Estimation of the value of using Monte-Carlo Method and Related Study of Errors*. https://www.academia.edu/1887423/

Singh, V. (2020). Prediction of COVID-19 corona virus pandemic based on time series data using Support Vector Machine. *Journal of Discrete Mathematical Sciences & Cryptography*.

Singhal, A. (2012). *Introducing the knowledge graph: things, not strings*. Official Google Blog.

Singla, N., Hall, M., Shands, B., & Chamberlain, R. D. (2008). Financial Monte Carlo simulation on architecturally diverse systems. *Workshop on High Performance Computational Finance, Supercomputing 08*, 1–7.

Sivakumar, P., Nagaraju, R., & Samanta, D. (2020). A novel free space communication system using nonlinear InGaAsP microsystem resonators for enabling power-control toward smart cities. *Wireless Networks*, *26*, 2317–2328. https://doi.org/10.1007/s11276-019-02075-7

Smith, A., & Anderson, J. (2014). *AI, Robotics, and the Future of Jobs*. Issue August.

Sodhi, P., Awasthi, N., & Sharma, V. (2019). Introduction to Machine Learning and Its Basic Application in Python. *SSRN Electronic Journal*, 1354–1375. doi:10.2139srn.3323796

Souag, A., Salinesi, C., & Comyn-Wattiau, I. (2012, June). Ontologies for security requirements: A literature survey and classification. In *International conference on advanced information systems engineering* (pp. 61-69). Springer. 10.1007/978-3-642-31069-0_5

Sparks, S., Embleton, S., Cunningham, R., & Zou, C. (2007, December). Automated vulnerability analysis: Leveraging control flow for evolutionary input crafting. In *Twenty-Third Annual Computer Security Applications Conference (ACSAC 2007)* (pp. 477-486). IEEE. 10.1109/ACSAC.2007.27

Speer, R., Chin, J., & Havasi, C. (2017, February). Conceptnet 5.5: An open multilingual graph of general knowledge. *Proceedings of the AAAI Conference on Artificial Intelligence, 31*(1).

Srimani, S. (2015). A Geometrical Derivation of π (Pi). *IOSR Journal of Mathematics (IOSR-JM), 11*(6), 19–22. doi:10.9790/5728-11611922

Srinivasan, A. (2002). Parallel and distributed computing issues in pricing financial derivatives through Quasi Monte Carlo. *Proceedings of the 16th International Parallel and Distributed Processing Symposium*, 14–19.

Srinivasan, A., Ceperley, D. M., & Mascagni, M. (1999). Random number generators for parallel applications. In D. M. Ferguson, J. I. Siepmann, & D. G. Truhlar (Eds.), Advances in Chemical Physics Series: Vol. 105. *Monte Carlo Methods in Chemical Physics* (pp. 13–36). John Wiley and Sons.

Sriram, V., & Kearney, D. (2007). High Throughput Multi-port MT19937 Uniform Random Number Generator. *Parallel and Distributed Computing Applications and Technologies, International Conference on*, 157–158.

Sritha, Z. D. B. (2020). Acting tools of ICT to tackle Covid19. *OmniScience: A Multi-disciplinary Journal, 10*(1), 1–9.

Stephen, L. C., Chancelier, J. P., & Nikoukhah, R. (2006). *Modeling and Simulation in Scilab/Scicos*. Springer.

Stolte, C., Tang, D., & Hanrahan, P. (2002). Polaris: A system for query, analysis, and visualization of multidimensional relational databases. *IEEE Transactions on Visualization and Computer Graphics, 8*(1), 52–65. doi:10.1109/2945.981851

Stramer, O. (1996). On the Approximation of Moments For Continuous Time Threshold ARMA Processes. *Journal of Time Series Analysis, 17*(2), 189–202. doi:10.1111/j.1467-9892.1996.tb00272.x

Subramanyam, R., & Krishnan, M. S. (2003). Empirical analysis of ck metrics for object-oriented design complexity: Implications for software defects. *IEEE Transactions on Software Engineering, 29*(4), 297–310. doi:10.1109/TSE.2003.1191795

Sun, J. T., Shen, D., Zeng, H. J., Yang, Q., Lu, Y., & Chen, Z. (2005). Web-page summarization using clickthrough data. In *Proceedings of the 28th Annual International ACM SIGIR Conference on Research and Development in Information Retrieval, SIGIR '05* (pp. 194–201). ACM. doi:10.1145/1076034.1076070

Sutton, M., Greene, A., & Amini, P. (2007). *Fuzzing: brute force vulnerability discovery*. Pearson Education.

Su, Y. S., & Wu, S. Y. (2021). Applying data mining techniques to explore user behaviors and watching video patterns in converged IT environments. *Journal of Ambient Intelligence and Humanized Computing*, 1–8. PMID:33425047

Tang, A., Liang, P., & Van Vliet, H. (2011, June). Software architecture documentation: The road ahead. In *2011 Ninth Working IEEE/IFIP Conference on Software Architecture* (pp. 252-255). IEEE. 10.1109/WICSA.2011.40

Tanne, J. H., Hayasaki, E., Zastrow, M., Pulla, P., Smith, P., & Rada, A. G. (2020). Covid-19: How doctors and healthcare systems are tackling coronavirus worldwide. *BMJ (Clinical Research Ed.), 368*, 1–5. doi:10.1136/bmj.m1090 PMID:32188598

Tao, K., Abel, F., Hauff, C., Houben, G. J., & Gadiraju, U. (2013). Groundhog Day: Nearduplicate Detection on Twitter. *Proc. Conference on World Wide Web (WWW)*.

Tao, R., Wu, H. Z., & Shan, T. (2010). Direct-path suppression by spatial filtering in digital television terrestrial broadcasting-based passive radar. *IET Radar, Sonar & Navigation, 4*(6), 791–805. doi:10.1049/iet-rsn.2009.0138

Thiruvinal, V. J., & Ram, S. P. (2017). Automated Blood Cell Counting and Classification Using Image Processing. *International Journal of Advanced Research in Electrical, Electronics and Instrumentation Engineering, 6*(1). doi:10.15662/IJAREEIE.2017.0601010

Thomas, D. B., & Luk, W. (2008). Multivariate gaussian random number generation targeting reconfigurable hardware. *ACM Trans. Reconfigurable Technol. Syst., 1*(2), 1–29.

Thompson, J. M. (1992). Visual Representation of Data Including Graphical Exploratory Data Analysis. In C. N. Hewitt (Ed.), *Methods of Environmental Data Analysis. Environmental Management Series.* Springer. doi:10.1007/978-94-011-2920-6_6

Thornton, M. A., & Chambers, M. J. (2017). Continuous time ARMA processes: Discrete time representation and likelihood evaluation. *Journal of Economic Dynamics & Control, 79*, 48–65. doi:10.1016/j.jedc.2017.03.012

Thrun, S., & Mitchell, T. M. (1995). Lifelong robot learning. *Robotics and Autonomous Systems, 15*(1-2), 25–46. doi:10.1016/0921-8890(95)00004-Y

Tigelaar, A.S., Opdenakker, R., & Hiemstra, D. (2010). Automatic summarisation of discussion fora. *Nat. Lang. Eng., 16*(2), 161–192. DOI doi:10.1017/S135132491000001X

Todorov, V., & Tauchen, G. (2006). Simulation methods for Lévy-driven CARMA stochastic volatility models. *Journal of Business & Economic Statistics, 24*, 455–469. doi:10.1198/073500106000000260

Traore, M., & Hill, D. R. C. (2001). The use of random number generation for stochastic distributed simulation: application to ecological modeling. *Proceedings of the 13th European Simulation Symposium*, 555-559.

Trautsch, F., Herbold, S., Makedonski, P., & Grabowski, J. (2016, May). Adressing problems with external validity of repository mining studies through a smart data platform. In *Proceedings of the 13th International Conference on Mining Software Repositories* (pp. 97-108). 10.1145/2901739.2901753

Tsay, R. S. (2012). *Analysis of Financial Time Series.* Wiley and Sons.

Tulshan, A., & Dhage, S. (2019). Survey on Virtual Assistant: Google Assistant, Siri, Cortana, Alexa. *4th International Symposium SIRS 2018.* 10.1007/978-981-13-5758-9_17

Typhoon Hagupit. (2014). https://en.wikipedia.org/wiki/TyphoonHagupit

Vanegue, J., & Lahiri, S. K. (2013, May). Towards practical reactive security audit using extended static checkers. In *2013 IEEE Symposium on Security and Privacy* (pp. 33-47). IEEE. 10.1109/SP.2013.12

Varin, C., Reid, N. M., & Firth, D. (2011). An overview of composite likelihoodmethods. *Statistica Sinica, 21*(1), 5–42.

Vijayalakshmi, M. N., & Senthilvadivu, M. (2016). Performance Evaluation of Object Detection Techniques for Object Detection. *2016 International Conference on Inventive Computation Technologies (ICICT).* 10.1109/Inventive.2016.7830065

Villano, M., Colone, F., & Lombardo, P. (2013). Antenna array for passive radar: Configuration design and adaptive approaches to disturbance cancellation. *International Journal of Antennas and Propagation.*

Viviani, E., Di Persio, L., & Ehrhardt, M. (2021). Energy Markets Forecasting. From Inferential Statistics to Machine Learning: The German Case. *Energies, 14*(2), 364. doi:10.3390/en14020364

Wald, A. (1949). Note on the consistency of the maximum likelihood estimate. *Annals of Mathematical Statistics*, *20*(4), 595601. doi:10.1214/aoms/1177729952

Wang, X., & Sloan, I. H. (2008). Low discrepancy sequences in high dimensions: How well are their projections distributed? *Journal of Computational and Applied Mathematics*, *213*(2), 366–386.

Wang, Z., Shou, L., Chen, K., Chen, G., & Mehrotra, S. (2015). On summarization and timeline generation for evolutionary tweet streams. *IEEE Transactions on Knowledge and Data Engineering*, *27*(5), 1301–1314. doi:10.1109/TKDE.2014.2345379

Wars, S., Sojourner, R., Pathfinder, M., Robots, U., Capek, K., Asimov, I., One, L., Two, L., Law, F., & Three, L. (2014). *Robotics : A Brief History Early Conceptions of Robots*. Academic Press.

Weaver, W., Jr., & Gere, J. M. (1986). Matrix Analysis of Framed Structures (2nd ed.). CBS Publishers and Distributors.

Weaver, B. C. (2020). Self-Driving Cars Learn to Read the Body Language of People on the Street. *IEEE Spectrum*, 1–5.

Wenjiang, L., Nanping, D., & TongShun, F. (2009). The application of Scilab / Scicos in the lecture of automatic control. *Open-source Software for Scientiifc Computation (OSSC), IEEE International Workshop*, 85–87.

Weyer, S., Schmitt, M., Ohmer, M., & Gorecky, D. (2015). Towards Industry 4.0 - Standardization as the crucial challenge for highly modular, multi-vendor production systems. *IFAC-PapersOnLine*, *48*(3), 579–584. doi:10.1016/j.ifacol.2015.06.143

WHO. (2020). *COVID-19*. WHO.

Wibawa, B., Siregar, J. S., Asrorie, D. A., & Syakdiyah, H. (2021, April). Learning analytic and educational data mining for learning science and technology. In AIP Conference Proceedings: Vol. 2331. *No. 1* (p. 060001). AIP Publishing LLC.

Wilson, W., Rosenberg, L., & Hyatt, L. (1997). Automated analysis of requirement specifications, *Proceedings of the 19th ACM international conference on Software engineering*, 161-171. 10.1145/253228.253258

Wisskirchen, G., Thibault, B., Bormann, B. U., Muntz, A., Niehaus, G., Soler, G. J., & Von Brauchitsch, B. (2017). Artificial Intelligence and Robotics and Their Impact on the Workplace. *IBA Global Employment Institute*.

Woo, M., Cha, S. K., Gottlieb, S., & Brumley, D. (2013, November). Scheduling black-box mutational fuzzing. In *Proceedings of the 2013 ACM SIGSAC conference on Computer & communications security* (pp. 511-522). 10.1145/2508859.2516736

Wu, Y., &Xu, J. (2021). 13 Statistical Problems with Planted Structures: Information-Theoretical and Computational Limits. *Information-Theoretic Methods in Data Science*, 383.

Wu,, P., & Huang, K. (2006). Parallel use of multiplicative congruential random number generators. *Computer Physics Communications*, *175*, 25–29.

Wuest, T., Weimer, D., Irgens, C., & Thoben, K.-D. (2016). Machine learning in manufacturing: Advantages, challenges, and applications. *Production & Manufacturing Research*, *4*(1), 23–45. doi:10.1080/21693277.2016.1192517

Xiaode, L., Jichuan, L., Kuan, L., Daojing, L., & Yi, Z. (2014, October). Range-Doppler NLMS (RDNLMS) algorithm for cancellation of strong moving targets in passive coherent location (PCL) radar. In *2014 International Radar Conference* (pp. 1-5). IEEE. 10.1109/RADAR.2014.7060444

Xin, Y., Kong, L., Liu, Z., Chen, Y., Li, Y., Zhu, H., Gao, M., Hou, H., & Wang, C. (2018). Machine Learning and Deep Learning Methods for Cybersecurity. *IEEE Access: Practical Innovations, Open Solutions*, *6*, 35365–35381. doi:10.1109/ACCESS.2018.2836950

Xu, W., Grishman, R., Meyers, A., & Ritter, A. (2013). A preliminary study of tweet summarization using information extraction. *Proc. NAACL 2013*, 20.

Yakubu, L. (2021). Africa's low COVID-19 mortality rate: A paradox. *International Journal of Infectious Diseases*, *102*, 18–122. PMID:33075535

Yamaguchi, F., Golde, N., Arp, D., & Rieck, K. (2014, May). Modeling and discovering vulnerabilities with code property graphs. In *2014 IEEE Symposium on Security and Privacy* (pp. 590-604). IEEE. 10.1109/SP.2014.44

Yang, W., Wang, X., Lu, J., Dou, W., & Liu, S. (2020). Interactive Steering of Hierarchical Clustering. *IEEE Transactions on Visualization and Computer Graphics*, 1. Advance online publication. doi:10.1109/TVCG.2020.2995100 PMID:32746252

Yan, L. (2011). Global exploratory analysis of massive neuroimaging collections using Microsoft Silverlight Pivot-Viewer. *Proceedings of the 2011 Biomedical Sciences and Engineering Conference: Image Informatics and Analytics in Biomedicine*, 1-4. 10.1109/BSEC.2011.5872323

Yan, Z., Schreiberhuber, S., Halmetschlager, G., Duckett, T., Vincze, M., & Bellotto, N. (2020). Robot perception of static and dynamic objects with an autonomous floor scrubber. *Intelligent Service Robotics*, *13*(3), 403–417. doi:10.100711370-020-00324-9

Yarowsky, D. (1995, June). Unsupervised word sense disambiguation rivaling supervised methods. In *33rd annual meeting of the association for computational linguistics* (pp. 189-196). 10.3115/981658.981684

Yauri-Machaca, M., Meneses-Claudio, B., & Vargas-Cuentas, N. (2018). Design of a Vehicle Driver Drowsiness Detection System through Image Processing using Matlab. IEEE.

Yew Lin, C. (2004). *Rouge: A package for automatic evaluation of summaries*. Academic Press.

Ye, X., Bunescu, R., & Liu, C. (2014, November). Learning to rank relevant files for bug reports using domain knowledge. In *Proceedings of the 22nd ACM SIGSOFT International Symposium on Foundations of Software Engineering* (pp. 689-699). 10.1145/2635868.2635874

Yoo, C., Ramirez, L., & Liuzzi, J. (2014). Big data analysis using modern statistical and machine learning methods in medicine. *International Neurourology Journal*, *18*(2), 50–57. doi:10.5213/inj.2014.18.2.50

Youngman, P. (2009). *Procyclicality and Value-at-Risk*. Bank of Canada Financial System Review Report.

YUIMA Project Team. (2018). *yuimaGUI: A Graphical User Interface for the 'yuima' Package. R package version 1.3.0.* CRAN.R-project.org/package=yuimaGUI

YUIMA Project Team. (2020, October 2). *YUIMAGUI: Web Application.* yuimaproject.com/yuimagui/

Zhang, C., Wang, X., & He, Z. (2021). Efficient Importance Sampling in Quasi-Monte Carlo Methods for Computational Finance. *SIAM Journal on Scientific Computing*, *43*(1), B1–B29. doi:10.1137/19M1280065

Zhao, Y. D., Zhao, Y. K., Lu, X. D., & Xiang, M. S. (2013). *Block NLMS cancellation algorithm and its real-time implementation for passive radar*. Academic Press.

Zhu, H., Shang, L., & Zhou, X. (2021). A Review of Statistical Methods for Identifying Trait-Relevant Tissues and Cell Types. *Frontiers in Genetics*, *11*, 1846. doi:10.3389/fgene.2020.587887 PMID:33584792

Zhu, N., Zhang, D., Wang, W., Li, X., Yang, B., Song, J., Zhao, X., Huang, B., Shi, W., Lu, R., Niu, P., Zhan, F., Ma, X., Wang, D., Xu, W., Wu, G., Gao, G. F., & Tan, W. (2020). A novel coronavirus from patients with pneumonia in China, 2019. *The New England Journal of Medicine*, *382*(8), 727–733. doi:10.1056/NEJMoa2001017 PMID:31978945

Zimmermann, T., & Nagappan, N. (2009, October). Predicting defects with program dependencies. In *2009 3rd international symposium on empirical software engineering and measurement* (pp. 435-438). IEEE. 10.1109/ESEM.2009.5316024

Zimmermann, T., & Nagappan, N. (2008, May). Predicting defects using network analysis on dependency graphs. In *Proceedings of the 30th international conference on Software engineering* (pp. 531-540). 10.1145/1368088.1368161

Zubiaga, A., Spina, D., Amigo, E., & Gonzalo, J. (2012). Towards Real-Time Summarization of Scheduled Events from Twitter Streams. Hypertext(Poster). doi:10.1145/2309996.2310053

About the Contributors

Debabrata Samanta is presently working as Assistant Professor, Department of Computer Science, CHRIST (Deemed to be University), Bangalore, India. He obtained his B.Sc. (Physics Honors), from Calcutta University; Kolkata, India. He obtained his MCA, from the Academy of Technology, under WBUT, West Bengal. He obtained his PhD in Computer Science and Engg. from National Institute of Technology, Durgapur, India, in the area of SAR Image Processing. His areas of interest are SAR Image Analysis, Video surveillance, Heuristic algorithm for image classification, Deep Learning Framework for Detection and Classification, Blockchain, Statistical Modelling, Wireless Adhoc Network, Natural Language Processing, V2I Communication. He is the owner of 17 Indian Patents. He has authored and coauthored over 134 research papers in international journal (SCI/SCIE/ESCI/Scopus) and conferences including IEEE, Springer and Elsevier Conference proceedings. He has received "Scholastic Award" at 2nd International conference on Computer Science and IT application, CSIT-2011, Delhi, India. He has published 9 books, available for sale on Amazon and Flipkart.

* * *

Yakup Arı is an Assistant Professor in the Department of Economics at Alanya Alaaddin Keykubat University. Having graduated with a bachelor's degree in Mathematics, he pursued an MBA degree in Finance and a PhD degree in Financial Economics – all of which at Yeditepe University with full scholarships. He worked as a statistical consultant at several private consultancy firms in Istanbul. He teaches courses in time series analysis, mathematical economics, technical analysis, probability and statistics, biostatistics and econometrics. His primary research interest lies in the area of time series analysis, Lévy driven stochastic processes, the Bayesian approach in statistics and econometrics, in addition to statistical methods in Engineering and Social Sciences.

Abhishek Bhattacharya received his M.Tech degree in Computer science from Birla Institute of Technology, Mesra. He is registered in Ph. D degree in Computer Science and Engineering at Birla Institute of Technology, Mesra. After finishing his undergraduate work in Dept. of the Computer science, he started his career in R S software India Pvt. Ltd. He has over 2 years of industry experience and on innovative technologies and also has the experience of working as the General Manager, Technical of a medium-sized IT company. In 2005, he had shifted to full-time academics and is presently serving as 'Assistant Professor and HOD' in the BCA and M.Sc (Information Sc) dept. of Institute of Engineering & Management, Kolkata. He has about 9 publications in international journals. His current areas of interest in research include Network Security, Social Networking, Data Analysis, Recommendation Algorithms, VLSI Physical Design Algorithms.

Nagadevi D. is Assistant Professor, Dept. of ECE, CBIT, Gandipet, Hyderabad.

Venu D. was born in India in 1982. He has a total experience of 14 years in teaching electronics related courses and is currently working in Kakatiya Institute of Technology and Science, Warangal, India. He is pursuing a Ph.D. from Osmania University in the area of Radar Signal Processing. He did his M.Tech in Embedded Systems and B.Tech in electronics and communication engineering from Jawaharlal Nehru Technological University(JNTU), Hyderabad. He is a life member of IAENG (International Association of Engineers), a life member of ISTE (Indian Society for Technical Education). He has five technical publications in various journals/conferences.

Pankaj Dadheech received his Ph.D. degree in Computer Science & Engineering from Suresh Gyan Vihar University (Accredited by NAAC with 'A' Grade), Jaipur, Rajasthan, India. He received his M.Tech. degree in Computer Science & Engineering from Rajasthan Technical University, Kota and he has received his B.E. in Computer Science & Engineering from University of Rajasthan, Jaipur. He has more than 15 years of experience in teaching. He is currently working as an Associate Professor & Dy. HOD in the Department of Computer Science & Engineering (NBA Accredited), Swami Keshvanand Institute of Technology, Management & Gramothan (SKIT), Jaipur, Rajasthan, India. He has published 15 Patents at Intellectual Property India, Office of the Controller General of Patents, Design and Trade Marks, Department of Industrial Policy and Promotion, Ministry of Commerce and Industry, Government of India. He has published 4 Australian Patents at Commissioner of Patents, Intellectual Property Australia, Australian Government. He has also Registered & Granted Research Copyright at Registrar of Copyrights, Copyright Office, Department for Promotion of Industry and Internal Trade, Ministry of Commerce and Industry, Government of India. He has presented 50 papers in various National & International conferences. He has 35 publications in various International & National Journals. He has published 2 Books & several Book Chapters. He is a member of many Professional Organizations like the IEEE Computer Society, CSI, ACM, IAENG & ISTE. He has appointed as a Ph.D. Research Supervisor in the Department of Computer Science & Engineering at SKIT, Jaipur (Recognized Research Centre of Rajasthan Technical University, Kota). He has also guided various M.Tech. Research Scholars. He has Chaired Technical Sessions in various International Conferences & Contributed as Resource Person in various FDP's, Workshops, STTP's, etc. He is also acting as a guest editor of the various reputed journal publishing houses and Bentham Ambassador of Bentham Science Publisher. His area of interest includes High Performance Computing, Cloud Computing, Cyber Security, Big Data Analytics, Intellectual Property Right and Internet of Things.

L. Nirmala Devi received her B.E, M.E and Ph.D degrees in Electronics and Communication Engineering from the Department of Electronics and Communication Engineering, University College of Engineering (Autonomous), Osmania University, Hyderabad, India. She is currently working as an Associate Professor and chairperson Board of studies (CBOS) in the Department of Electronics and Communication Engineering, Osmania University and also Director, CHW Osmania University. She has teaching experience of more than 20 Years in subjects like, Signals and Systems, Digital Signal Processing, Adaptive Signal Processing, Neural Networks, AI,ML,IoT and Wireless Communications, Wireless

Sensor Network, Data and Computer Communications, Analog Communication, Digital Communication, and Research Methodology for Engineering. Her research interests include Ad-hoc networks, wireless communication, wireless sensor networks, IoT and Signal Processing Machine Learning and AI. Currently, she is working on various research projects sponsored by Ministry of Electronics and Information Technology (MeiTY), Government of India, New Delhi, Department of Science & Technology (DST) and UGC. Recently she has received a grant of 1crore to work on smart cities project from RUSA 2.0 sponsored by MHRD, Govt of INDIA. She delivered keynote and invited talks at several international conferences and workshops, she has served as TPC member organizing committee member of numerous international conferences and conducted two GAIN programs in the year 2017.She has presented her paper in IEEE flagship conference International Conference on Communications (ICC-2019) at Shanghai. She has been selected for "Outstanding Women in Engineering Award "during Annual women's meet (AWM 2018) in the year 2018. She has published many papers in various national & international journals IEEE conferences. She is also a member of IEEE, IEI and OSA.

Arijit Ghosal is working as Associate Professor in the Department of Information Technology at St. Thomas' College of Engineering & Technology, Kolkata.

Salah-ddine Krit received the Hability Physics-Infromatics from the Faculty of Sciences, University Ibn Zohr Agadir morocco in 2015, the B.S. and Ph.D degrees in Software Engineering from Sidi Mohammed Ben Abdellah university, Fez, Morroco in 2004 and 2009, respectively. During 2002-2008, he worked as an engineer Team leader in audio and power management Integrated Circuits (ICs) Research, Design, simulation and layout of analog and digital blocks dedicated for mobile phone and satellite communication systems using Cadence, Eldo, Orcad, VHDL-AMS technology. He is currently a professor of Informatics with Polydisciplinary Faculty of Ouarzazate, Ibn Zohr university, Agadir, Morroco. His research interests include Wireless Sensor Networks (Software and Hardware), computer engineering and wireless communications, Genetic Algorithms, Gender and ICT.

Muralidhar Kurni is an Independent Consultant for Pedagogy Refinement, EduRefine, India. He is currently working as an Assistant Professor in the Department of Computer Science & Engineering at Anantha Lakshmi Institute of Technology & Sciences, Anantapuramu, Andhra Pradesh, India. He is an IUCEE & IGIP certified International Engineering Educator & researcher. He has several scholarly publications to his credit. He has been a reviewer for various International Conferences & Journals, including SCIE & Scopus indexed Journals. His research interests include Mobile Ad Hoc Networks, Cloud Computing, Ad Hoc Mobile Cloud Computing, Learning Analytics, Learning Strategies, Digital Peda-gogy, Design Thinking, Pedagogy refinement & Engineering Education Research.

Vinay Kumar Nassa is working as a Professor(CSE)/Director Principal at South Point Group of Institutions(Sonepat -Hr).He has about 30+ years of teaching, research & industry. He has experience of NAAC & NBA accreditation process of institutes. He has 30+ Research papers in SCI/SCopus/UGC-Care journals as research papers and in other international/national research journals. He had Australian and Indian patents to his credit. He is a reviewer of international/national journals, conference proceedings.

Ahmed Obaid is from the Department of Computer Science, Faculty of Computer Science and Mathematics, University of Kufa, Iraq.

Binay Kumar Pandey is currently working as an Assistant Professor in Department of Information Technology of Govind Ballabh Pant University of Agriculture and Technology Pantnagar Uttrakhand, India. He obtained his M. Tech with Specialization in Bioinformatics from Maulana Azad National Institute of Technology Bhopal M. P. India, in 2008 . He obtained his First Degree B. Tech at the IET Lucknow (Uttar Pradesh Technical University, Uttar Pradesh and Lucknow) India, in 2005. In 2010, he joined Department of Information Technology of College of Technology in Govind Ballabh Pant University of Agriculture and Technology Pantnagar as an Assistant Professor and worked for various UG and PG projects till date. He has more than ten years of experience in the field of teaching and research. He has more than 40 publications in reputed peer journal reputed journal Springer, Inderscience (sci and scopus indexed journal and others) and 3 patent. He has many awards such PM Scholarship, etc. He session chair in IEEE International Conference on Advent Trends in Multidisciplinary Research and Innovation (ICATMRI-2020) on December 30, 2020 organized by Pankaj Laddhad Institute of Technology and Management Studies; Buldhana, Maharashtra, India.

Digvijay Pandey is a lecturer in the Department of Technical Education, Research Scholar, IET Lucknow, India.

Dharmendra Patel received Bachelor Degree in Industrial Chemistry- BSc. (Industrial Chemistry) from North Gujarat University, Gujarat, India and Master Degree in Computer Applications (MCA) from North Gujarat University, Gujarat, India. He completed his Ph.D., in the field of Web Usage Mining. He is the professor of MCA Department at Smt. Chandaben Mohanbhai Patel Institute of Computer Applications, under Charusat University of Science and Technology (CHARUSAT), Changa, Gujarat, India.

Sabyasachi Pramanik is a Professional IEEE member. He obtained a Ph.D. in Computer Science and Engineering from the Sri Satya Sai University of Technology and Medical Sciences, Bhopal, India. Presently, he is an Assistant Professor, Department of Computer Science and Engineering, Haldia Institute of Technology, India. He has many publications in various reputed international conferences, journals, and online book chapter contributions (Indexed by SCIE, Scopus, ESCI, etc). He is doing research in the field of Artificial Intelligence, Data Privacy, IoT, Network Security, and Machine Learning. He is also serving as the editorial board member of many international journals. He is a reviewer of journal articles from IEEE, Springer, Elsevier, Inderscience, IET, and IGI Global. He has reviewed many conference papers, has been a keynote speaker, session chair and has been a technical program committee member in many international conferences. He has authored a book on Wireless Sensor Network. Currently, he is editing 6 books from IGI Global, CRC Press, EAI/Springer and Scrivener-Wiley Publications.

A. Nageswar Rao received B.E and M.E degrees in Electronics and Communication Engineering in 2000 and 2004 from Nagarjuna University and IIT Kahragpur and Submitted PhD under Visveswaraiah Scholarship sponsored by DeitY and Media Asia Labs in Wireless Sensor Networks at Osmania University.

N. V. Koteswara Rao was born in India in 1966. He has a vast experience of 28 years and has been working in Chaitanya Bharathi Institute of Technology(A), Hyderabad, India, since 1992. He did his Ph.D. from Osmania University in the area of "Microstrip Antennas", M.Techin field of Microwave Electronics from University of Delhi and B.Tech in field of electricity and communication engineering from Nagarjuna University. He is a Fellow of IETE, life member of ISTE and IEEE. Presently, he is the Professor and 'Director-Academics'. Under his supervision, two scholars have been awarded Ph.D. and nine Ph.D. research scholars are pursuing. He has been awarded the 'Best Teacher' of the institute and 'Distinguished Teacher' of the department. He has published sixty four technical publications in various journals/conferences and has filed one patent to his credit. He has successfully completed two R&D projects, one MODROB project and two FDP programs which are sponsored by AICTE. He has completed one in-house project and two consultancy projects from RCI/ISRO. Presently, he is associated with two R&D projects.

Raghavendra Rao Althar has obtained his Bachelor's in Mechanical Engineering from VTU and MBA in Operations Management from Indira Gandhi Open University. His area of interest is in application of Data Science in Software Development Processes. He is Currently researching as Quality Management Specialist in a Software Development team of Insurance domain-based company. For last 15 years, he has been researching on building Quality Management Systems for various domains like manufacturing, Retail, Telecom and Software industries, based on International Standards like ISO, Six sigma and best Global practices for industry. Adopting best practices of Data Science to optimize Software Development processes and connected Business Processes has been the recent focus area. He is Six sigma Black Belt certified and Quality Management & Information Security Audit Standards certified practitioner. He is pursuing Doctor of Philosophy in Data Science from Christ (Deemed to be University), Bangalore, India since June 2020. ORCIDID: 0000-0001-5859-0662.

K. Saritha is currently working as Principal, S.V. Degree & P.G. College, Anantapur, Andhra Pradesh. She has received Ph.D. in Computer Science from Sri Padmavati Mahila Visvavidyalayam (Women's University), Tirupati, Andhra Pradesh, India. She has 20+ years of teaching experience. She has presented 15 papers at various national & international conferences and journals. Her research interests are Data Mining, Blockchain, Cloud Computing, Machine Learning and Engineering education research.

Venkat Narayana Rao T. is a professor, Department of Computer Science and Engineering, Sreenidhi Institute of Science and Technology, Hyderabad, Telangana, India.

Manogna Thumukunta completed her B.Tech in Computer Science at Sreenidhi Institute of Science and Technology in 2020. She lives in Hyderabad and is currently working in a Software job at a multinational company. Machine Learning has been her key area of interest, and aims to dive deeper and do further research on the subject.

Index

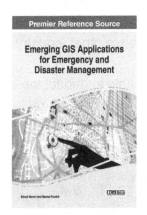
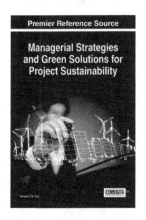

IGI Global Author Services

Providing a high-quality, affordable, and expeditious service, IGI Global's Author Services enable authors to streamline their publishing process, increase chance of acceptance, and adhere to IGI Global's publication standards.

Benefits of Author Services:

- **Professional Service:** All our editors, designers, and translators are experts in their field with years of experience and professional certifications.

- **Quality Guarantee & Certificate:** Each order is returned with a quality guarantee and certificate of professional completion.

- **Timeliness:** All editorial orders have a guaranteed return timeframe of 3-5 business days and translation orders are guaranteed in 7-10 business days.

- **Affordable Pricing:** IGI Global Author Services are competitively priced compared to other industry service providers.

- **APC Reimbursement:** IGI Global authors publishing Open Access (OA) will be able to deduct the cost of editing and other IGI Global author services from their OA APC publishing fee.

Author Services Offered:

English Language Copy Editing
Professional, native English language copy editors improve your manuscript's grammar, spelling, punctuation, terminology, semantics, consistency, flow, formatting, and more.

Scientific & Scholarly Editing
A Ph.D. level review for qualities such as originality and significance, interest to researchers, level of methodology and analysis, coverage of literature, organization, quality of writing, and strengths and weaknesses.

Figure, Table, Chart & Equation Conversions
Work with IGI Global's graphic designers before submission to enhance and design all figures and charts to IGI Global's specific standards for clarity.

Translation
Providing 70 language options, including Simplified and Traditional Chinese, Spanish, Arabic, German, French, and more.

Hear What the Experts Are Saying About IGI Global's Author Services

"Publishing with IGI Global has been **an amazing experience** for me for sharing my research. The **strong academic production** support ensures quality and timely completion." – **Prof. Margaret Niess, Oregon State University, USA**

"The service was **very fast, very thorough, and very helpful** in ensuring our chapter meets the criteria and requirements of the book's editors. I was **quite impressed and happy** with your service." – **Prof. Tom Brinthaupt, Middle Tennessee State University, USA**

Learn More or Get Started Here:

For Questions, Contact IGI Global's Customer Service Team at cust@igi-global.com or 717-533-8845

www.igi-global.com

Publisher of Peer-Reviewed, Timely, and
Innovative Academic Research Since 1988

IGI Global's Transformative Open Access (OA) Model:
How to Turn Your University Library's Database Acquisitions Into a Source of OA Funding

Well in advance of Plan S, IGI Global unveiled their OA Fee Waiver (Read & Publish) Initiative. Under this initiative, librarians who invest in IGI Global's InfoSci-Books and/or InfoSci-Journals databases will be able to subsidize their patrons' OA article processing charges (APCs) when their work is submitted and accepted (after the peer review process) into an IGI Global journal.

How Does it Work?

Step 1: **Library Invests in the InfoSci-Databases:** A library perpetually purchases or subscribes to the InfoSci-Books, InfoSci-Journals, or discipline/subject databases.

Step 2: **IGI Global Matches the Library Investment with OA Subsidies Fund:** IGI Global provides a fund to go towards subsidizing the OA APCs for the library's patrons.

Step 3: **Patron of the Library is Accepted into IGI Global Journal (After Peer Review):** When a patron's paper is accepted into an IGI Global journal, they option to have their paper published under a traditional publishing model or as OA.

Step 4: **IGI Global Will Deduct APC Cost from OA Subsidies Fund:** If the author decides to publish under OA, the OA APC fee will be deducted from the OA subsidies fund.

Step 5: **Author's Work Becomes Freely Available:** The patron's work will be freely available under CC BY copyright license, enabling them to share it freely with the academic community.

Note: This fund will be offered on an annual basis and will renew as the subscription is renewed for each year thereafter. IGI Global will manage the fund and award the APC waivers unless the librarian has a preference as to how the funds should be managed.

Hear From the Experts on This Initiative:

"I'm very happy to have been able to make one of my recent research contributions *freely available* along with having access to the *valuable resources* found within IGI Global's InfoSci-Journals database."

— **Prof. Stuart Palmer,**
Deakin University, Australia

"Receiving the support from IGI Global's OA Fee Waiver Initiative *encourages me to continue my research work without any hesitation.*"

— **Prof. Wenlong Liu**, College of Economics and Management at Nanjing University of Aeronautics & Astronautics, China

For More Information, Scan the QR Code or Contact:
IGI Global's Digital Resources Team at eresources@igi-global.com.

IGI Global
PUBLISHER of TIMELY KNOWLEDGE

Printed in the United States
by Baker & Taylor Publisher Services